Jürgen Schlabbach
Karl-Heinz Rofalski
Power System Engineering

Related Titles

Krause, P., Wasynczuk, O., Sudhoff, S.D.

Analysis of Electric Machinery and Drive Systems, Second Edition

2 Edition
2002
Print ISBN: 978-0-471-14326-0

Stoft, S.

Power System Economics

Designing Markets for Electricity

2002
Print ISBN: 978-0-471-15040-4

Rothwell, G., Gomez, T.

Electricity Economics

Regulation and Deregulation

2003
Print ISBN: 978-0-471-23437-1

Anderson, P.M., Fouad, A.A.

Power System Control and Stability, Second Edition

2 Edition
2003
Print ISBN: 978-0-471-23862-1

Webster, J.G. (ed.)

Encyclopedia of Electrical & Electronics Eng Online

2000
Print ISBN: 978-0-471-34608-1

Yamayee, Z.A., Bala, J.L.

Electromechanical Energy Devices and Power Systems

1994
Print ISBN: 978-0-471-57217-6

Sabonnadière, J., Hadjsaïd, N.

Smart Grids

2011
Print ISBN: 978-1-848-21261-9

Jürgen Schlabbach
Karl-Heinz Rofalski

Power System Engineering

Planning, Design and Operation of
Power Systems and Equipment

Second, Updated and Enlarged Edition

Verlag GmbH & Co. KGaA

The Authors

Professor Jürgen Schlabbach
Bielefeld College of Further Education
Schneidemühler Str. 88b
33605 Bielefeld
Germany

Dipl.-Ing. Karl-Heinz Rofalski
Freelance Engineer and Consultant
Im Winkel 15
61352 Bad Homburg
Germany

Library of Congress Card No.:
applied for

British Library Cataloguing-in-Publication Data
A catalogue record for this book is available from the British Library.

Bibliographic information published by the Deutsche Nationalbibliothek
The Deutsche Nationalbibliothek lists this publication in the Deutsche Nationalbibliografie; detailed bibliographic data are available on the Internet at <http://dnb.d-nb.de>.

Print ISBN: 978-3-527-41260-0
ePDF ISBN: 978-3-527-67905-8
ePub ISBN: 978-3-527-67904-1
Mobi ISBN: 978-3-527-67903-4
oBook ISBN: 978-3-527-67906-5

Cover Design Adam-Design, Weinheim, Germany
Typesetting Toppan Best-set Premedia Limited
Printing and Binding Markono Print Media Pte Ltd., Singapore

Printed on acid-free paper

Contents

Foreword, 2nd Edition

Since the first edition of this book in 2008, significant changes have taken place with respect to the generation and network structures of electric power supply systems. Under the term "Energiewende," Germany acts as a pioneer regarding the conversion and restructuring of electrical energy generation. In the past, electrical energy generation was based mainly on nuclear energy, coal, and gas power plants as against the new technologies pertaining to renewable energy sources, such as photovoltaic, wind energy, and biomass plants. The actual development by interconnection of photovoltaic and biomass plants with low-voltage (LV) and medium-voltage (MV) networks, of wind turbines with MV and high-voltage (HV) networks, likewise the development of offshore wind farms, indicated that a careful and thorough planning, design, and implementation of transmission and distribution networks is an absolutely necessary condition for the development of renewable energy sources and the successful completion of the "Energiewende."

This second edition of the book takes account of these developments, such as smart grids, HVDC-transmission, technical conditions for grid connection of generation plants, and application of new technologies, such as high temperature overhead conductors.

All standards and technical regulations have been brought up to date. The book reflects the actual status of the techniques, norms, and standards. All comments stated in this book are given to the authors' best knowledge, based on comprehensive technical knowledge and experience. Actual issues of the standards, available from the national standards organization, are only to be applied.

The authors like to thank all companies and organizations for supporting the book by providing technical data and other information, especially Prof. Dr. Stamminger from University of Bonn/Germany for information on smart domestic appliances. Special thanks are addressed to the staff of Wiley-VCH for their ongoing excellent cooperation in preparing the second edition. Our heartfelt thanks are with our families for their patience and support; otherwise, this book could never have been revised and extended.

Bielefeld, Bad Homburg, November 2013
juergen.schlabbach@fh-bielefeld.de rofalski-hg@gmx.de

Foreword, 1st Edition

The supply of electrical energy at competitive prices, in sufficient quantity and quality, and under the aspect of safe supply through reliable equipment, system structures and devices is of crucial importance for the economic development of countries and for the well-being of each individual. When planning power systems different boundary conditions must be considered, which are based on regional, structural, technical, environmental and financial facts, having a considerable impact on the technical design in many cases. Each investment decision requires a particularly careful planning and investigation, to which power system engineering and power system planning contribute substantially.

This book deals with nearly all aspects of power system engineering starting from general approach such as load estimate and the selection of suitable system and substation topology. Details for the design and operational restrictions of the major power system equipment, like cables, transformers and overhead lines are also dealt with. Basics for load-flow representation of equipment and short-circuit analysis are given as well as details on the grounding of system neutrals and insulation coordination. A major chapter deals with the procedures of project definition, tendering and contracting.

The purpose of this book is to serve as a reference and working book for engineers working in practice in utilities and industry. However, it can also be used for additional information and as a hand-book in post-graduate study courses at universities. The individual chapters include theoretical basics as far as necessary but focus mainly on the practical application of the methods as presented in the relevant sections. Carrying out engineering studies and work moreover requires the application of the latest edition of standards, norms and technical recommendations. Examples are given based on projects and work carried out by the authors during the last years.

The preparation of this book was finalised in *March 2008* and reflects the actual status of the techniques, norms and standards. All comments stated in this book are given to the best of knowledge, based on the comprehensive technical experience of the authors.

Extracts from IEC 60905:1978 and DIN-norms with VDE-classification are permitted for this edition of the book by licensee agreement 252.007 of DIN (Deutsches Institut für Normung e.V.) and VDE (Verband der Elektrotechnik Elektronik

Informationstechnik e.V.) on 08.11.2007. An additional permission is required for other usages and editions. Standards are only to be applied based on their actual issues, available from VDE-Verlag GmbH, Bismarckstr. 33, D-10625 Berlin, Beuth-Verlag GmbH, Burggrafenstr. 6, D-10787 Berlin or the national standard organisation.

The authors would like to thank Dipl.-Ing. Christian Kley and cand. ing. Mirko Zinn, University of Applied Sciences in Bielefeld, for spending time and efforts to draw the figures and diagrams. Special thanks are addressed to the staff of Wiley-VCH for the very good cooperation during the preparation of this book. Our thanks are also extended to our families for their patience during the uncounted hours of writing the book.

Bielefeld, Bad Homburg, March 2008

juergen.schlabbach@fh-bielefeld.de rofalski-hg@gmx.de

1 Introduction

1.1 Reliability, Security, Economy

Power system engineering is the central area of activity for power system planning, project engineering, operation and rehabilitation of power systems for electrical power supply. Power system engineering comprises the analysis, calculation and design of electrical systems and equipment, the setup of tender documents, the evaluation of offers and their technical and financial assessment and contract negotiations and award. It is seen as an indispensable and integral part of the engineering activities for feasibility studies, for planning and operating studies, for project engineering, for the development, extension and rehabilitation of existing facilities, for the design of network protection concepts and protective relay settings and also for clearing up of disturbances e.g. following short-circuits.

The supply of electricity – as for other sources of energy – at competitive unit price, in sufficient quantity and quality, and with safe and reliable supply through reliable equipment, system structures and devices is of crucial importance for the economic development of industries, regions and countries. The planning of supply systems must take into account different boundary conditions, which are based on regional and structural consideration that in many cases have a considerable impact on the technical design. Given that, in comparison with all other industries, the degree of capital investment in electric utilities takes the top position, not only from the monetary point of view but also in terms of long-term return of assets, it becomes clear that each investment decision requires particularly careful planning and investigation, to which power system engineering and power system planning contribute substantially.

The reliability of the supply is determined not only by the quality of the equipment but also by careful planning and detailed knowledge of power systems, together with a consistent use of relevant standards and norms, in particular IEC standards, national standards and norms as well as internal regulations. Furthermore, the mode of system operation must conform to the conditions specified by standards, including the planning process, manufacturing of equipment and commissioning. Just as faults in equipment cannot be totally excluded because of technical or human failure, likewise the equipment and installations cannot be

Power System Engineering: Planning, Design and Operation of Power Systems and Equipment, Second Edition.
Jürgen Schlabbach and Karl-Heinz Rofalski.

designed to withstand any kind of fault: accordingly, the effects of faults must be limited. Thus, violation of or damage to other equipment must be prevented in order to ensure undisturbed system operation and reliable and safe supply to the consumers.

The security of the electrical power supply implies strict adherence to the conditions specified in standards, norms and regulations concerning the prevention of accidents. In low-voltage systems the protection of individuals is seen of primary importance; at higher voltage levels the protection of equipment and installations must also be considered.

1.2
Legal, Political and Social Restrictions

Electrical power systems are operated with certain restrictions imposed by legal requirements, technical standards, political issues, financial constraints and social, political and environmental parameters which have a strong influence on the system structure, the design and the rating of equipment and thus on the cost of investment and cost of energy, without any justification in terms of aspects of security, reliability and economy. Some general areas pertaining to regulations, guidelines and laws for electrical power supply are simply stated below, without any elaboration at this stage.

- Concession delivery regulations
- Market guidelines for domestic electricity supply
- Electrical power industry laws
- Energy taxation
- Laws supporting or promoting "green-energy"
- Environmental aspects
- Safety and security aspects
- Right-of-way for overhead-line and cable routing.

Such regulations, laws and guidelines will have an impact on planning, construction and operation of power systems, likewise on the reliability of the power supply, the cost structure of equipment, the cost of electrical energy and finally on the attractiveness of the economic situation within the particular country.

- Generating plants will be operated in merit order, that is, the generator with lowest production cost will be operated in preference to operating generation with the highest efficiency.
- Criteria of profitability must be reevaluated in the light of laws supporting "green-energy."
- Reduced revenues from energy sales will lead to a decrease in the investments, personnel and maintenance costs, with consequences of reduced availability and reliability.
- Increasing the proportion of "green-energy" generation plants that have low availability leads to an increase in the running reserve of conventional power

stations, with consequences of reduced efficiency of these plants and thus higher costs.

- Reduction of investment for the construction of new power stations leads to a decrease in reserve capabilities and thus to a decrease in the reliability of the power supply.
- Expenditures for coordination during normal operation and during emergency conditions are increased with rising numbers of market participants, with the consequence of an increased risk of failures.
- Power systems of today are planned for the generation of electrical energy in central locations by large power stations with transmission systems to the load centers. A change of the production structure, for example, by increase of "green-energy" production plants and development of small co-generation plants, mainly installed in distribution systems, requires high additional investment for the extension of the power system, resulting in rises in energy prices as well as reduced usage of existing plants.
- The power system structure up to now has been determined by connections of the load centers with the locations of power stations, which were selected on the basis of the availability of primary energy (e.g. lignite coal), the presence of cooling water (e.g. for nuclear power stations) or hydrological conditions (e.g. for hydro power stations). The construction of offshore wind energy parks requires substantial investment in new transmission lines to transmit the generated energy to the load centers.
- Increase of "green-energy" production plants, in particular photovoltaic, wind energy and fuel-cells, reduces the quality of the power supply ("Power quality") due to the increased requirement for power electronics.
- The long periods for planning and investment of power stations and high-voltage transmission systems do not allow for fast and radical changes. Decisions on a different development, for example, away from nuclear power generation towards "green-energy" production, are to a certain extent irreversible if these decisions are not based on technical and economic background and detailed knowledge but are predominantly politically and ideologically motivated.

As an example, the structure of public tariffs for electrical energy in the Federal Republic of Germany is characterized by numerous measures initiated by the government. These taxes, concessionary rates, expenditures occasioned by the "green-energy" law, and so on amounted in the year 2011 to nearly 23.2 billion euro (€) according to data of the Bundesnetzagentur. Included in this are 0.2% for the support of combined cycle plants, 6.7% for concessionary rates for use of public rights of way, 14.1% for expenditures for the "green-energy" laws and 8.2% for energy taxes. Additionally, VAT (Value Added Tax) of 19% is added for private households. For the average electricity consumption of a private household of 3500 kWh per year, these costs as a result of governmental actions amount to almost 35€ per household per month.

1.3
Needs for Power System Planning

Power system planning must take due consideration of the restrictions mentioned above and must develop concepts and structures which are technically and economically sound. This includes the planning and project engineering of generation systems, transmission and distribution networks, and optimization of systems structures and equipment, in order to enable flexible and economic operation in the long as well as the short term. Power system planning also has to react to changes in the technical, economic and political restrictions. Key activities are the planning and construction of power stations, the associated planning of transmission and distribution systems, considerations of long-term supply contracts for primary energy, and cost analysis.

The systematic planning of power systems is an indispensable part of power system engineering, but it must not be limited to the planning of individual system components or determination of the major parameters of equipment, which can result in suboptimal solutions. Power system engineering must incorporate familiar aspects regarding technical and economic possibility, but also those that are sometimes difficult to quantify, such as the following:

- Load forecast for the power system under consideration for a period of several years
- Energy forecast in the long term
- Standardization, availability, exchangeability and compatibility of equipment
- Standardized rated parameters of equipment
- Restrictions on system operation
- Feasibility with regard to technical, financial and time aspects
- Political acceptance
- Ecological and environmental compatibility.

Power system engineering and power system planning require a systematic approach, which has to take into account the financial and time restrictions of the investigations as well as to cope with all the technical and economic aspects for the analysis of complex problem definitions. Planning of power systems and project engineering of installations are initiated by:

- Demand from customers for supply of higher load, or connection of new production plants in industry
- Demand for higher short-circuit power to cover requirements of power quality at the connection point (point of common coupling)
- Construction of large buildings, such as shopping centers, office buildings or department stores
- Planning of industrial areas or extension of production processes in industry with requirement of additional power
- Planning of new residential areas
- General increase in electricity demand.

- Connection of "green-energy" generation, such as photovoltaic and biomass generation to LV and MV systems and wind energy plants in MV and HV systems.

Power system planning is based on a reliable load forecast which takes into account the developments in the power system mentioned above. The load increase of households, commercial and industrial customers is affected by the overall economic development of the country, by classification by land development plans, by fiscal incentives and taxes (for example, for the use or promotion of "green-energy") and by political measures. Needs for power system planning also arise as a result of changed technical boundary conditions, such as the replacement of old installations and equipment, introduction of new standards and regulations, construction of new power stations and fundamental changes in the scenario of energy production, for example, by installation of photovoltaic generation. The objective of power system planning is the determination and justification of system topologies, schemes for substations and the main parameters of equipment considering the criteria of economy, security and reliability.

Further aspects must be defined apart from the load forecast:

- The information database of the existing power system with respect to geographical, topological and electrical parameters
- Information about rights-of-way, right of possession and space requirements for substations and line routes
- Information about investment and operational costs of installations
- Information about the costs of losses
- Knowledge of norms, standards and regulations.

The fundamental relations of power system planning are outlined in Figure 1.1.

1.4 Basic, Development and Project Planning

Load forecast, power system planning and project engineering are assigned to special time intervals, defining partially the tasks to be carried out. Generally three steps of planning are to be considered – basic planning, development planning and project planning – which cover different time periods as outlined in Figure 1.2.

1.4.1 Basic Planning

For all voltage levels the fundamental system concepts are defined: standardization of equipment, neutral earthing concepts, nominal voltages and basics of power system operation. The planning horizon is up to 10 years in low-voltage systems and can exceed 20 years in high-voltage transmission systems.

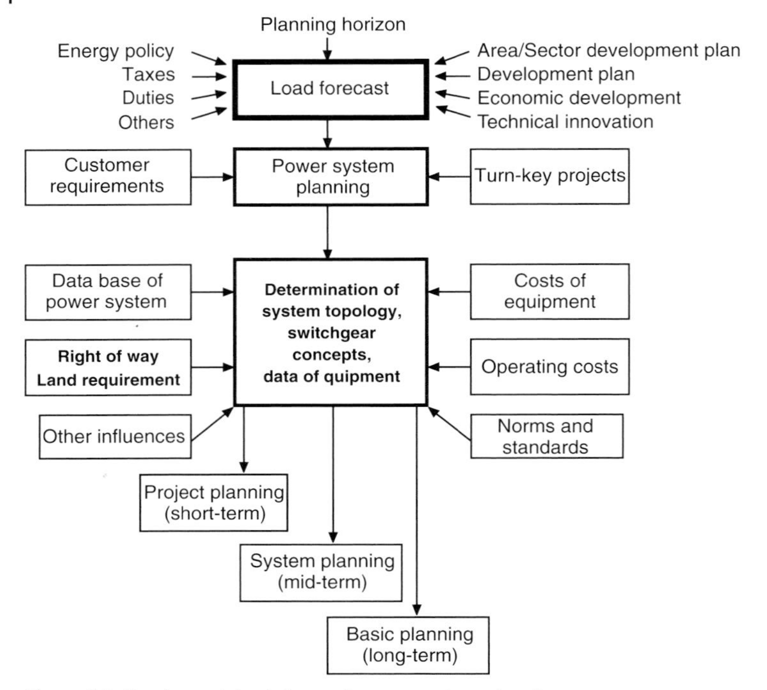

Figure 1.1 Fundamental relations of power system planning.

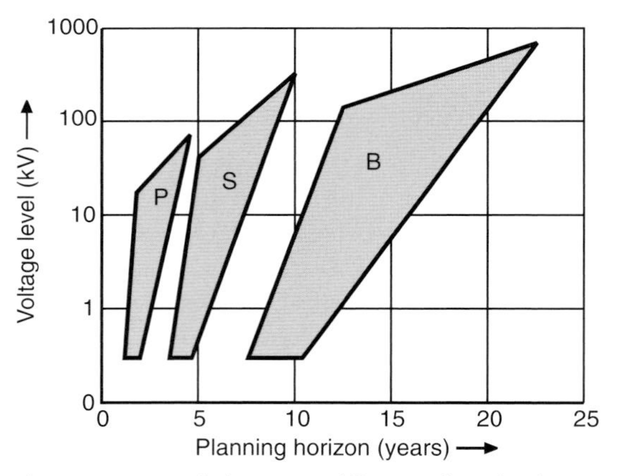

Figure 1.2 Steps of planning at different voltage levels. P, project planning; S, system development planning; B, basic planning.

1.4.2
System Development Planning

Detailed planning of the system topology is carried out based on the load forecast. Alternative concepts are analyzed technically by load-flow calculations, short-circuit analysis and stability computations. Cost estimates are also carried out. Disturbance and operational statistics are evaluated and locations for installations are determined. The main parameters of equipment, such as cross-section of overhead lines and cables, short-circuit impedance of transformers are defined. The planning horizon is approximately five years in a low-voltage system and up to 10 years for a high-voltage transmission system.

1.4.3
Project Planning

The projects defined in the system development planning stage are implemented. Typical tasks of the project engineering are the connection types of new customers, connection of new substations to the power system, restructuring measures, evaluation of information on system loading, preparation of tender documents and evaluation of offers, supervising construction contracts, cost calculation and cost control. Project planning covers a time range of one year in the low-voltage system and up to four years in the high-voltage system.

1.5
Instruments for Power System Planning

The use of computer programs as well as the extent and details of the investigations are oriented at the desired and/or required aim of the planning process. The fundamental investigations that must be accomplished by power system planning are explained below.

The load-flow analysis (also named power-flow calculation) is a fundamental task for planning and operation of power systems. It serves primarily to determine the loading and the utilization of the equipment, to calculate the active and reactive power-flow in the branches (lines, transformers, etc.) of the power system, to determine the voltage profile and to calculate the power system losses. Single or multiple outages of equipment can be simulated in the context of the investigations for different preloading conditions. The required setting range of the transformer tap-changer and the reactive power supply by generators or compensation devices are determined.

Short-circuit current calculations are carried out for selected system configurations, defined by load-flow analysis. For special applications, such as protection coordination, short-circuit current calculation should consider the preloading conditions as well. Symmetrical and unsymmetrical faults are simulated and the results are taken as a basis for the assessment of the short-circuit strength. Calculations of short-circuit current for faults between two systems are sometimes necessary to clarify system disturbances. Faults between two systems may occur in cases of multiple-circuit towers in overhead-line systems.

The results obtained by calculation programs are as exact as the main parameters of the equipment. If those data are not available, the parameters must be determined by calculation. In the case of overhead-lines and cables, the reactances, resistances and capacitances in the positive-sequence and zero-sequence component are calculated from the geometrical arrangement of the conductors and from the cable construction. Subsequent calculation may determine the permissible thermal loading, the surge impedance, natural power and, in case of overhead lines, additionally the electric field strength at the conductor as well as the electric and magnetic field strengths in the surrounding of the line for certain applications.

The permissible thermal loading of equipment under steady-state conditions and under emergency conditions is based on ambient conditions, for example, ambient temperature, thermal resistance of soil, wind velocity, sun exposure and so on. The calculation of the maximum permissible loading plays a larger role with cables than with overhead lines because of the poorer heat dissipation and the lower thermal overload capability.

The investigation of the static and in particular transient stability is a typical task when planning and analyzing high-voltage transmission systems. Stability analysis is also important for the connection of industrial plants with their own generation to the public supply system. Stability analysis has to be carried out for the determination of frequency- and voltage-dependent load-shedding schemes. The stability of a power system depends on the number and type of power stations, the type and rating of generators, their control and excitation schemes, devices for reactive power control, and the system load as well as on the voltage level and the complexity of the power system. An imbalance between produced power and the system load results in a change of frequency and voltage. In transient processes, for example, short-circuits with subsequent disconnection of equipment, voltage and frequency fluctuations might result in cascading disconnections of equipment and subsequent collapse of the power supply.

In industrial power systems and auxiliary supply systems of power stations, both of which are characterized by a high portion of motor load, the motors must start again after short-circuits or change-overs with no-voltage conditions. Suitable measures, such as increase of the short-circuit power and time-dependent control of the motor starts, are likewise tasks that are carried out by stability analyses.

The insulation of equipment must withstand the foreseeable normal voltage stress. It is generally economically not justifiable and in detail not possible to design the insulation of equipment against every voltage stress. Equipment and its overvoltage protection, primarily surge arresters, must be designed and selected with regard the insulation and sensitivity level, considering all voltage stresses that may occur in the power system. The main field of calculation of overvoltages and insulation coordination is for switchgears, as most of the equipment has non-self-restoring insulation.

Equipment in power systems is loaded, apart from currents and voltages at power-frequency, also by those with higher frequencies (harmonics and interharmonics) emitted by equipment with power electronics in common with the industrial load, in the transmission system by FACTS (flexible AC transmission systems)and by generation units in photovoltaic and wind-energy plants. Higher frequencies in current cause additional losses in transformers and capacitors and can lead to maloperation of any equipment. Due to the increasing electronic load and application of power electronics in generation plants, the emission of harmonics and interharmonics is increasing. Using frequency-dependent system parameters, the statistical distribution of the higher-frequency currents and the voltage spectrum can be calculated as well as some characteristic values, such as total harmonic distortion (THD), harmonic content, and so on.

Equipment installations, communication circuits and pipelines are affected by asymmetrical short-circuits in high-voltage equipment due to the capacitative, inductive and conductive couplings existing between the equipment. Thus, inadmissible high voltages can be induced and coupled into pipelines. In power systems with resonance earthing, unsymmetry in voltage can occur due to parallel line routing with high-voltage transmission lines. The specific material properties and the geometric outline of the equipment must be known for the analysis of these interference problems.

Electromagnetic fields in the vicinity of overhead lines and installations must be calculated and compared with normative specified precaution limit values, to assess probable interference of humans and animals exposed to the electric and magnetic fields.

Earthing of neutrals is a central topic when planning power systems since the insulation coordination, the design of the protection schemes and other partial aspects, such as prospective current through earth, touch and step voltages, depend on the type of neutral earthing.

In addition to the technical investigations, questions of economy, loss evaluation and system optimization are of importance in the context of power system planning. The extension of distribution systems, in particular in urban supply areas, requires a large number of investigations to cover all possible alternatives regarding technical and cost-related criteria. The analysis of all alternative concepts for distribution systems cannot normally be carried out without using suitable programs with search and optimization strategies. Optimization strategies in high-voltage transmission systems are normally not applicable because of restrictions, since rights of way for overhead lines and cables as well as locations of substations cannot be freely chosen.

The conceptual design of network protection schemes determines the secure and reliable supply of the consumers with electricity. Network protection schemes must recognize incorrect and inadmissible operating conditions clearly and separate the faulty equipment rapidly, safely and selectively from the power system. An expansion of the fault onto other equipment and system operation has to be avoided. Besides the fundamental design of protection systems, the parameters of voltage and current transformers and transducers must be defined and the settings of the protective devices must be determined. The analysis of the protection concept represents a substantial task for the analysis of disturbances.

1.6 Further Tasks of Power System Engineering

Project engineering is a further task of power system engineering. Project engineering follows the system planning and converts the suggested measures into defined projects. The tasks cover

- The evaluation of the measures specified by the power system planning
- The design of detailed plans, drawings and concept diagrams
- The description of the project in form of texts, layout plans, diagrams and so on
- The definition of general conditions such as test provisions, conditions as per contract, terms of payment and so on
- The provision of tender documents and evaluation of offers of potential contractors
- The contacts with public authorities necessary to obtain permission for rights of way and so on.
- The setup of project team to carry out the specific tasks of the project.

2
Power System Load

2.1
General

The forecasting of power system load is an essential task and forms the basis for planning of power systems. The estimation of the load demand for the power system must be as exact as possible. Despite the availability of sophisticated mathematical procedures, the load forecast is always afflicted with some uncertainty, which increases the farther the forecast is intended to be projected into the future. Power systems, however, are to be planned in such a way that changing load developments can be accommodated by the extension of the system. Long-term planning is related either to principal considerations of power system development or to the extra-high voltage system, so that no irrevocable investment decisions are imposed. These investment decisions concern the short term, as they can be better verified within the short-term range, for which the load forecast can be made with much higher accuracy. One thinks here, for example, of the planning of a medium- and a low-voltage system for a new urban area under development or the planned connection of an industrial area.

If the three stages of the planning process, explained in Chapter 1, are correlated with the required details and the necessary accuracy of the load forecast, it is clear that planning procedures are becoming more detailed within the short-time range and less detailed within the long-time range. Accordingly, different methods of load forecast have to be applied, depending on the planning horizon and thus on the voltage level and/or task of planning. From a number of different load forecasting procedures, five methods are described below.

- Load forecast with load increase factors
- Load forecast based on economic characteristic data
- Load forecast with estimated values
- Load forecast based on specific load values and extend of electrification
- Load forecast with standardized load curves.

The precise application of the different methods cannot be determined exactly and combinations are quite usual.

Power System Engineering: Planning, Design and Operation of Power Systems and Equipment, Second Edition.
Jürgen Schlabbach and Karl-Heinz Rofalski.

2.2 Load Forecast with Load Increase Factors

This method is based on the existing power system load and the increase in past years and estimates the future load increase by means of exponential increase functions and trend analyses. The procedures therefore cannot consider externally measured variables and are hardly suitable to provide reliable load and energy predictions. On the basis of the actual system load P_0 the load itself in the year n is determined by an annual increase factor of $(1 + s)$ according to Equation 2.1.

$$P_n = P_0 \cdot (1+s)^n \tag{2.1}$$

Assuming a linear load increase instead of exponential growth, the system load in the year n is given by Equation 2.2.

$$P_n = P_0 \cdot \left(1 + n \cdot \frac{\Delta P}{P_0}\right) \tag{2.2}$$

An increase in accuracy is obtained if the load forecast is carried out separately for the individual consumption sectors, such as households, trade, public supply and so on. The individual results are summed for each year to obtain the total system load.

Another model for load forecasting is based on the phenomenological description of the growth of electrical energy consumption [1]. The appropriate application for different regions must be decided individually for each case. The change of the growth of system load P with time is calculated from Equation 2.3.

$$\frac{\mathrm{d}P}{\mathrm{d}t} = c \cdot P^k \cdot (B-P)^l \tag{2.3}$$

where

k = growth exponent
c = growth rate
B = saturation level of the growth process as standardization value
l = saturation exponent.

With this model, adjustments can be combined with the process of load development of the past with different increases and saturation effects for the future. Experience indicates that, if l can be set $l > k$, the load increase follows the growth processes in the saturation phase at the limiting external conditions. Figure 2.1 shows typical load developments, calculated with the load development model [2]. The load development was standardized at the saturation level $B = 1$ at the end of the period under investigation; the growth rate was set equal to unity.

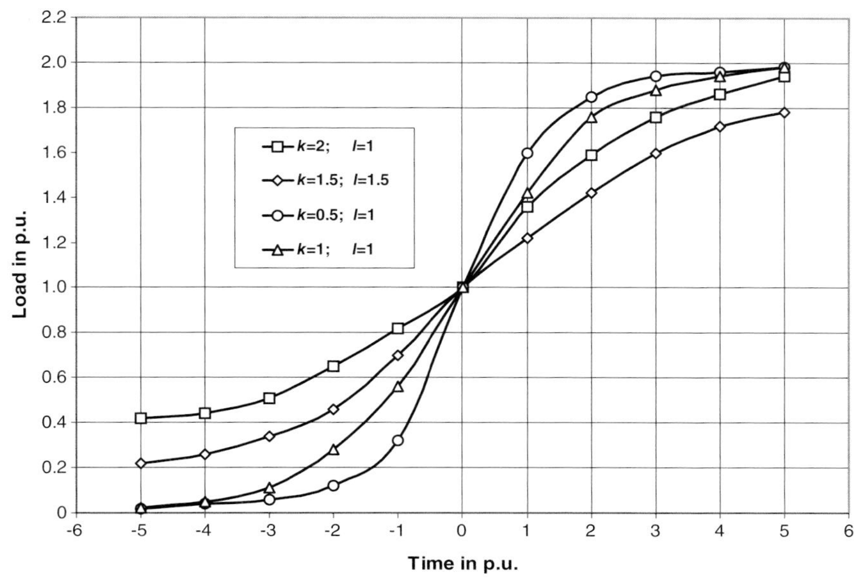

Figure 2.1 Load forecast calculated with the load development model (curves for various values of k and l).

2.3
Load Forecast with Economic Characteristic Data

Load forecast with economic characteristic data obtained from energy statistics assumes different relations between economic growth, availability of energy resources, energy consumption and requirements in general, such as the increase in energy consumption due to growth of population, and in special applications, such as energy requirements of industry. The requirement for electrical energy per capita of the population is determined to a large extent by the standard of living and the degree of industrialization of a country. However, it has to be considered that high consumption of energy can be also an indicator of the waste of energy, for example, in the case of high numbers of buildings with air conditioning or where there are comparatively low energy prices arising from differences in generation structure, as seen in countries with a high proportion of electrical heating because of cheap energy production in hydro plants.

In the past the increase of electrical energy consumption in industrialized countries was less affected by the growth of population and predominantly by the growth of the gross domestic product (GDP) and/or the gross national product (GNP). The economic ascent of Germany in the years 1950 to 1980 resulted in a rise of primary energy needs and demand for electrical energy. Growth rates of the so-called gross electrical consumption (GEC) amounted to about 7% per year until the end of the 1960s. Thus, doubling of the annually generated electricity every ten years was observed. The average rise of the GEC amounted to approximately 2%

in the 30 years upto 2004, during which economic recession caused decreases in the years 1975, 1982 and 1992. In this context it is not to be assumed that higher energy consumption automatically leads to an increase in the economic indicators GDP or GNP. Uncoupling of economic growth and electrical power requirement for industrialized countries today appears possible. We do not in this book discuss energy predictions based on various economic characteristics in more detail.

2.4
Load Forecast with Estimated Values

The aim of power system planning is to develop structures and concepts for the secure and reliable supply of electricity to the various types of consumers. For the load forecast, land development plans and land registers of town and regional planning authorities can be used. In the case of connection of bulk loads and industrial customers, the system load to be supplied must be determined via the owner of the industrial installation on the basis of the industrial processes operated and the installed number and types of devices and machinery.

Land development plans contain general information about the area development and use of land, and the size, location and types of residential, industrial and commercial areas, without allowing one to be able to derive detailed individual measures from them. The plans are suitable for the preliminary estimation of the future power system load, however. The need for construction of new substations and transformer stations, for example, from the 110-kV system to feed the distribution system, and area requirements needed for it can be justified using them. Estimated values for load densities of different type of land usage are illustrated in Table 2.1.

Larger industrial plants and special large consumers, such as shopping centers, are usually considered with their actual load based on internal planning.

2.5
Load Forecast with Specific Loads and Degrees of Electrification

More exact planning is possible using development plans available from town planning authorities, from which data can be taken concerning the structural use of the areas. Land usage by houses of different types and number of storeys, infrastructure facilities such as schools, kindergartens and business centers, as well as roads and pathways are included in the development plans. Thus a more exact determination of the system load can be achieved. The bases of the load forecast are the loads of typical housing units, which may vary widely depending on the degree of electrification. For the calculation of the number of housing units N_{WE}, the relationship of floor space (floor area) A_W to the housing estate surface A_G, the so-called floor space index G indicated in the development plan is used, see Equation 2.4 [3].

Table 2.1 Estimated values of load densities for different types of land usage (European index).

Type of usage	Load density	Remarks
Individual/single plot	$1\,\mathrm{MW\,km^{-2}}$	Free-standing single-family houses, two-family houses
Built-up area	$3\,\mathrm{MW\,km^{-2}}$	Terrace houses, small portion of multiple-family houses with maximum of three stories
Dense land development	$5\,\mathrm{MW\,km^{-2}}$	Multiple-story buildings, multifamily houses
Business	$5\,\mathrm{MW\,km^{-2}}$	Manufacturing shops, small business areas
	$0.2\,\mathrm{kW\,m^{-2}}$	Warehouses
	$0.3\,\mathrm{kW\,m^{-2}}$	Supermarkets and shopping malls
Industry	Up to $15\,\mathrm{MW\,km^{-2}}$	Medium-size enterprises, not very spatially expansive
General consumption	$2\,\mathrm{MW\,km^{-2}}$	Schools, kindergartens, street lighting

$$G = \frac{A_{\mathrm{W}}}{A_{\mathrm{G}}} \tag{2.4}$$

If the sizes of the housing estates are not yet well known, the total area must be reduced by about 25–40% for roads, pathways and green areas. The inhabitant density E (capita per km^2) is derived from the empirical value of a gross floor space of 20–22 m^2 per inhabitant (German index) according to Equation 2.5.

$$E \approx 25.000 \times G \tag{2.5}$$

In the case of only two persons per housing unit the housing density D is derived from Equation 2.6.

$$D \approx 13.000 \times G \tag{2.6}$$

The number of the housing units N_{WE} for given land development surface A_{G} is derived from Equation 2.7.

$$N_{\mathrm{WE}} \approx 13.000 \times A_{\mathrm{G}} \times G \tag{2.7}$$

The increase of gross floor space per inhabitant as well as the trend to more one-person households in industrialized countries leads to a reduced number of housing units N_{WE} for the area under investigation. The type of residential area must therefore be considered in the estimation of load forecast.

Table 2.2 Degree of electrification and load assumptions for households, authors' index.

Degree of electrification	Peak load of one household (kW)	Portion of peak load per household P_W (kW)	Degree of simultaneous usage g_∞	Remarks
EG1	5	0.77–1.0	0.15–0.2	Low electrification (old buildings, lighting only), today of less importance
EG2	8	1.0–1.2	0.12–0.15	Partial electrification (lighting, cooking)
EG3	30	1.8–2.1	0.06–0.07	Complete electrification (without electrical heating or air-conditioning
EG4	15	10.5–12	0.7–0.8	Total electrification (with electrical heating and air-conditioning)

Households have different grades of usage of electrical appliances. Usage depends on differing attitudes of individual groups within the population to the use of electrical energy, on the age of the house and also on the ages and the incomes of the inhabitants. As indicated in Table 2.2, one can divide the different household appliances in terms of the degree of electrification of the household or building. As not all appliances within one household are in operation at the same time and as not all households have the same consumption habits at the same time, the respective portion of the peak load P_W has to be set for the total load determination, which for increasing number of housing units reaches the limit value g_∞.

The degree of simultaneous usage g_n for the number of households N_{WE} (denoted n in Equation 2.8) can be calculated according to Equation 2.8, whereas values for g_∞ are taken from Table 2.2.

$$g_n = g_\infty + (1 - g_\infty) \cdot n^{-0.75} \tag{2.8}$$

The proportion of peak load to be taken for the load forecast P_{tot} for households with different degrees of electrification ($I = 1, \ldots, 4$) can be calculated using Equation 2.9.

$$P_{tot} = \sum_{i=1,4} g_{ni} \cdot P_{Wi} \cdot n_i \tag{2.9}$$

The loads for commercial, industrial and common consumers are added to the load of the housing units.

2.6
Load Forecast with Standardized Load Curves

Another possibility for the determination of the system load is based on the annual energy consumption of the individual consumer or consumer groups, which can be taken from the annual electricity bill. The system load can be determined by means of standardized load curves or load profiles [4] for different consumer groups:

- Household consumers
- Commercial consumers of different kinds (24-hour shift, shop or manufacturing enterprise, opened or closed at weekends, seasonal enterprise, etc.)
- Agricultural enterprises of different kinds (dairy farming or water pumping)
- Other customers (schools, public buildings, etc.).

Individual load profiles of special customers and bulk loads can be obtained by measurement. As consumption profiles of the particular customer groups not only change with time of day but also show day-of-week and seasonal changes, characteristic days are defined, such as working-day, Saturday, Sunday (or Friday in Islamic countries) and holiday as well as seasonal differences in winter, summer and transition periods. It is to be noted that the consumption of households is subject to substantially stronger seasonal fluctuations than is the case with other consumer groups. The load profiles of characteristic days are adjusted dynamically as explained in the example below.

Typical household load profiles are presented in Figure 2.2. The calculation proceeds from a standardized diagram of the load, which corresponds to an annual consumption of 1000 kWh. Small deviations, for example, due to changes of holidays or leap years, can normally be neglected in the context of power system planning.

On the basis of the allocation of the respective load profiles to the different days in the different seasons, the individual daily load curves are adjusted dynamically [4] by means of a factor F_d according to Equation 2.10.

$$F_d = -3.92 \times 10^{-10}\, d^4 + 3.2 \times 10^{-7}\, d^3 - 7.02 \times 10^{-5}\, d^2 + 2.1 \times 10^{-3}\, d + 1.24 \qquad (2.10)$$

The quantity d is the current day of the year, with $d=1$ for 1 January of each year. The time dependence of the dynamization factor F_d during a year is presented in Figure 2.3.

After the daily load curves have been corrected with the dynamization factor according to Figure 2.3, the standardized consumer load curve of the year is determined. Taking account of the annual consumption of the consumers, the load of each customer and thus of the total customer group can be determined for each time. This load can also be used for load forecast after the number of expected new consumers is known. Furthermore, the data can be used for the management of load balance circle. By adding up the yearly load diagrams of all consumers or consumer groups to be supplied via the same feeder or transformer

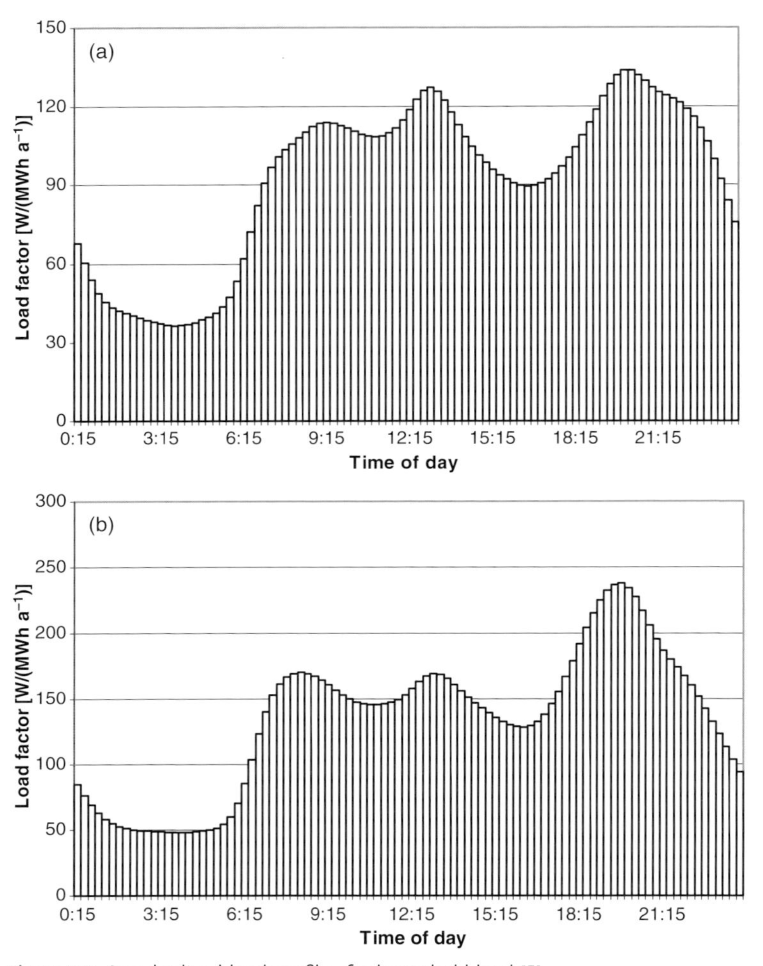

Figure 2.2 Standardized load profiles for household load [5]. (a) Summer working-day; (b) winter working-day; (c) summer Sunday; (d) winter Sunday.

station, the yearly peak and low load can be determined as the basis for planning. Two more examples of load profiles are presented in Figures 2.4 and 2.5. The differences in the load profile for the bakery shop are significant, whereas the load profiles for the dairy farming are nearly the same on working-days and on weekends.

In principle the load profile of other consumer groups, such as hotels and small business enterprises, are quite different from those of household consumers [6]. Differences can obviously be seen due to the work time and/or utilization periods, and the daily variations can be very large, but the seasonal variations are comparatively small. As an example, the load diagrams of a metal-working enterprise

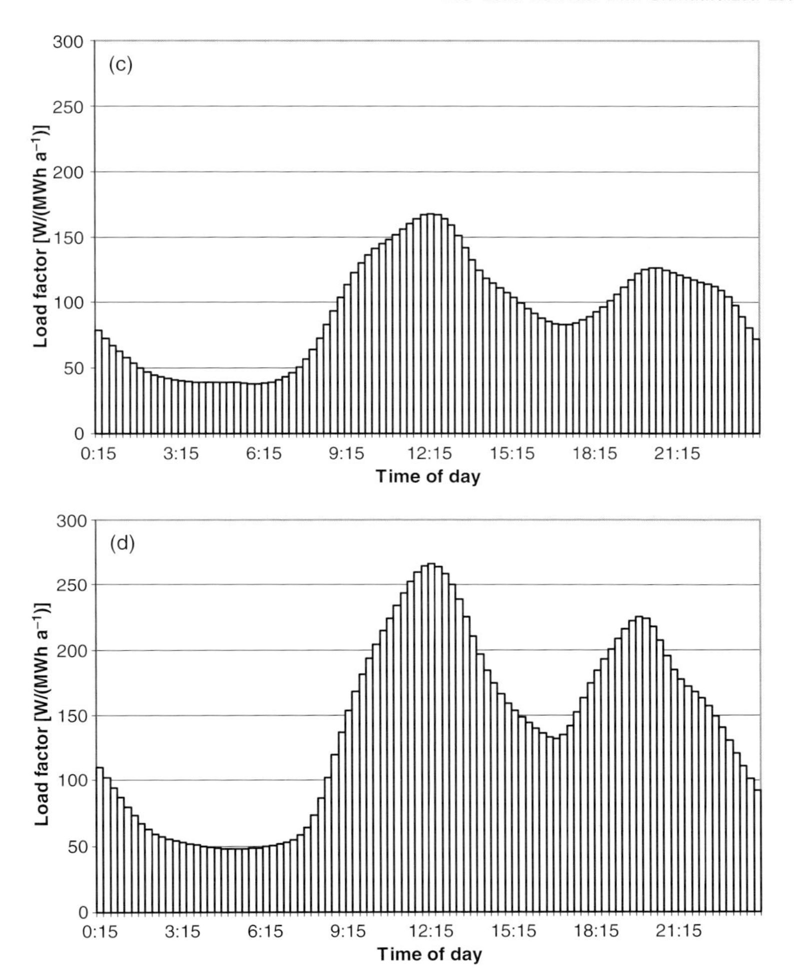

Figure 2.2 *Continued.*

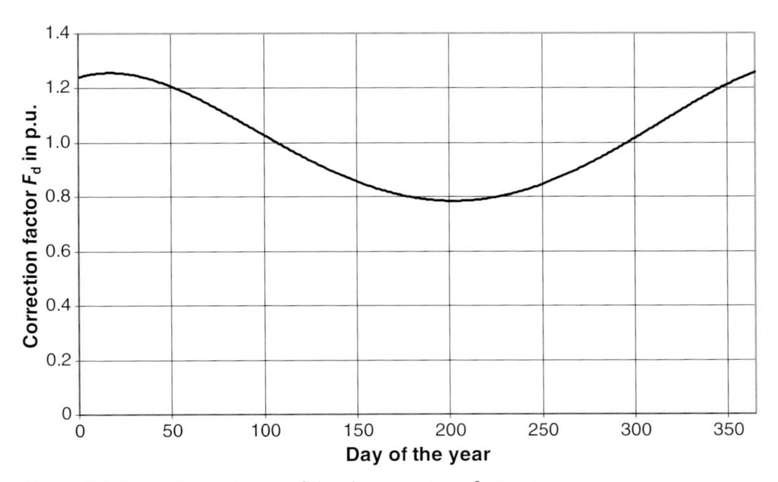

Figure 2.3 Time dependence of the dynamization factor F_d.

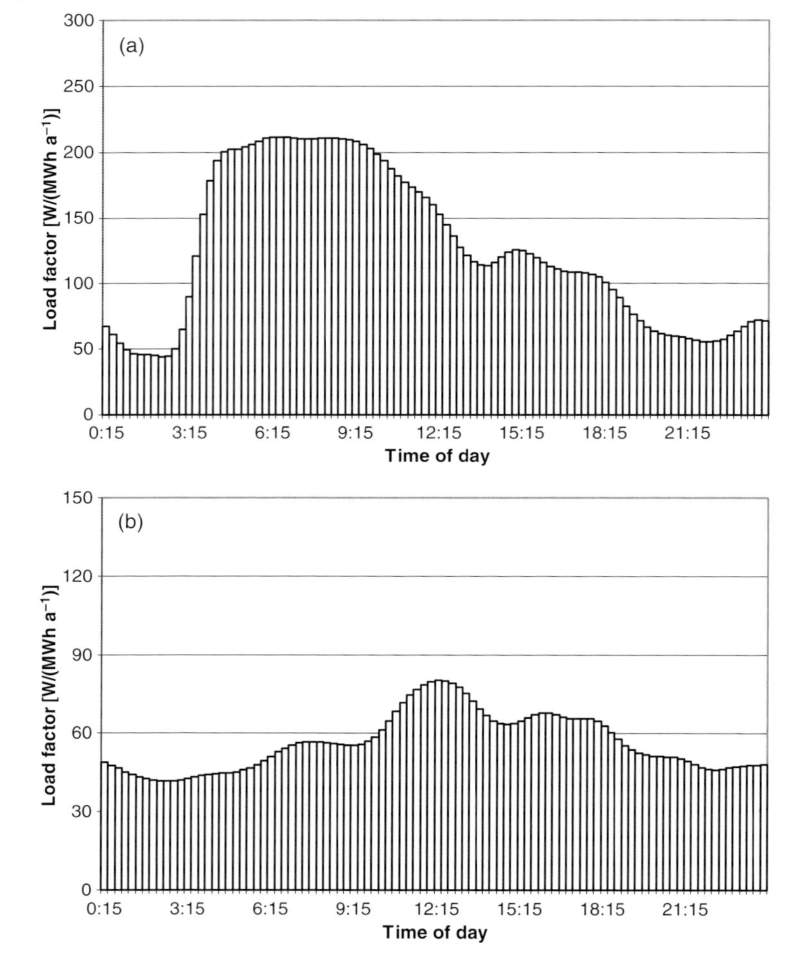

Figure 2.4 Standard load profile of a bakery shop [5]. (a) Summer working-day; (b) summer Sunday.

(Figure 2.6a) and of a hotel (Figure 2.6b) are presented for one week in summer in each case. The diagrams for the winter period differ only marginally from those in the summer.

2.7
Typical Time Course of Power System Load

Electricity can be stored only in small quantities and at high technical and financial expense. The power and energy demand of each second, hour, day and season must be covered by different types of power stations. The variation of the power

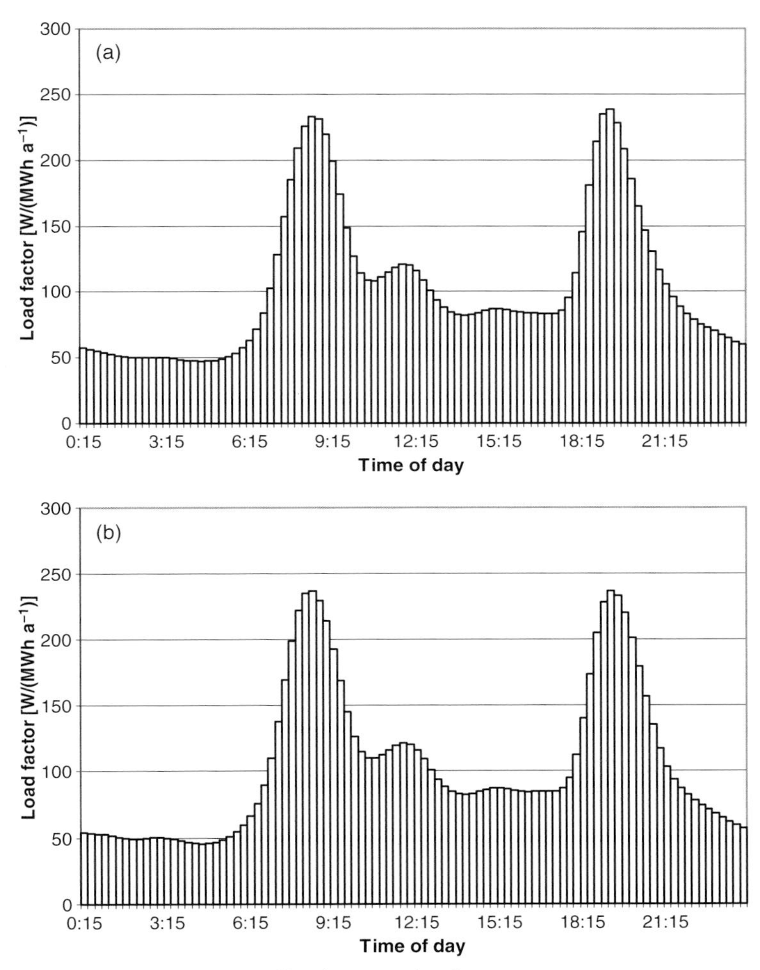

Figure 2.5 Standard load profile of an agricultural enterprise with dairy farming [5]. (a) Summer working-day; (b) summer Sunday.

system load in Germany in the year 2006 was between 61.5 GW and 77.8 GW (ratio of low load to high load equal 0.79). This ratio can be completely different in other countries, especially if there are no or only few industrial consumers. Considering, for example, a power system in an Arabian country with only 7.4% industrial load, the system load is determined mainly by air-conditioning devices and lighting. Taking a weekend day in August, the load varied between 704 MW and 1324 MW (ratio between low load and peak load equal 0.53). On the day of the yearly low load, a weekday in February, the load varied between 274 MW and 450 MW (ratio of low load to peak load equal 0.61).

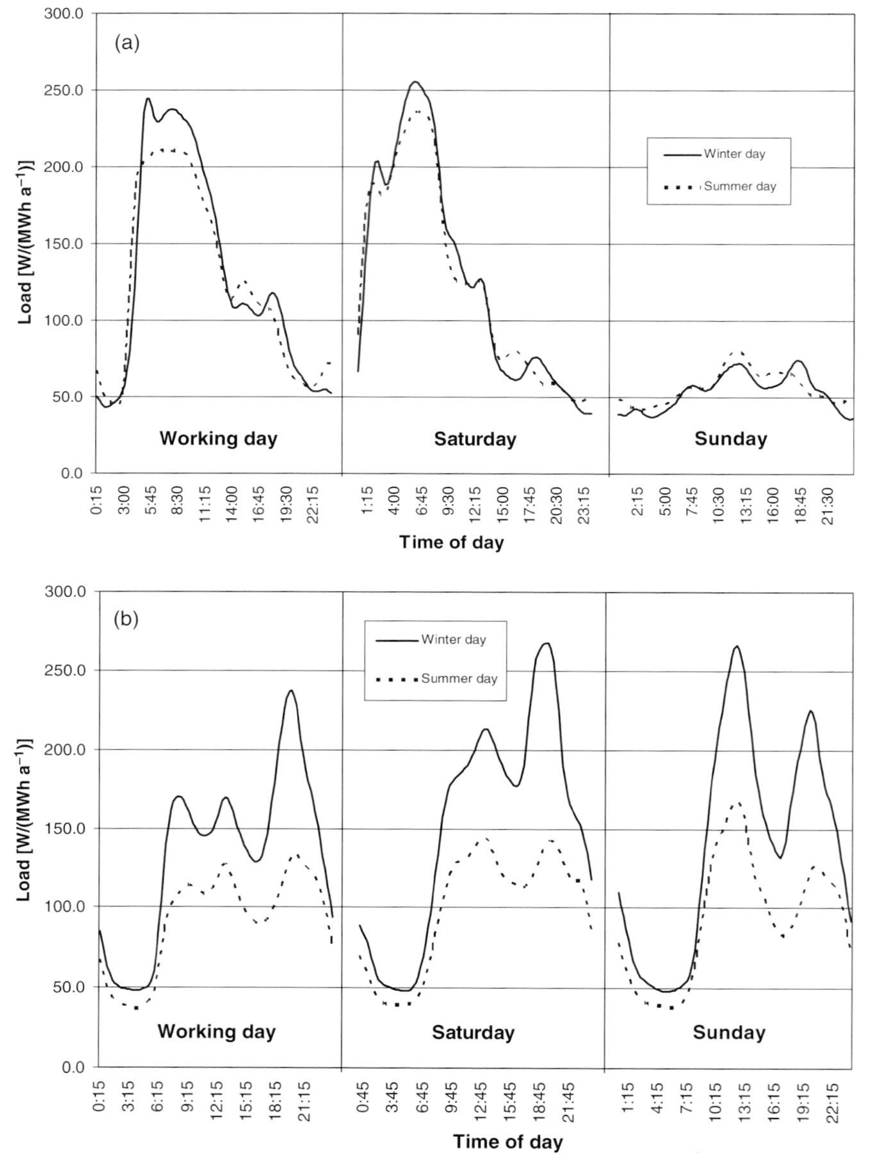

Figure 2.6 Weekly load diagrams of business enterprises [6, 7]. (a) Metal-processing enterprise; (b) hotel.

2.8
Smart Grid and Smart Domestic Appliances

One of the important items to realize the so-called "Energiewende" in Germany and in future years in other countries is the introduction of smart energy grids or smart grids.

The term smart grid (smart energy system) includes the coordination and intelligent control of generation and storage facilities, consumers, and power system equipment in transmission and distribution networks with the help of information and communication technology (ICT). The target is to maintain a sustainable and environmentally friendly energy supply based on transparency, energy efficiency, cost efficiency, as well as safe and reliable system operation. Smart grid is an intelligent power supply system, not to be mixed up with intelligent network. It includes the operation of power distribution and transmission networks with new ICT-based technologies for network automation as well as the integration of centralized and decentralized power generation facilities and controllable and intelligent consumer equipment including domestic appliances to achieve an improved control and operation of the entire power system.

A study carried out with support by the European Commission through the IEE program (contract no. EIE/06/185/SI2.447477) indicated a high potential for load shifting and load control within the European community by smart domestic appliances [8]. By this means the daily load curves can be smoothed and partly adjusted to the generation curve dominated in the future by the stochastic generation by wind turbines and photovoltaic installations.

Figure 2.7 indicates the estimated daily load curve of an average European household. As can be seen from the diagram, the peak demand with more than 800 W (mainly caused by heating appliances such as water heater, tumble dryer, washing machine and air conditioning) at approximately 8 h p.m. is followed by a sharp decline to less than 200 W in the early morning hours.

The load curves in the various areas of Europe differ from each other, due to differences in market penetration and time duration of usage of the domestic appliances. Figure 2.8 indicates the estimated daily load curve of an average household in Germany/Austria in the years 2010 and 2025.

The daily energy consumption is reduced from approximately 5.9 kWh in 2010 to 5.5 kWh in 2025 due to the usage of appliances with increased power efficiency, despite a higher market penetration of all types of domestic appliances. The load curve in 2010 is characterized by two peak patterns mainly caused by cooking (around 11 h a.m.) and by the use of dish washers and washing machines in the evening hours. The ratio of peak load to low load at 11 h a.m. is around 2.6. In 2025, the evening peak is increased as compared with the peak at noon due to the increased use of water heating, air conditioning, and tumble dryer; the ratio of peak to low load at 19 h a.m. is around 3.4.

The comparison of the two load curves indicate the high potential of smoothing the load curve by the use of intelligent domestic appliances, mainly by shifting of operation hours of washing machines, dish washers, and tumble dryers to late

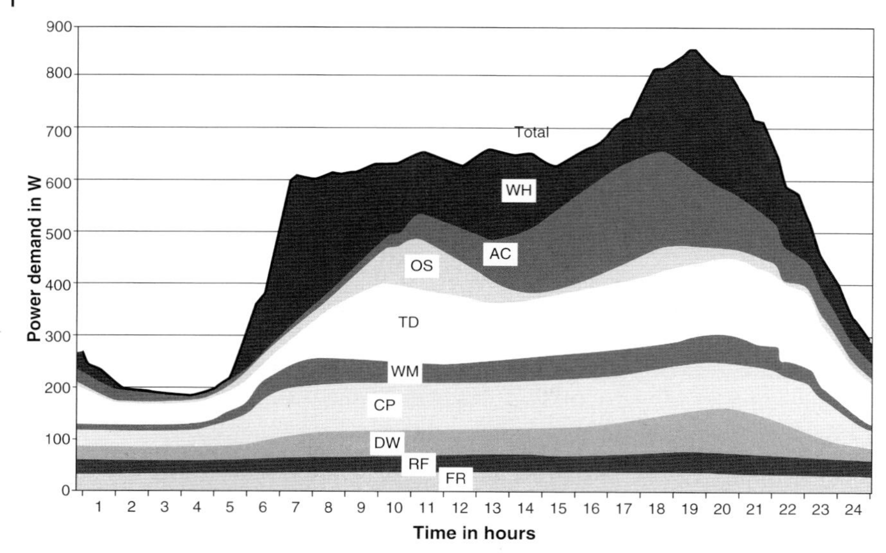

Figure 2.7 Estimated load curve of an average European household, market penetration of all household appliances 100% except electrical heating [8].
Total: total power demand of all appliances; WH: water heater; AC: air conditioning; OS: oven and stove; TD: tumble dryer; WM: washing machine; CP: circulating pump for heating; DW: dish washer; RF: refrigerator; FR: freezer.

evening hours. Figure 2.9 indicates the simulated results for the year 2025 for an average household in Germany/Austria and the shifting potential [8].

The total amount of load shifted is in the range of 190 Wh per household, the reduction of peak power is approximately 31 W per household. The potential of reduction of peak power demand for all regions in Europe is about 98 W per household, resulting in a total power demand of 18 GW, which is equal to the installed power of nearly 13 nuclear power stations (1.4 GW each) or 30 coal-fired power stations (600 MW each).

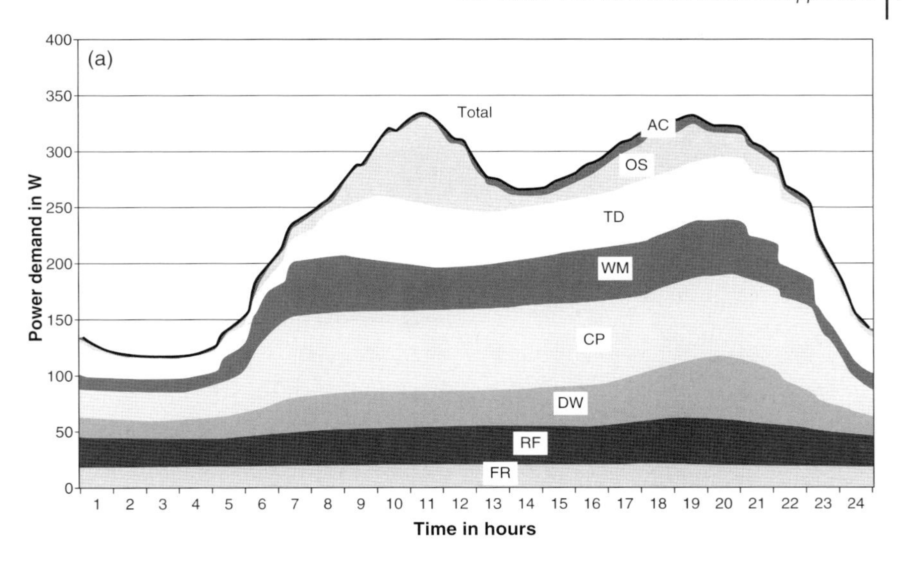

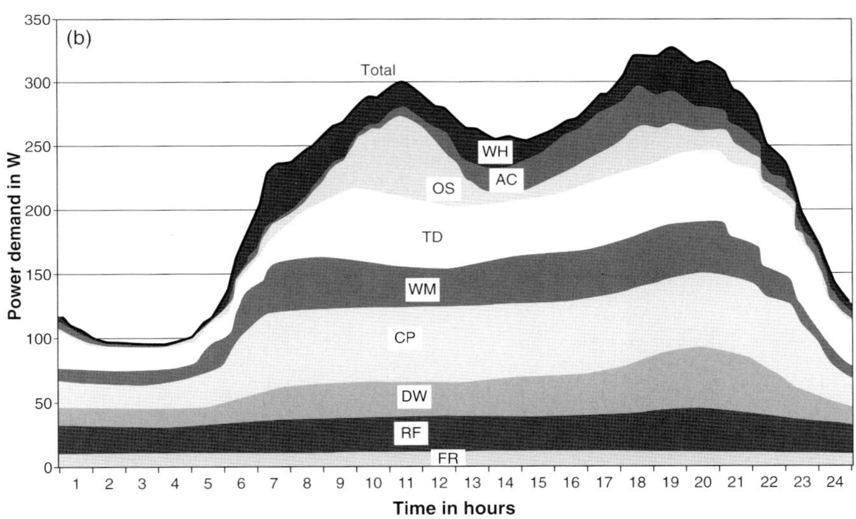

Figure 2.8 Estimated load curve of an average household in Germany/Austria [8]. Legend: see Figure 2.7.
(a) year 2010
(b) year 2025

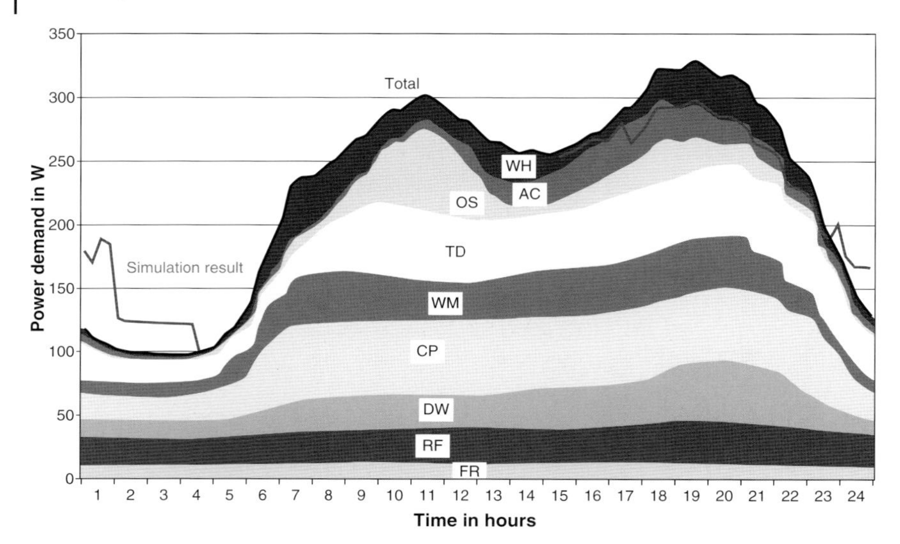

Figure 2.9 Simulated load curve of an average household in Germany in the year 2055 with shifting potentials [8]. Legend: see Figure 2.7.

3
Planning Principles and Planning Criteria

3.1
Planning Principles

The aim of planning electrical power systems is to fully serve the interests of the consumers to be supplied with electricity. The active and reactive power of the supply area to be expected in the long-range planning period are taken as basic parameters. In order to determine the configuration of power system in terms of technical, operational, economic, legal and ecological criteria, planning principles have to be defined and used. High priority is to be given to the supply of consumers with a defined need for supply reliability, which can be accomplished if sufficient data are available on system disturbances (faults, scheduled and unscheduled outages) or by means of quantitative and if necessary additional qualitative criteria.

The reliability of the electrical power supply system (power station, transmission and distribution system, switchgear, etc.) is influenced by:

- The fundamental structure of the power system configuration (topology)
  Example: The consumer is supplied only via one line (overhead line or cable) forming a radial supply system. In case of failure of the line, the supply is interrupted until the line is repaired.
- The selection of equipment
  Qualified and detailed specification and tendering of any equipment, consistent use of international norms for testing and standardization of equipment guarantee high-quality installations at favorable costs on an economic basis.
- The operational mode of the power system
  The desired reliability of supply can be guaranteed only if the power system is operated under the conditions for which it was planned.
- Earthing of neutral point
  A single-phase fault with earth connection (ground fault) in a system with resonance earthing does not lead to a disconnection of the equipment, whereas a single-phase earth fault in a system with low impedance neutral grounding (short-circuit) leads to a disconnection of the faulted equipment and in some cases to interruption of supply.

Power System Engineering: Planning, Design and Operation of Power Systems and Equipment, Second Edition.
Jürgen Schlabbach and Karl-Heinz Rofalski.

- Qualification of employees
  Apart from good engineering qualifications, continuing operational training of personnel obviously leads to an increase of employees' competence and through this to an increase of supply reliability.
- Regular maintenance
  Regular and preventive maintenance according to specified criteria is important to preserve the availability of equipment.
- Uniformity of planning, design and operation
  Operational experience must be included in the planning of power systems and in the specification of the equipment.
- Safety standards for operation
  The low safety factor for "human failure" can be improved by automation and implementation of safety standards, thus improving the supply reliability.

It is axiomatic that 100% security and reliability of electrical power supply cannot be achieved and does not have to be achieved. In each case a compromise between supply reliability, the design of the system and any equipment and the operational requirements must be agreed, and of course the interests of the consumers are to be considered. The planning should be guided by the following precept:

Reliability as high as necessary,
design and operation as economical as possible!

Prior to the definition of planning principles, agreement must be obtained concerning acceptable frequency of outages, their duration up to the reestablishment of the supply and the amount of energy not supplied and/or the loss of power due to outages. Outages include both planned or scheduled outages due to maintenance and unplanned or unscheduled outages due to system faults. Unscheduled outages result from the following:

- The equipment itself, the cause here being the reduction of insulation strength, leading to short-circuits and flash-over
- Malfunctioning of control, monitoring and protection equipment (protection relays), which can cause switch-off of circuit-breakers
- External influences, such as lightning strokes or earthquakes, which lead to the loss of equipment and installations
- Human influences, such as crash-accidents involving installations (overhead towers) or cable damage due to earthworks, followed by disconnection of the overhead line or the cable.

Even with careful design and selection of equipment, loading and overloading sequences in normal operation cycles, detailed monitoring of the system operation and preventive or regular maintenance have to be considered. Faults and outages are difficult to foresee. Thus the frequency and duration of outages can hardly be predicted and can only be estimated on the basis of evaluation and assessment of disturbance statistics.

The duration of outages up to the reestablishment of the supply can be estimated as a maximum value and is determined by the following:

- Power system configuration and planning criteria
  If the power system is planned in such a way that the outage of one item of equipment or power system element does not lead to overloading of the remaining equipment, safe power supply is secured in case of failure of any piece of equipment, independently of the repair and reconnect duration.
- Design of monitoring, protection and switching equipment
  If switchgear in a power system can only be operated manually and locally, then the duration of the supply interruption is longer and thus the energy not supplied is larger than if the switches are operated automatically or from a central load dispatch center.
- Availability of spare parts
  A sufficient number of spare parts reduces the duration of supply interruption and the amount of energy not supplied, as the repair can be carried out much more quickly.
- Availability of personnel (repair)
  The timely availability of skilled and qualified personnel in sufficient number reduces the repair time significantly.
- Availability of personnel (fault analysis)
  The causes of failures and faults in the power system have to be analyzed and assessed carefully prior to any too-hasty reestablishment of the supply after outages, in order to avoid further failures due to maloperation and erroneous switching.
- Availability of technical reserves
  A sufficient and suitable reserve is needed to cover the outage of any equipment. This need not imply the availability of equipment of identical designed to the faulty equipment; for example, after the outage of a HV/MV-transformer the supply can be ensured temporarily by a mobile emergency power generator.

The amount of energy not supplied and/or the loss of power must be established taking account of the importance of the consumers. One might accept, say, a range for the energy not supplied between 500 kWh for urban supply and major customers (for example, a poultry farm) and 2000 kWh for rural supply or less important consumers (for example, skating-rink). The results of these assumptions are shown diagrammatically in Figure 3.1 (average value of the energy not supplied equals 1000 kWh).

Figure 3.1 can be interpreted with the examples (a) to (d) below:

(a) The loss of a 0.4 kV cable loaded with 100 kW is acceptable for a period between approximately 5 and 12 hours. The faulty cable must be repaired during this time or restitution of supply has to be achieved by other measures, such as local switch-over of the supply to other LV-systems or by mobile emergency power generators.

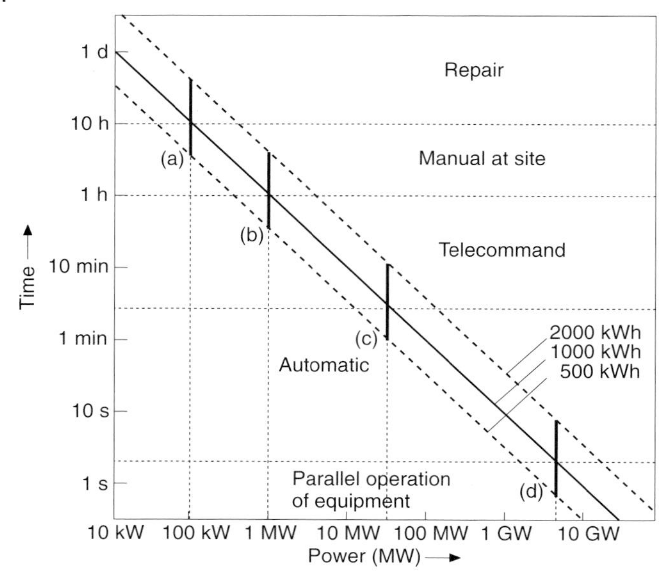

Figure 3.1 Relation between energy not supplied and loss of power for a failure frequency of 0.1 outages per year; (a) to (d) refer to the examples given in the text.

(b) The outage of a 10 kV cable supplying a load of 1 MW is acceptable for a period between 1.5 and 2 hours. Within this time the supply can be restored by switch-over locally or remote switching to other supply systems.
(c) The outage of a MV-transformer loaded with 30 MW can be accepted only for ~5–10 minutes. For the reestablishment of the supply, an automatic switch-over or a remote controlled switch-over to other supply systems are sufficient.
(d) Outage of equipment or power systems with load in the range of some gigawatts requires countermeasures within milliseconds to restore the supply. Equipment operated in parallel (transformers, transmission lines, cables or hot-running reserve generation units) are needed. Automatic switch-over can be adequate in exceptional cases.

3.2 Basics of Planning

Power systems for electrical power supply must be planned and operated considering the loading of the equipment in such a way as to achieve the following:

- A reasonable and/or suitable relation between the maximal thermal stress (acceptable load current) and the actual load in the final stage of system voltage and/or until restructuring measures become effective and
- No inadmissible thermal loadings (load current) arise, except those which are permitted under certain operating and ambient site conditions.

Thermal loading of equipment beyond the permissible values can lead to a reduction of the insulation strength in case of non-self-healing insulating media (oil-impregnated paper, XLPE, PVC, etc.) with consequent reduction of the lifetime, an increase in frequency of insulation faults and thus an increase of the failure rate of the equipment. The mechanical strength and the elasticity of overhead line conductors, bars and other metallic connections can be reduced if the temperature exceeds the permissible temperature for a certain duration (see Chapter 9). This may result in irreversible deformations, which can lead, for example, with overhead lines, to a reduction of the insulation clearances between conductor and earth and consequently to an increase of the flash-over frequency and to an increased failure rate of the equipment. Mechanical damage is also possible. The thermally permissible load of equipment depends on

- Material properties
- Ambient temperature
- Other site conditions, for example, wind and sun exposure
- Number of load cycles
- Preloading conditions of the equipment in case of variable load
- Duration of the additional load arising after the preloading conditions
- Past total actual time under operation.

The permissible loads are to be taken from standards or manufacturers' data or can be determined with suitable computer programs.

Power systems must be planned and operated with regard to short-circuit currents in such a way that

- The thermal strength of equipment and installations is always higher than the prospective thermal effects of the short-circuit currents.
- The electromagnetic effects of short-circuit currents are lower than the associated mechanical strength of the equipment and installations.
- The short-circuit and fault currents through earth do not cause any impermissible step or touch voltages or impermissible voltages at earthing electrodes.

The conditions mentioned must be fulfilled for symmetrical and unsymmetrical short-circuits. Short-circuit currents as well as their thermal and electromechanical effects and the thermal and electromechanical strength of equipment and installations must be determined in accordance with international standards, most suitably using computer programs. The voltage at earthing resistances or reactances and step and touch voltages are to be determined on the basis of unsymmetrical short-circuit currents with earth connection. For pipelines and other metallic circuits running in parallel, induced voltages need to be calculated.

Power systems must be planned and operated with regard to the generation, transmission and distribution of electrical power and energy in such a way that

- Sufficient generation capacity is available to supply the expected (forecast) load as well as the power system losses and to cover the internal consumption under normal operating conditions and in case of outages of power stations and any other equipment in the power system.

- Transmission and distribution systems have sufficient capacity to supply the power system load under normal operating conditions and under defined outage conditions.

In view of typical realization times (planning period, tender period, construction time) for the construction of power systems, very careful and comprehensive planning is necessary. The range of realization times, for example, in Germany, is from 2 years (distribution system), to 5 years to 10 years (medium-voltage power systems) and up to 15 years (power stations and high-voltage transmission systems).

Power systems must be planned and operated with regard to the system voltage in such a way that

- A suitable and internationally standardized voltage level is selected for transmission and distribution systems.
- The voltage is within a suitable bandwidth as defined by international standards or by planning criteria under normal operation and under outage conditions.
- The power factor of the system is on the lagging side and the generators can run in the over-excited mode.

IEC 60038 "IEC standard voltages" stipulates standardized voltages and the permissible voltage bandwidth or voltage range is also defined for some voltage levels. In the extra-high-voltage range only the highest voltage for equipment is defined, see Tables 3.1– 3.4. The permissible voltage band must be defined in the planning criteria in cooperation with other supply companies.

Power systems must be planned and operated with regard to frequency control and transient behavior in such a way that

- The frequency in the steady-state condition remains within a permissible and agreed range for the entire power system.
- Deviations from the nominal frequency (50 Hz or 60 Hz) are to be compensated within a fixed time period.

Table 3.1 Nominal system voltages according to IEC 60038 for LV three-phase and single-phase AC systems.

Nominal voltage		Voltage bandwidth	Remarks (Table I of IEC 60038)
Three-phase systems (V)	Single-phase systems (V)		
230/400	120/240	$\Delta U_{max} < \pm 10\% \times U_n$	277/480 V
277/480 400/690 1000		Normal operating conditions	Not to be used together with 400/690 V

Table 3.2 Nominal system voltages according to IEC 60038 for MV systems, AC voltage, 1 kV < $U_n \leq$ 35 kV.

Highest voltage for equipment, U_m (kV)	Nominal voltage, U_n (kV)		Tolerance ΔU	Remarks (Table III of IEC 60038) reference U_m
3.6	3.3	3	$\Delta U_{max} < \pm$ 10% $\times U_n$	Not for public supply systems
7.2	6.6	6	Normal operating conditions	Not for public supply systems
12	11	10		–
17.5		15		Not recommended for new installations
24	22	20		–
36	33	–		–
40.5	–	35		–

Table 3.3 Nominal system voltages according to IEC 60038 for HVAC-systems, 35 kV < $U_n \leq$ 230 kV.

Highest voltage for equipment, U_m (kV)	Nominal voltage, U_n (kV)		Remarks (Table IV of IEC 60038)	Remarks U_m
52	45		Not recommended for new installations	123 and 145 kV not to be used both in one country. 245 kV not to be used in one country, if 300 or 363 kV are present (Table 3.4)
72.5	66	69	–	
123	110	115	–	
145	132	138	–	
170	150		Not recommended for new installations	
245	220	230	–	

- No inadmissible frequency fluctuations shall be initiated due to disconnections of loads or due to short-circuits in the power system. (Load-shedding by frequency relays is seen only as the last measure to secure the stability of the power system).
- The frequency range shall remain within the limits defined for the operation of synchronous and asynchronous machines.

A stable frequency under steady-state conditions represents an essential condition for the regulation of the exchange of electricity between different supply

Table 3.4 Highest voltages for equipment according to IEC 60038 for EHVAC-systems, $U_m > 245$ kV.

Highest voltage for equipment, U_m (kV)	Remarks (Table V of IEC 60038)	Remarks U_m
300	Not recommended for new installations	245, 300 and 363 kV
363	Not recommended for new installations	or 363 and 420 kV or
420	–	420 and 525 kV not
525	Instead of 525 kV, 550 kV are to be used also	be used together in
765	Values between 765 kV and 800 kV to be used	one geographical area
1200	–	(see also Table 3.3)

partners. The conditions (criteria) for the analysis of transient stability, such as load conditions to be considered, time delay of the protection, automatic reclosing sequences, types of faults and short-circuits in the power system and so on, must be defined in planning criteria.

Power systems must be planned and operated with regard to flexibility and economy in such a way that

- The type and topology of the system allow supply to some extent even for load developments different from those forecast.
- The system losses are minimal under normal operating conditions.
- Different schedules of operation of power stations are possible.
- The generation of energy is possible in economic priority sequence (merit order), and ecological and environmental conditions are taken into account.
- A suitable and favorable relation between design and rating of equipment and the actual load, in particular their thermal permissible loading, is achieved in the final system development stage.
- Standardization of the equipment is possible, without impairment of operational flexibility.

3.3 Planning Criteria

3.3.1 Voltage Band According to IEC 60038

Planning criteria are understood to be objectively verifiable conditions, parameters and data, which are fixed in a quantitative way for planning and operation of equipment and the power system defined in standards or other agreed documents. The nominal voltage and the acceptable voltage range cannot be specified arbitrarily, since this is fixed for some voltage levels in IEC 60038. For low-voltage systems (LV systems) the data of Table 3.1 are to be applied.

Power systems having nominal voltage above 1 kV and below 35 kV are termed medium-voltage systems (MV systems). The specified voltages according to Table 3.2 are to be assured. It has to be considered that the highest voltage for equipment U_m does not correspond in some cases with the indicated voltage tolerance ΔU related to the nominal voltage; for example, for the nominal voltage $U_n = 3.3\,kV$ the highest voltage for equipment $U_m = 3.6\,kV$ is lower than the upper tolerance value of the voltage ($1.1 \times 3.3\,kV = 3.63\,kV$). Therefore, the upper tolerance of the voltage is to be taken as always smaller than or equal to the highest voltage for equipment.

Power systems with voltages above 35 kV up to 230 kV (inclusive) are termed high-voltage systems (HV systems). IEC 60038, Table IV, specifies only the highest voltage for equipment, but no voltage tolerance. The respective data are given in Table 3.3.

Power systems having voltages above 245 kV are termed extra-high-voltage systems (EHV systems). IEC 60038 specifies only the highest voltage for equipment as outlined in Table 3.4. No voltage tolerance and no nominal voltage are defined; nevertheless, the term nominal voltage is also used in loosely in this voltage range. Thus one calls a power system having $U_m = 420\,kV$ (highest voltage for equipment) a 400 kV system (sometimes also 380 V system) with nominal voltage $U_n = 400\,kV$.

The conditions indicated in Tables 3.1–3.4 define the minimal requirements. Stronger conditions can be defined and may be needed especially for high- and extra-high-voltage systems, for which no voltage tolerances are defined. In addition, voltage tolerances should also be specified for operation after loss of equipment or under emergency conditions.

3.3.2 Voltage Criteria

Criteria for voltage tolerance are to be defined for power systems under normal operating conditions and under single outage conditions, in some case also for multiple outages, as outlined below.

3.3.2.1 Low-Voltage Systems

If one uses the values in low-voltage systems according to Table 3.1 (Table I of IEC 60038) at the point of connection of the customer to the utility (point of common coupling, PCC), then the voltage drop in the customer's installation remains unconsidered; connected low-voltage consumers are supplied with a system voltage below the permissible limit. Similar considerations apply to voltage increase, which occurs mainly while connecting local generation in LV systems, such as photovoltaic generation. Therefore the following criteria are recommended:

- In low-voltage systems the voltage is to be held in a tolerance of −5% and +10% in relation to the nominal system voltage during normal operation at the PCC.
- In case of single outage in the low-voltage system, the voltage at the PCC is to be kept within a tolerance of −8% and +10% in relation to the nominal system voltage. If necessary, corrective measures are to be considered, such as change-over to another supply point in the low-voltage system or in the medium-voltage system, by adjusting the tap-changer of the MV/LV-transformer or by shifting the section point in the ring main feeders (see Section 5.3.2).
- In low-voltage systems without redundant supply, for example, in a radial power system, a longer outage time up to the repair has to be accepted (see Figure 3.1).

3.3.2.2 Medium-Voltage Systems

In medium-voltage power systems with $U_n < 35\,kV$, similar criteria as for low-voltage systems can be applied. The voltage is regulated by the automatic tap-changer of the feeding transformer. Note that in the medium-voltage power system increases of voltage may have effects on the LV system as well and also have to be considered. The following criteria are recommended:

- In medium-voltage systems the voltage is to be kept within a tolerance of ±5% in relation to the nominal system voltage during normal operation.
- With single outage in the medium-voltage system without supply interruptions, the voltage is to be held within a tolerance of ±10% in relation to the nominal system voltage. After carrying out corrective measures such as switch-on of a reserve transformer or operating the tap-changer of the feeding transformer, the voltage should remain in a tolerance of ±8% in relation to the nominal system voltage.
- In medium-voltage power systems without redundant supply, for example, in radial power systems, a longer outage time up to the repair must be accepted (see Figure 3.1). After restoring the supply, without replacement or repair of the failed equipment, the voltage is to be held within a tolerance of ±10% in relation to the nominal system voltage.

3.3.2.3 High- and Extra-High-Voltage Systems

High- and extra-high-voltage systems with $U_n > 110\,kV$ are usually planned and operated as meshed power systems. IEC 60038 does not define criteria for the voltage tolerance; the tolerance can be freely specified. The voltage is regulated by the automatic tap-changer of the feeding transformers or by any other measures influencing the voltage, such as generator voltage control, reactive power compensation equipment and so on. Voltage drops and voltage increase are to be compensated and controlled. As the nominal system voltage is not defined for extra-high voltage systems, a value of $0.9\,U_m$ (90% of the highest voltage for equipment) is to be taken as nominal system voltage. The following criteria are recommended:

- In high-voltage systems the voltage is to be held in a tolerance band of ±5% in relation to the nominal system voltage during normal operation.
- With single outage in the high-voltage system and without switching or corrective measures, the voltage is to be kept within a tolerance of ±10% in relation to the nominal system voltage. After execution of corrective measures, for example, by operating the tap-changer of feeding transformers or changing the generation schedule of power stations, the voltage is to be kept within a tolerance of ±8% in relation to the nominal system voltage.

If the high- and/or extra-high-voltage system is planned in such a way that several independent outages do not lead to supply interruptions, then graded criteria for single and multiple outages can be defined.

- With single outage the voltage is to be held within a tolerance of ±8% in relation to the nominal system voltage. After execution of corrective measures, for example, by operating the tap-changer of feeding transformers or changing the generation schedule of power stations, a tolerance of ±6% in relation to the nominal system voltage is to be maintained.
- If multiple outages, that is, two independent outages, are permitted as a planning principle, a voltage tolerance of ±10% in relation to the nominal system voltage is allowed in case of multiple outage. After execution of corrective measures, for example, by operating the tap-changer of feeding transformers or changing the generation schedule of power stations, the voltage is to be held within a tolerance of ±8% in relation to the nominal system voltage.

3.3.3 Loading Criteria

Criteria for the permissible loading of equipment in power systems under normal operating conditions and/or with single outage or multiple outages can be specified.

- The loading of any equipment may not exceed the values as defined in standards, norms and regulations, data sheets of manufacturers or by means of computer programs during normal operating conditions. For details see the appropriate Chapters 7 to 9.
- In case of outage of any equipment, the loading of the remaining ones (still in operation in the power system) may not exceed the values defined for a given period as specified in standards, norms and regulations, data sheets of manufacturers or determined by means of computer calculations. For reference see the appropriate Chapters 7 to 9.
- For the determination of load criteria such as duration and height of the load, preloading conditions and so on, application-oriented standards are specified, which are to be used in planning and operation. Details can be seen in the appropriate Chapters 7 to 9.

3.3.4
Stability Criteria

Power systems have to be operated in a stable manner in the event of transient disturbances without subsequent faults. Such faults may include load shedding by frequency relays, switch-off of generators, isolating of subsystems and so on. The power system frequency has to fulfill tolerance criteria. Oscillations of power in case of defined scenarios most not be allowed to lead to loss of stability:

- Three-phase faults or single-phase faults on any equipment with subsequent switch-off (fault clearing) of the faulted equipment within a specified time determined by the operating time of the protection
- Three-phase faults on any overhead-line with subsequent successful fault clearing by auto-reclosing with a specified time sequence
- Three-phase faults on any overhead-line with subsequent unsuccessful fault clearing by auto-reclosing with a specified time sequence
- Single-phase faults on any overhead-line with subsequent successful fault clearing by three-phase or single-phase auto-reclosing with a specified time sequence
- Three-phase faults on any overhead-line with subsequent unsuccessful three-phase or single-phase fault clearing by auto-reclosing with a specified time sequence
- Loss of load in the system, for example, by switch-off of a HV-transformer or any other fault
- Loss of generation in the system, for example, by switch-off of a power station or any generator

To ensure normal frequency control in the power system, sufficient generation reserve under primary control (primary reserve) must be available and activated in timely fashion. The amount of primary reserve depends on the typical power station schedule.

The recommendations of the UCTE (Union for the Co-ordination of Transmission of Electricity) for the European transmission system specify for primary reserve an amount of 2.5% of the total generation feeding the system. Half of it is to be available within 5 seconds, the remaining part is to be activated within 30 seconds. In subsystems isolated from the UCTE system, for example, after system separation following severe faults, the primary reserve is to be doubled as compared with the first case. It should be noted that the actual primary reserve and the control speed in the UCTE system are higher than mentioned. Figure 3.2 outlines some typical relations of primary reserve and activating time in the UCTE system.

If drop in the frequency, for example, after large system disturbances, cannot be avoided by load frequency control or if during transition periods the frequency remains below certain limit values for a longer time, then load-shedding by frequency relays has to be applied. In the UCTE system with nominal frequency 50 Hz the agreed grading plan is

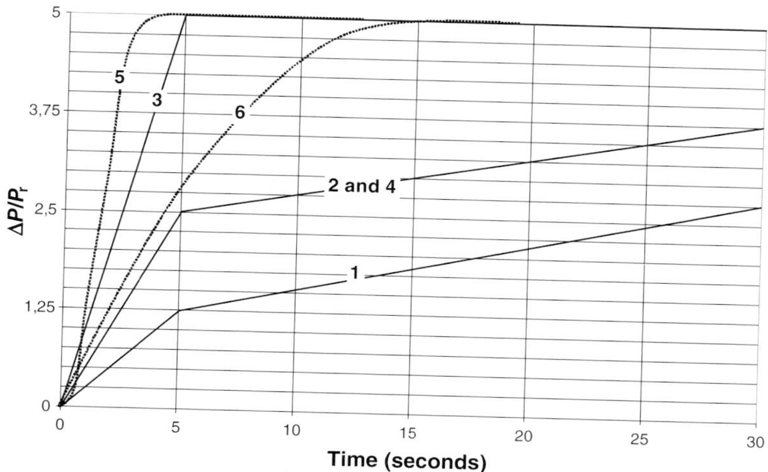

Figure 3.2 Relation between primary reserve and activating time as agreed in the UCTE system. (1) Operation without subsystems; (2) operation of sub-systems isolated from the total UCTE system; (3) reserve capability of a nuclear power station (P_r = 1300 MW); (4) primary reserve capability of a coal-fired power generator unit (P_r = 600 MW); (5) primary reserve capability of a pumped-storage generator unit (P_r = 220 MW); (6) primary reserve capability of a pumped-storage generator unit (P_r = 55 MW).

- Frequency drops to 49.8 Hz: activation of reserves according to a coordinated plan
- Frequency drops to 49.0 Hz: instantaneous disconnection of 10–15% of the power system load
- Frequency drops to 48.7 Hz: instantaneous disconnection of a further 10–15% of the power system;
- Frequency drops to 48.4 Hz: third load-shedding stage, a further 15–20% of the power system load to be switched off; French power system: 48 Hz
- Frequency drops to 47.5 Hz: power stations to disconnect from the power system immediately

Intermediate steps with special countermeasures are specified by different utilities in different ways.

4
Economic Consideration and Loss Evaluation

4.1
Present Value and Annuity Method

In the context of power system planning and project engineering, economic aspects must be considered, compared and assessed for different alternatives and scenarios, such as investment costs of the project as well as an evaluation of the losses resulting from system operation. Various methods are available; the most common ones – present value and annuity method – are explained in the context of this book [9].

The present value takes account of all incomes and expenditures of the period under review, which are referred to one reference time instant t_0, usually the time of project commissioning. Payments K_{Bi} resulting during the project engineering, building and commissioning phase of the project will be cumulated to the time instant t_0 in accordance with Equation 4.1a.

$$K_{B0} = \sum_i K_{Bi} \cdot q^i \tag{4.1a}$$

The present value method for investment cost is well-suited for the comparison of different financing scenarios of projects and also for the comparison of different prices of equipment. The present value method is also suitable for the comparison of loss costs during the foreseeable operating time and for comparison of annual costs of different project concepts and/or equipment resulting from the loss costs. One includes the costs K_{Ri} (e.g. for maintenance and repair), arising during the operation time, into the present value method according to Equation 4.1b and the expected incomes K_{Vi} according to Equation 4.1c.

$$K_{R0} = \sum_i \frac{K_{Ri}}{q^i} \tag{4.1b}$$

$$K_{V0} = \sum_i \frac{K_{Vi}}{q^i} \tag{4.1c}$$

Costs are to be set as negative values, incomes as positive values. The project is profitable if the total present value becomes positive over the expected lifespan.

Power System Engineering: Planning, Design and Operation of Power Systems and Equipment, Second Edition.
Jürgen Schlabbach and Karl-Heinz Rofalski.

Among other things, the annual costs of the project, financing expenses and costs of losses can be calculated with the annuity method. The minimal necessary current sales can also be determined by this method. The annuity method is appropriate if the annual costs are continuous in order to finance an investment, for example, with the determination of the annual costs. The investment costs, set equal to the present value of the investment cost K_0 can be calculated with Equation 4.2.

$$K = r_n \cdot K_0 \tag{4.2}$$

The method supposes that the entire investment costs are to be financed as credit and that the credit is repaid at constant annual rates K, the annuity, over the stipulated lifespan of n years. The annuity factor r_n for the lifespan of n years is given by Equation 4.3, with the interest factor $q = 1 + p$ and the interest rate p.

$$r_n = \frac{p \cdot (1+p)^n}{(1+p)^n - 1} = \frac{q^n \cdot (q-1)}{q^n - 1} \tag{4.3}$$

The specific costs per unit of installed power k_0 as given by Equation 4.4 are to be converted into annual specific costs per unit of installed power, including the annuity r_n, the expenditures Δk for repair, maintenance, taxes, insurance and so on.

$$k_0 = \frac{K_0}{P_{\max}} \tag{4.4}$$

The annuity r_n and the expenditures per year Δk are summed to the fixed service costs f according to Equation 4.5.

$$f = r_n + \Delta k \tag{4.5}$$

The specific costs per kW and year k_P are calculated according to Equation 4.6.

$$k_P = f \cdot \frac{K_0}{P_{\max}} \tag{4.6}$$

4.2 Evaluation of Losses

For the calculation, evaluation and assessment of losses in the context of power system planning, simplifications may have to be made particularly regarding the expected load shape of the equipment and the power system. Generally the losses consist of a load-sensitive part (current-dependent) and a part independent from the load (voltage-dependent). Load-sensitive losses are mainly ohmic losses, which means losses due to the thermal effects of the current. They are proportional the square of the current and can also be taken as approximately proportional the square of the apparent power S.

The losses independent of the load are the no-load losses of transformers and electrical machines, corona losses of overhead lines and insulation losses and dielectric losses of cables, transformers and so on, that is, in equipment with no self-restoring insulation. They are proportional in the first approximation to the square of the operating voltage. The relevance of the losses to the loss evaluation in the context of system planning and design have to be estimated and assessed individually for each project.

4.2.1 Energy Losses

The losses of equipment and power systems are generated in the power station and transferred from the power station (power-source) through other equipment up to the equipment under investigation (loss-sink). Losses anywhere in the system must be generated in the power station. With "green-energy" production plants without fuel inventory, the loss portion reduces the energy available to the final consumer. This part of the loss is called energy loss. Energy losses reduce the current available to be transferred to the consumers through overhead lines, cables and transformers between power-source and loss-sink. Their importance depends on the time for which the energy losses have to be transported and whether thereby the maximum permissible current loading of equipment is limited or not. As long as the maximal permissible current loading of the equipment is not reached, this consideration may be of less importance.

For description of the importance of the losses the maximum load portion h according Equation 4.7 can be used. The maximum load portion is defined as the correlation of the load P_{i} of the equipment concerned or part of the power system at peak load time t_{Pmax} of the entire system with the maximum load P_{imax} of the equipment or a part of the power system.

$$h = \frac{P_{\mathrm{i}}|_{t=t_{\mathrm{Pmax}}}}{P_{\mathrm{imax}}} \tag{4.7}$$

The energy losses P_{Vi} of equipment at peak load time are calculated based on the maximal losses of the equipment according to Equation 4.8.

$$h_{\mathrm{Vi}} = h^2 \cdot P_{\mathrm{Vi\,max}} \tag{4.8}$$

For the losses independent of load, the maximum load portion is equal to unity in each case. The relation between the factors and the system load is outlined in Figure 4.1.

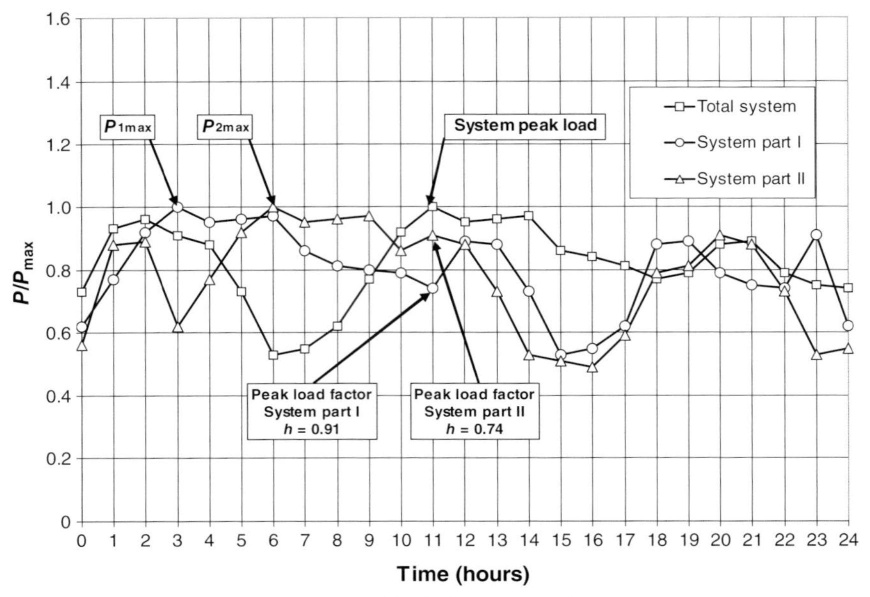

Figure 4.1 Correlation of system load and load of one piece of equipment with regard to the maximum load portion h [9].

4.2.2
Power Losses

Meeting the losses caused by other equipment necessitates the supply of extra capacity from power in the power stations. This depends on whether the losses arise during a larger period (e.g. one year) or only for a short time within the year under investigation. As an example, the typical load curve of a HV/MV transformer is outlined in Figure 4.2. For the evaluation of losses and for the determination of production costs the load curve is described by the period of use T_m and the degree of utilization m. The period of use is defined according to Equation 4.9 as the ratio of the energy W during the period considered T_Z, for example, one year, to the peak load P_{max}:

$$T_m = \frac{W}{P_{max}} = \frac{\int_0^{T_Z} P(t)dt}{P_{max}} \tag{4.9}$$

The degree of utilization or load rate m, also termed the utilization rate as given by Equation 4.10 is the relation of the energy actually produced to the maximal possible energy during the period T_Z.

$$m = \frac{T_m}{T_Z} = \frac{\int_0^{T_Z} P(t)dt}{P_{max} \cdot T_Z} \tag{4.10}$$

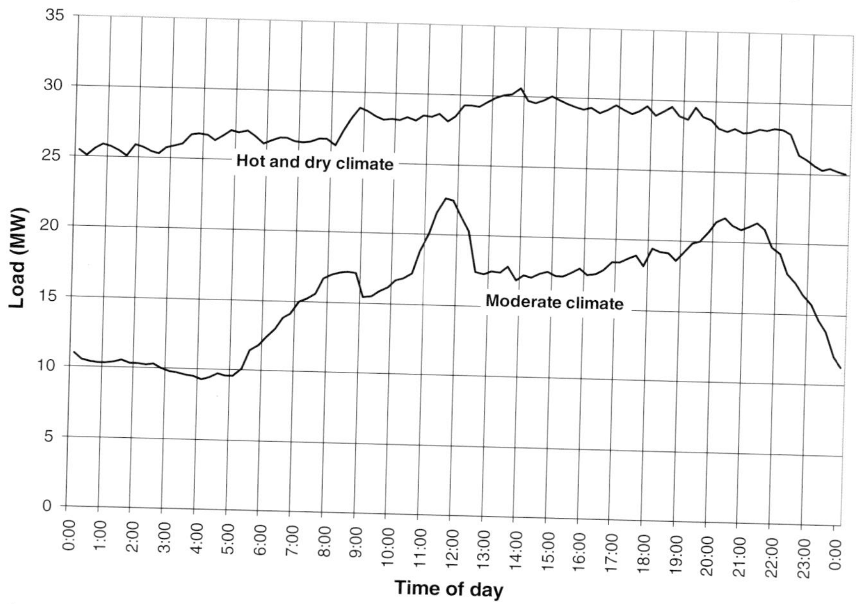

Figure 4.2 Daily load curve of a 31.5 MVA transformer 110 kV/10 kV on the day of yearly peak load.

The period of use T_m, also termed number of full-load hours, represents that time interval in which the equipment would produce the same energy running with rated power (full load) as generated actually during the specified period T_Z, see also Figure 4.3.

The energy loss factor ϑ and the period of energy losses T_V are defined similarly to the definitions of load factor m and the period of use T_m. The energy loss factor ϑ as given by Equation 4.11a is defined as the ratio of the energy losses P_V to the energy losses at peak load (maximal current) P_{Vmax} of the equipment.

$$\vartheta = \frac{P_V}{P_{Vmax}} = \frac{R \cdot \int_0^{T_Z} I(t)^2 \cdot dt}{R \cdot I(t)_{max}^2 \cdot T_Z} \tag{4.11a}$$

If the current is set proportional to the apparent power S, Equation 4.11b is obtained.

$$\vartheta = \frac{\int_0^{T_Z} S(t)^2 \cdot dt}{S(t)_{max}^2 \cdot T_Z} \tag{4.11b}$$

The energy loss factor ϑ is difficult to determine due to lack of data in the planning phase. The energy loss factor can be estimated from the load factor m which

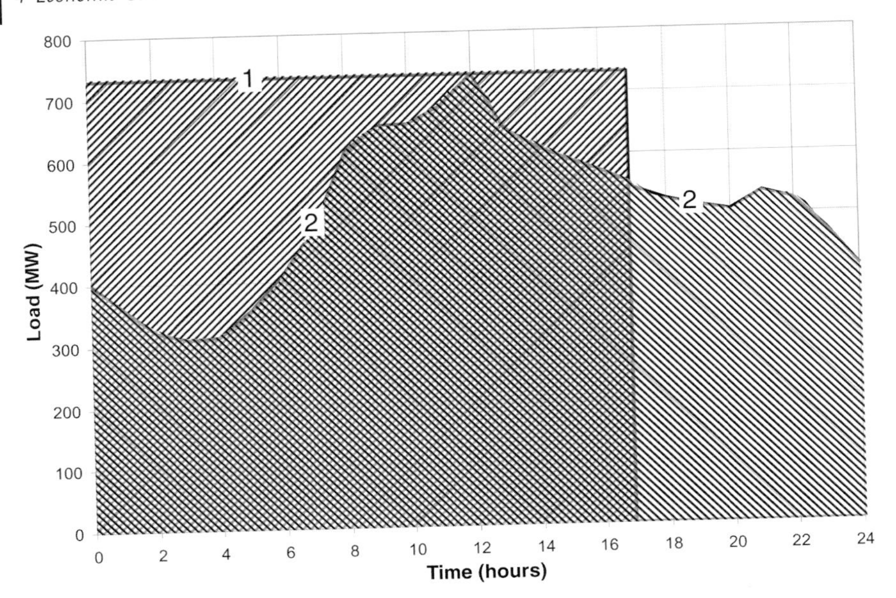

Figure 4.3 Explanation of period of use T_m and load factor m: **2**, load curve; **1**, equivalent area of generated energy.

can be determined comparatively easily using Equation 4.12 for load factors $m >$ 0.25 which is a reasonable assumption for power systems.

$$\vartheta \approx 0.17 \cdot m + 0.38 \cdot m^2 \tag{4.12}$$

The period of energy hours, sometimes called loss hours T_V is defined according to Equation 4.13 as the ratio of energy losses W_V during the period concerned to the maximal loss P_{Vmax}.

$$T_V = \frac{W_V}{P_{Vmax}} = \frac{\int_0^{T_Z} I(t)^2 \cdot dt}{I(t)^2_{max}} \tag{4.13}$$

Increase in the equipment load leads to an increase of the power losses and of the energy losses, which are called "added losses." Apart from the questions of loss evaluation described here in the context of power system planning, knowledge of the "added losses" is necessary for clarifying the costs of energy transfer through other power systems. The "added losses," consisting of additional power and energy losses, are defined as in Equation 4.14a for constant load increase during the total planning period.

$$\Delta P_V = P_V \cdot \left[2 \cdot \frac{\Delta P}{P} + \left(\frac{\Delta P}{P} \right)^2 \right] \tag{4.14a}$$

With small load changes ΔP in Equation 4.14b and the additional energy losses in Equation 4.14c the added losses can be evaluated with knowledge of the additional cost of the equipment taking account of the load-dependent losses and the maximal load portion h.

$$\Delta P_V \approx P_V \cdot \left(2 \cdot \frac{\Delta P}{P}\right) \tag{4.14b}$$

$$\Delta W_V = W_V \cdot \left(2 \cdot \frac{\Delta P}{P}\right) \tag{4.14c}$$

It is difficult to determine the additional cost of equipment because the investment (for example, construction of a new line or substitution of a transformer to increase the transfer capability) must be determined and justified by detailed and careful planning. The additional costs of equipment have to be assigned to the various system voltage levels (transmission, distribution level). Further, the losses resulting from transmission of energy must be included in an increase of transmission capacity and thus in an increase of investment cost for the respective voltage level.

For the evaluation of the energy costs, the production costs of electricity must be determined. With external supply, the contractual energy prices can be taken for this. If energy is produced in the user's own power station the production costs will be calculated from the energy cost and the cost per installed power. Energy costs of a power station comprise first of all the fuel costs. The costs for transmission losses can be taken into account by a fixed loss factor or the transmission efficiency η_N. The energy costs are calculated from Equation 4.15

$$k_W = \frac{A}{H \cdot \eta_{PS} \cdot \eta_N} \tag{4.15}$$

with

A = unit price of fuel
H = caloric value of the fuel
η_{PS} = efficiency of the power station
η_N = transmission efficiency of the power system.

The total generation costs K in the period considered T_z, referred to the energy unit, are given by Equation 4.16.

$$K = k_p \cdot P_{max} \cdot T_Z + k_W \cdot W \tag{4.16}$$

If one refers the total costs K to the energy produced W, one obtains the production costs k_E as in Equation 4.17.

$$k_E = \frac{k_P}{m} + k_W \tag{4.17}$$

For the evaluation of the losses in the context of power system planning over the entire long-range planning period, the losses are usually determined for each year and will be represented as present value, discounted on the time of the investment decision or commissioning. The losses in one year i are calculated using Equation 4.18.

$$K_{Vi} = W_{Vi} \cdot \left(\frac{k_P}{\vartheta_i \cdot T_Z} + k_{wi} \right) \tag{4.18}$$

The present value of the costs of losses are calculated from Equation 4.19

$$K_{V0} = \sum_{i=1}^{n} \frac{K_{Vi}}{q^i} \tag{4.19}$$

with the interest factor $q = 1 + p$. If future load increase with an annual growth rate g is taken into account, the annual losses and the annual cost of losses are increased in accordance with Equation 4.20a,

$$K_{V0} = \sum_{i=1}^{n} \frac{K_{Vi}}{q^i} \cdot \frac{1}{(1+g)^{2(i-1)}} \tag{4.20a}$$

and for continuous annual costs of losses K_V according to Equation 4.20b with the increase factor $r = 1 + g$,

$$K_{V0} = K_V \cdot \frac{q^n - r^{2n}}{q^n \cdot (q - r^2)} \tag{4.20b}$$

If the load of the equipment under investigation or the power system does not increase continuously from the initial loading S to the final loading $k \times S$ over the long-range planning period, but rather experiences several load cycles (number j) of S on $k \times S$ during n years, then the present value of the cost of losses is given by Equation 4.21.

$$K_{V0} = K_V \cdot \frac{q^{jn} - 1}{q^{jn} \cdot (q-1)} \cdot \frac{(q^n - r^{2n}) \cdot (q-1)}{(q - r^2) \cdot (q^n - 1)} \tag{4.21}$$

with

K_V = cost of losses
j = number of load cycles
n = number of years
q = interest factor: $q = 1 + p$
r = load increase factor $r = 1 + g$
p = interest rate
g = load increase rate.

The second term in Equation 4.21 indicates the annuity factor for discounting of the constant cost of losses K_V and the last term takes account of the increase of load in the load cycles described.

5
Topologies of Electrical Power Systems

5.1
Development of Power Systems

The first three-phase alternating current transmission system was built in the year 1891 in Germany on the occasion of the International Electrotechnical Fair. The electrical energy of a hydro power plant with rated power of 140 kW, located in Lauffen on the river Neckar, was transmitted over 175 km by a single-circuit three-phase overhead transmission line, voltage around 14 kV, 40 cycles per second, to the Electrotechnical Fair exhibition in Frankfurt/Main.

Starting from small isolated systems in urban areas, the three-phase AC system became generally accepted throughout the world. Different voltage levels are coupled by transformers from the low-voltage system to an extra-high-voltage level up to 1500 kV. Electrical power systems spread across continents and through DC links also connect different continents or power systems with different frequencies.

The power system of the Federal Republic of Germany, as represented in Figure 5.1, included the following at the end of the year 2011 [10]:

- 1 714 000 km electric circuit length, comprising
  - 1 123 000 km with $U_n = 0.4\,kV$
  - 479 000 km with $U_n = 6\,kV–60\,kV$
  - 77 000 km with $U_n > 60\,kV–110\,kV$
  - 35 000 km with $U_n = 220\,kV–380\,kV$
- 566 300 transformers with total apparent power 839 200 MVA (data as per 2006).

The power system in Germany can be characterized into three different sub-systems according to its tasks.

- The high-voltage transmission grid, the UCTE grid, in the whole of Europe connects the large power stations and the main consumer centers. Its major task is the transportation of electrical energy over long distances under normal and emergency conditions, for example, in case of major outages of power stations. The nominal system voltage is 400 kV (also referred to as 380 kV); some 220 kV lines still exist.
- The sub-transmission systems or regional distribution systems are formed in Germany by the 110 kV network (supplied from the 380 kV system and partly

Power System Engineering: Planning, Design and Operation of Power Systems and Equipment, Second Edition.
Jürgen Schlabbach and Karl-Heinz Rofalski.

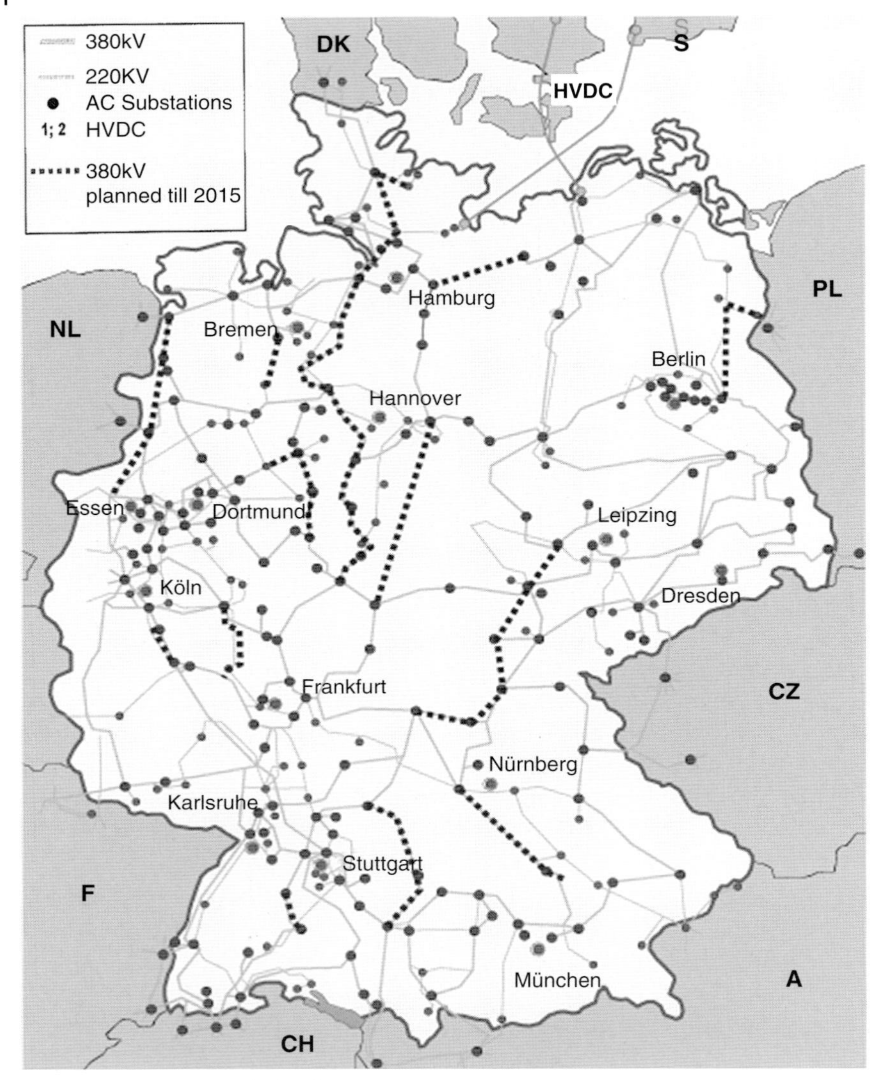

Figure 5.1 German extra-high-voltage power system, planning status 2015. Source: Reproduced by permission of VDN, Berlin.

also from the 220 kV system), which serves as a network of distribution in urban population centers and additionally as the regional transportation system in rural areas; large industrial consumers are supplied from the 110 kV system as well.

- Distribution systems with nominal voltages of 10 kV up to 35 kV, in some countries up to 60 kV, serve the regional and local distribution of electrical energy at different voltage levels depending upon load density and type of consumer. Special customers such as industrial installations are supplied at higher voltage levels, for example, 10 kV, than standard tariff customers who are supplied uniformly with the rated voltage of 400 V.

Due to the increasing utilization of "green-energy," power stations are connected to different voltage levels from 0.4 kV (small photovoltaic installations) up to high voltages (wind power farms), the tasks of the different voltage levels cannot be defined so clearly in the future.

5.2 Recommended Voltage Levels

Nominal voltages in power systems are recommended in IEC 60038. Table 5.1 outlines the appropriate voltage levels as applicable in Germany. In addition to the nominal voltage, typical application and supply tasks are mentioned as well.

Table 5.1 Recommended system voltages according to IEC 60038 as applicable in German power systems.

Common name	Nominal system voltage	Supply task	Remarks
Low-voltage (LV)	400 V/230 V	Household customers Small industrial consumers	IEC 60038 Table I
	500 V	Supply of motors in industry	Not mentioned in IEC 60038
Medium-voltage (MV)	6 kV	HV-motors in industry and power stations	IEC 60038 Table III
	10 kV	Urban supply, industrial power systems	
	20 kV	Rural supply, industrial power systems	
	30 kV	Industrial supply (electrolysis, thermal processes) Rural power supply	Not mentioned in IEC 60038
High-voltage (HV)	110 kV	Urban transport and sub-transmission systems	IEC 60038 Table IV
	220 kV	Transmission systems (decreasing importance)	
	380 kV (400 kV)	UCTE transmission system	IEC 60038 Table V Definition of highest voltage of equipment $U_{bmax} = 420$ kV

It should be mentioned in this respect that the existence of the different voltage levels is due partially to historical development. Economic studies indicate that a suitable grading of the individual voltage levels in MV and HV systems should be in the order of 1 : 3 to 1 : 7, that is, 10 kV or 20 kV in MV systems, 110 kV and 400 kV in HV systems and 1200 kV in EHV systems, as far as applicable.

5.3
Topology of Power Systems

Power systems are constructed and operated as

- Radial systems or
- Ring-main systems or
- Meshed systems.

Additional criteria for distinction can be defined, such as the number and kind of feeders from supplying system level, the number and arrangement of lines and the reserve capability of the system to cover loss of load. The three system topologies are constructed and operated at all voltage levels. In the context of the sections below, the following definitions are used:

Feeder	Outgoing connection of any overhead line or cable from a MV or LV substation
Gridstation	Switchyard including busbars, transformers and outgoing feeders to the EHV level
Substation	Switchyard including busbars, transformers and outgoing feeders to the HV level
Station	Switchyard including busbars, transformers and outgoing feeders to the MV level
Primary	Switchyard including busbars, transformers and outgoing feeders to the LV level
Line	Any overhead line or cable of any voltage level.

5.3.1
Radial Systems

The simplest system configuration, the radial system, can be found particularly at the low-voltage and the medium-voltage levels. The individual feeders or lines, connected to the primary or station, connect the primaries by radial feeders, as represented in Figure 5.2a. Branching of the lines is possible and in fact usual (Figure 5.2b). This network configuration is suitable for areas with low load density and is also used for the connection of bulk loads. In this case the system is called a connection system. The advantages of the simple topology and low capital investment cost have to be compared with the disadvantage that, in case of failure of lines, the load of the faulted lines cannot be supplied. The branch-off points in the low-voltage system are sometimes implemented in the form of branch-off

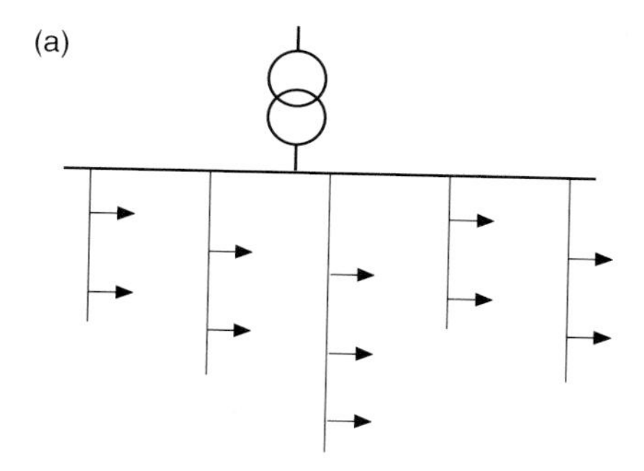

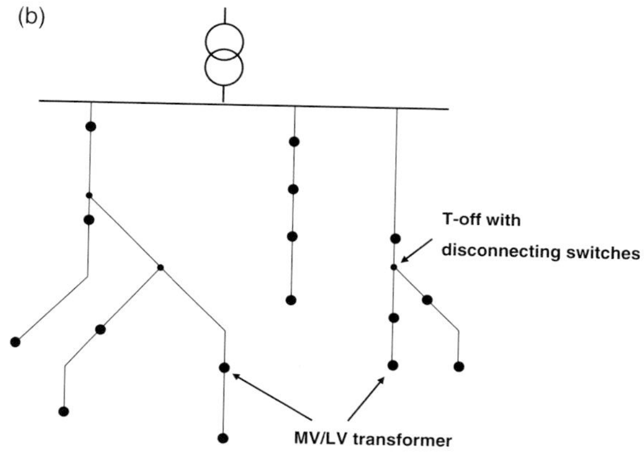

Figure 5.2 Topology of radial power systems. (a) Simple structure radial network (LV system); (b) radial network with branch lines (MV system).

joints or T-connections without switching capability. Radial systems in the medium-voltage level are usually built only in areas with low load density. The branch-off points can then be implemented both with and without disconnecting switches or load-break switches. The possibility of switch-off for each line exists in the feeding station or primary in the form of disconnecting switches, rarely as circuit-breakers. If only disconnecting switches are installed, the switch on the secondary side of the feeding transformer has to be implemented as circuit-breaker.

The loading of the lines can be selected such that the thermally permissible load at the sending end of the lines amounts to 100%.

The system is characterized by:

- Clear and simple structure
- Low planning expenditure
- Simple operation under normal operating conditions
- Loading of lines during normal operation up to 100%

- No reserve for loss of lines
- Low investment cost
- Maintenance cost rather small
- System losses comparatively high; losses cannot be minimized
- Voltage profile not very good; distinct voltage drop between the feeding and the receiving end of the lines
- Flexibility for changed load conditions is comparatively small
- Reserve for losses of the feeding MV transformer usually missing
- Standardization of cross-sections of lines possible, but not advisable
- Protection usually only with overcurrent relays at the feeder, in MV systems, sometimes also with circuit-breakers, in LV systems with fuses
- Typical application in MV systems up to 60 kV with small load densities up to approximately $1\,\mathrm{MW\,km^{-2}}$, typically in low-voltage systems.

The reliability and/or the reserve capability of radial systems can only be improved in principle if another concept is used: the pure radial system is converted into a ring-main system or even a meshed system.

5.3.2 Ring-Main Systems

Ring-main systems are common in the medium-voltage range. A large variety of ring-main systems are in operation with respect to permissible loading of the lines, reserve capability against outages, different arrangement of the feeding station and supply reliability.

5.3.2.1 Ring-Main System – Simple Topology

The simplest kind of ring-main system is obtained by connecting the line ends (starting from the radial system topology) back to the feeding station as outlined in Figure 5.3. Usually ring-main systems are operated with open disconnection points (load disconnecting switch) at defined locations on each line, which provides for simple operation including a switchable reserve capability, depending on the loading of the lines.

The loading of the lines must be selected in such a way that in case of failure of a line the total load of this line concerned can be supplied after closing the load-switch at the open disconnection point. This means that the loading of each feeder must be maintained at 50% of the thermally permissible loading as an average for normal operating conditions. Each feeder offers reserve in case of faults of the respective feeder itself.

The system is characterized by:

- Clear and simple structure
- Moderate planning expenditure
- Simple operation (similar to radial system) under normal operating conditions
- Loading of lines during normal operation 50% of the permissible loading, higher loading possible depending on load duration

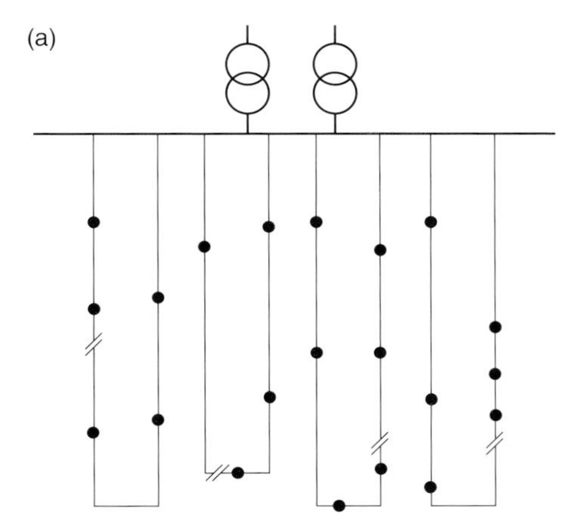

// Disconnection point (n.o.)
• MV/LV transformer connected through fuse or disconnecting switch

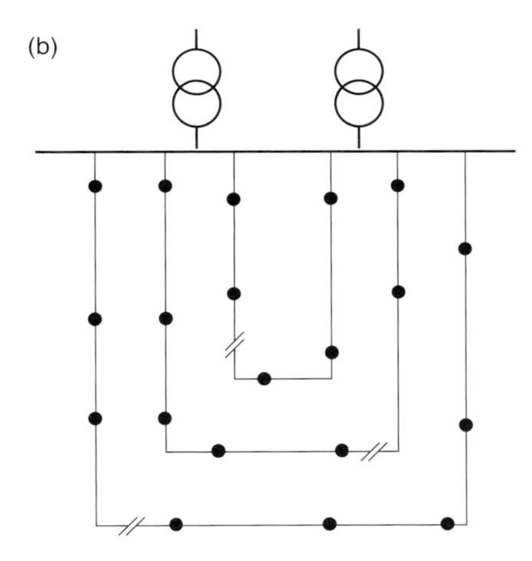

// Disconnection point (n.o.)
• MV/LV transformer connected through fuse or disconnecting switch

Figure 5.3 Ring-main system simple topology.
(a) Arrangement with limited reserve in feeding station;
(b) arrangement with reserve to cover outages in the feeding station.

- Reserve for outage of each line section given by the line itself
- Investment cost not very high; reduction possible, if circuit-breakers are omitted; in this case one circuit-breaker has to be installed on the secondary side of the feeding transformer
- Maintenance cost rather low
- System losses can be minimized by changing the location of the open disconnection point

- Voltage profile can be optimized, differences between feeding and receiving end of the lines depend on the location of the open disconnection point
- Flexibility to respond to changing load conditions
- Reserve for outage of feeding transformer or bus section usually available if an arrangement is selected as in Figure 5.3b
- Standardization of cross-sections of the lines is given (usually only one cross-section shall be used)
- Feeder protection can be realized with overcurrent protection
- Application in medium-voltage systems up to 35 kV, in case of high load density, in principle also in low-voltage systems.

5.3.2.2 Ring-Main System with Remote Station (Without Supply)

Connecting the individual lines of the system at the receiving end to a station without infeed from a higher voltage supply level forms a ring-main system as outlined in Figure 5.4. Normal operation condition is with open disconnection point, thus forming a radial system. Load-switches are also installed to select suitable disconnection point locations similarly to the ring-main system shown in Figure 5.3. The remote station should always be kept energized, so as to guarantee a quick switchover of line sections in case of outages.

The permissible loading of individual lines can be above 50% of the thermally permissible load of a line, depending on the loading of all lines. In case of outage of any feeder, the load will be supplied by separation of the faulted line section through the remote station by one or more lines connected to it. This type of ring-main system with remote station can be seen as an intermediate step of system development between a pure radial system and a ring-main system with a feeding remote station (see below).

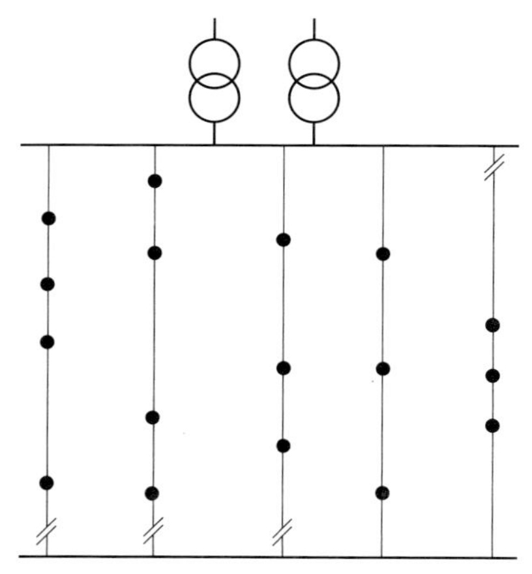

// Disconnection point (n.o.)

• MV/LV transformer connected through fuse or disconnecting switch

Figure 5.4 Ring-main system with remote station (without supply).

The system is characterized by:

- Clear and simple structure
- Moderate expenditure for planning
- Simple operation under normal operating conditions
- Loading of lines under normal operating conditions more than 50% of permissible thermal loading
- Reserve for outages available depending on the preloading of the remaining lines
- Investment cost in the medium range; reduction possible if a topology without circuit-breakers for the feeders is selected. In this case one circuit-breaker has to be installed on the secondary side of the feeding transformer
- Maintenance cost rather low
- Power system losses cannot be minimized
- Voltage profile in the system not optimal, differences between feeding station and remote station are significant
- Flexibility to respond to changed load conditions
- Reserve for outage of the feeding MV transformer usually available
- Standardization of cross-sections of all lines given (usually one cross-section used)
- Feeder protection can be realized with overcurrent protection
- Application in medium-voltage systems up to 35 kV, also in case of medium load density, in principle also in low-voltage systems.

5.3.2.3 **Ring-Main System with Reserve Line**

If a separate line between the feeding station and the remote station is constructed, without supplying load through this line under normal operating conditions, the system performance is significantly improved by this reserve line. The general topology is indicated in Figure 5.5. If the cross-section of the reserve line is chosen identical to the other lines, the loading of the lines can be increased to as much as 100%. The outage of one line is covered under the system topology. The second line outage is no longer covered, however; to achieve this, the reserve line must have a larger cross-section.

The system is characterized by:

- Clear and simple structure
- Moderate planning expenditure
- Simple operation under normal operating conditions
- Loading of lines under normal operating conditions up to 100%, resulting in outage reserve for one line
- Reserve for outages of more than one line depending on preloading conditions of the feeders or on the cross-section of the reserve line
- Investment cost within the medium range but extra cost for reserve line; reduction possible if circuit-breakers for the feeders are omitted. In this case one circuit-breaker has to be installed on the secondary side of the feeding transformer

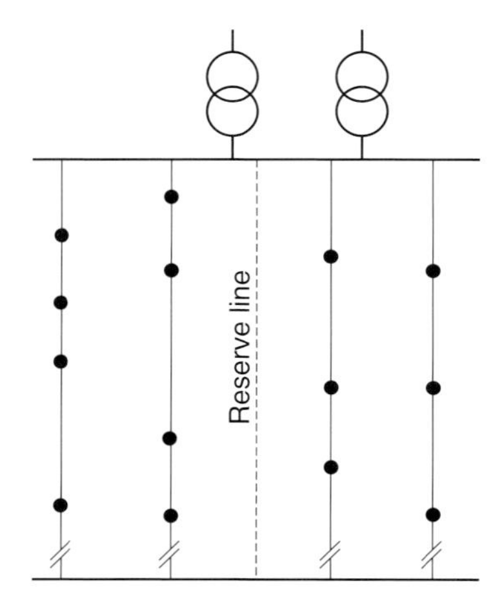

// Disconnection point (n.o.)

• MV/LV transformer connected through fuse or disconnecting switch

Figure 5.5 Ring-main system with reserve line.

- Maintenance cost rather small
- System losses cannot be minimized
- Voltage profile in the system not optimal; differences between feeding station and remote station are significant
- Flexibility to respond to changed load conditions
- Reserve for outage of the feeding MV transformer usually available
- Standardization of cross-sections of all lines given (usually one or two cross-sections sufficient)
- Feeder protection can be realized with overcurrent protection
- Application in medium-voltage systems up to 35 kV, also in case of medium load density, in principle also in low-voltage systems.

Ring-main systems with reserve line can also be designed and operated so that the reserve line acts as reserve for one or two lines. However, this seems acceptable only as a temporary solution or for the supply of important consumers who have their own redundant supply. Examples are represented in Figure 5.6.

The system is characterized by:

- Not very clear structure
- Moderate planning expenditure
- Simple operation under normal operating conditions
- Loading of lines for normal operating conditions up to 100% for those lines, which can be supplied by means of reserve line
- Reserve for outages given for single outage, for double outage only for special lines

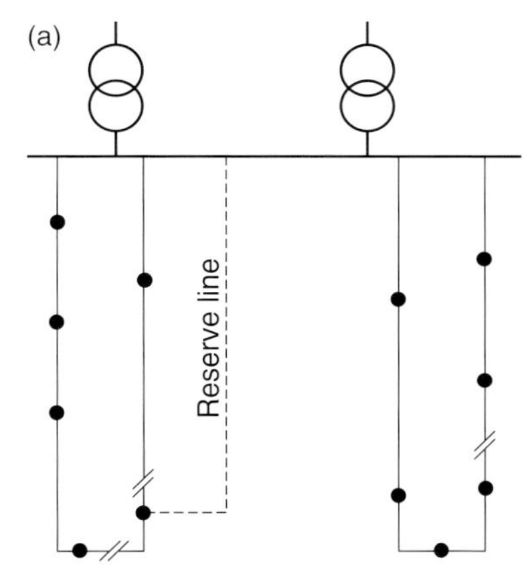

// Disconnection point (n.o.)

• MV/LV transformer connected through fuse or disconnecting switch

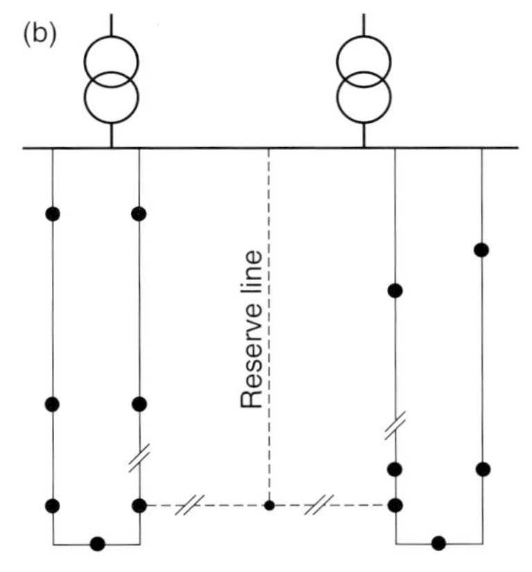

// Disconnection point (n.o.)

• MV/LV transformer connected through fuse or disconnecting switch

Figure 5.6 Ring-main system with reserve lines. (a) Reserve for outage of one line/two feeders; (b) reserve for outage of several lines/feeders.

- Investment cost within the medium range but extra cost for reserve lines. Reduction possible if circuit-breakers in the lines are omitted
- Maintenance costs rather small
- System losses can be reduced, but this must be accomplished by an increased number of switching actions in case of outages
- Voltage profile sufficient; differences between feeding and receiving end of the lines depend on the location of the open disconnection point
- Flexibility to respond to changed load conditions

- Reserve for loss of the feeding MV transformer usually available
- Standardization of cross-sections given (usually one or two cross-sections are sufficient)
- Feeder protection can be realized with overcurrent protection
- Application in medium-voltage systems up to 35 kV, also in case of medium load density, in principle also in low-voltage systems.

5.3.2.4 Ring-Main System with Feeding Remote Station

An improvement of the ring-main system with remote station shown in Figure 5.4 can be achieved if an additional infeed from the higher voltage level can be installed in the remote station. Figure 5.7 indicates the general structure.

The system covers outages of any feeder, up to a load of 100% of the maximal permissible thermal loading. The system is characterized by:

- Clear and simple structure
- Moderate planning expenditure
- Simple operation under normal operating conditions
- Loading of lines under normal operating conditions up to 100%
- Reserve for outages of more than one line depending on preloading conditions
- Investment cost within the high range
- Maintenance costs rather small
- System losses can be minimized
- Voltage profile in the system optimal; small differences between feeding station and remote station
- High flexibility for changed load conditions

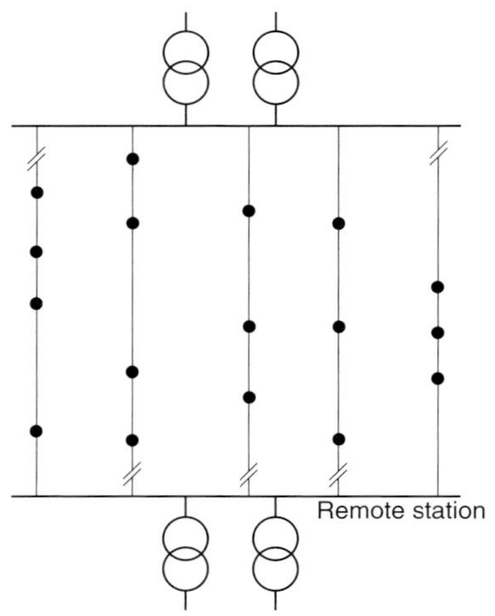

Figure 5.7 Ring-main system with feeding remote station.

- Reserve for outage of the feeding MV transformer and station is available
- Standardization of cross-sections of all lines given (usually one cross-section is sufficient)
- Feeder protection can be realized with overcurrent protection
- Application in medium-voltage systems up to 35 kV, also in case of high load density, in principle also in low-voltage systems.

5.3.2.5 Ring-Main System as Tuple System

Modifications and combinations of the ring-main systems described above are quite common as temporary solutions, but are sometimes operated for longer periods. Ring-main systems with tuples are to be mentioned, which can be described as a ring-main system with additional switching stations *en route* of the lines. An example is outlined in Figure 5.8, the so-called triple-system.

The reserve in case of outages depends on the preloading conditions. Full outage reserve is achieved if the loading of the lines is set on average to 67% of the permissible thermal loading.

The system is characterized by:

- Somewhat unclear system structure
- Moderate planning expenditure
- Simple operation under normal operating conditions
- Loading of lines under normal operating conditions up to 67%
- Reserve for single outage of lines; second outage cannot be covered under all conditions
- Investment cost in the medium range; reduction possible if circuit-breakers of the feeders are omitted
- Maintenance cost in the medium range
- System losses can be minimized

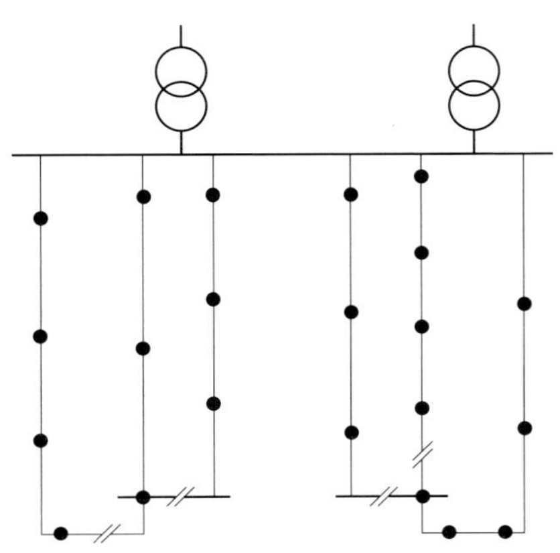

// Disconnection point (n.o.)

• MV/LV transformer connected through fuse or disconnecting switch

Figure 5.8 Ring-main system as triple-system.

- Voltage profile sufficient; differences between feeding and receiving end of the lines depend on the location of the open disconnection point
- Flexibility for changed load conditions
- Reserve for outage of the feeding MV transformer usually available
- Standardization of cross-sections of all lines given (usually one or two cross-sections)
- Feeder protection can be realized with overcurrent protection
- Application in medium-voltage systems up to 35 kV with medium load density.

5.3.2.6 Ring-Main System with Cross-Link

Ring-main systems with cross-link as outlined in Figure 5.9 are mainly built as an intermediate stage for system reinforcement and are still in operation as remaining parts of temporary solutions. They likewise offer reserve possibilities, which can be used as required.

The reserve for outages depends on the preloading of the loaded lines, the location of the fault causing the outage and the location of the cross-link in the system. With skillful choice of the connection points of the cross-link a preloading of the lines of up to 100% can be covered, depending on the location of the fault. On average 60–70% loading in relation to the permissible thermal loading can be accepted. The system is characterized by:

- Somewhat unclear system structure
- Comparatively high planning expenditure
- Simple operation under normal operating conditions
- Loading of lines under normal operating conditions up to 60–70%, sometimes up to 100% for those parts of the lines connected through cross-link

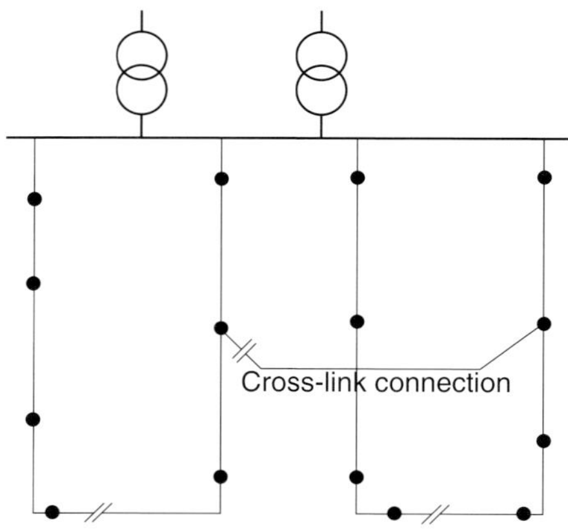

// Disconnection point (n.o.)

• MV/LV transformer connected through fuse or disconnecting switch

Figure 5.9 Ring-main system with cross-link.

- Reserve for single outage of lines given, depending on the preload; second outage cannot be covered under all conditions
- Investment cost within the medium range, but extra cost for the cross-link; reduction possible if circuit-breakers in the feeders are omitted
- Maintenance cost in the medium range
- System losses can be minimized
- Voltage profile sufficient; differences between feeding and receiving end of the lines depend on the location of the open disconnection point
- Flexibility for changed load conditions not very high
- Reserve for outage of the feeding MV transformer usually available
- Standardization of cross-sections of all lines given (usually one or two cross-sections)
- Feeder protection can be realized with overcurrent protection
- Application in medium-voltage systems up to 35 kV with medium load density.

5.3.2.7 Ring-Main System with Base Station

The ring-main system with base station, as can be found in urban power systems, is a simple ring-main system with connection of the base station by means of a double circuit line to the feeding station. If a reserve line is connected between different base stations having a larger cross-section than the lines between the base station and the feeding station, a reserve capability is obtained for outages of feeding cables and the feeding substation to some extent (Figure 5.10).

The system is characterized by:

- Somewhat unclear system structure
- Comparatively high planning expenditure

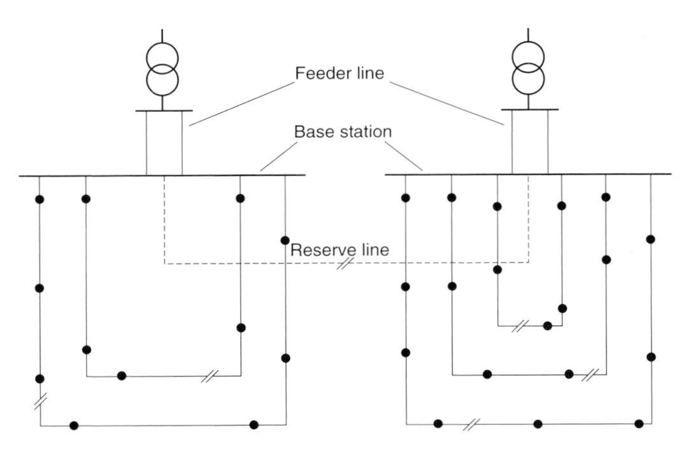

Figure 5.10 Ring-main system with base station.

- Simple operation under normal operating conditions
- Loading of lines under normal operating conditions up to 67%, sometimes up to 100%
- Reserve for single outage of lines given by each line itself; reserve for the feeding line given by parallel line
- Investment cost in the high range due to the additional base station
- Maintenance cost high
- System losses can be minimized
- Voltage profile sufficient; differences between feeding and receiving end of the lines depend on the location of the open disconnection point
- Flexibility for changed load conditions
- Reserve for outage of the feeding MV transformer is available
- Standardization of cross-sections of lines (usually two cross-sections are sufficient)
- Feeder protection can be realized with overcurrent protection
- Application in medium-voltage systems up to 35 kV with high load density.

5.3.2.8 **Special-Spare Cable System**

The name "special-spare cable system" is widely known, though the system can be realized with cables or overhead lines. The special-spare cable system is found in urban areas of densely populated areas in Asia but also as overhead line systems in North and South America. In European countries this system configuration is rather uncommon. The system can be described as a combination of ring-main system and radial system. Figure 5.11 indicates the fundamental concept.

The load is supplied through radial lines, called tap-lines, connected to main-lines having larger cross section. Connection is realized by means of T-connectors

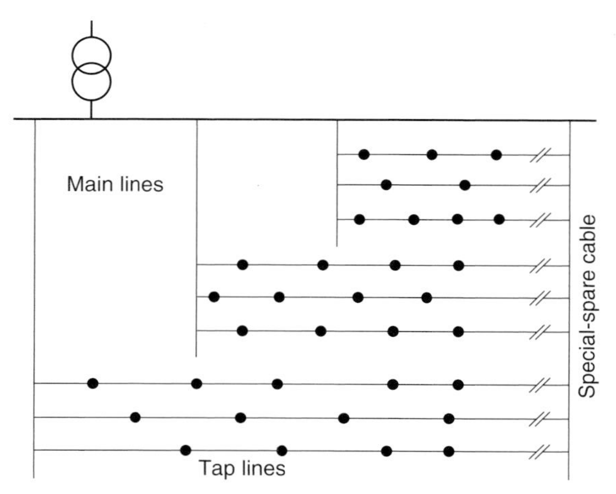

// Disconnection point (n.o.)

• MV/LV transformer connected through fuse or disconnecting switch

Figure 5.11 Special-spare cable system.

in overhead line systems and special T-joints in cable systems. A spare cable, called the special-spare cable, with the same cross-section as the main cable is connected to the other end of the tap lines by T-connectors or T-joints. The special-spare cable does not supply load under normal operating conditions but is kept energized. Reserve for outages by the special-spare cable is only given for outages of the tap lines and only to a minor extent for outage of a main line. Generally one can assume the loading of the tap lines as up to 100% of the thermal permissible loading.

The system is characterized by:

- Simple and clear structure
- Moderate planning expenditure
- Easy operation under normal operating conditions
- Loading of the lines under normal operating conditions 100% for tap lines and up to 70% for the main lines, dependent on the preload conditions and the cross-section of main lines and special-spare cable
- Reserve for line outage depends on cable layout (cross-section) and loading conditions
- Investment cost comparatively low if disconnecting load-switches are omitted
- Maintenance cost rather small
- System losses comparatively high and cannot be minimized
- Voltage profile only moderate, partially large differences between feeding and receiving end of the tap-lines
- Flexibility for changed load conditions is small
- Reserve for losses of the feeding MV transformer usually available
- Standardization of cross-sections of line possible (at least two or three different cross-sections are needed)
- Protection usually only with overcurrent protection for the main lines
- Advisable application in medium-voltage systems up to 35 kV with small load densities; however, the system is widely used in urban power systems.

5.3.2.9 **Double-T Connection**

The double-T connection concept in MV systems is found especially in centers of large cities with high load density. The primaries, feeding the LV systems, are connected to two independent lines which are connected to different substations. High reliability is achieved by this scheme compared with the ring-main systems described so far (Figures 5.4 and 5.5). The system can cover any outage of a MV line. A line fault leads to the outage of the complete feeder. If the primaries are connected to the line by three load-switches, the outage of one line results only in the outage of a section of the line. The system can be operated either in open-circuit mode or fed from both primaries, in this case it is more a meshed system than a ring-main system. Figure 5.12 indicates the system structure.

The system is characterized by:

- Simple and clear system structure
- Moderate planning expenditure

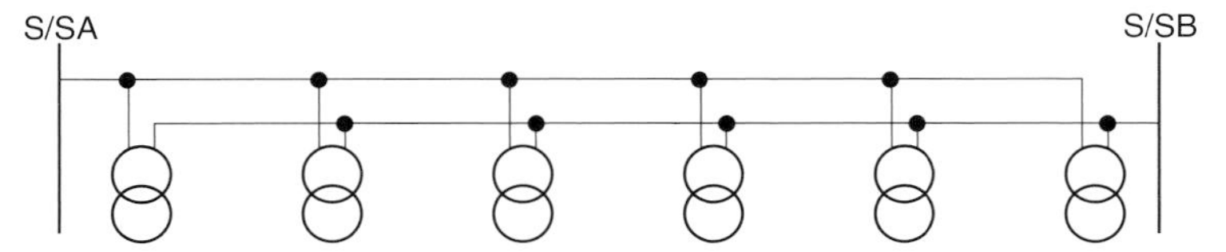

Figure 5.12 System with double-T connection.

- Simple operation for normal operating and emergency conditions
- Loading of lines under normal operating conditions up to 60% of thermal permissible loading
- Reserve for outage of MV line
- Investment cost moderate
- Maintenance cost comparatively low
- System losses can be minimized by suitable selection of open disconnection point
- Voltage profile flat
- Flexibility to respond to change in load conditions
- Reserve for loss of substation available
- Standardization of cross-sections of all lines
- Protection of circuits with overcurrent relays
- Application in medium-voltage systems with cables up to 35 kV with medium to high load densities in urban areas.

5.3.3
Meshed Systems at HV and MV Levels

5.3.3.1 HV Transmission Systems

Meshed systems used for the medium- and high-voltage level are planned and operated in such a way that outage of any equipment, such as overhead lines, cables, transformers, compensation or busbar sections, will not cause a loss of supply to any load. Increase of loading of the remaining equipment is obvious and can be accepted for a specified period up to the emergency rating of the equipment. The voltage profile in the system will worsen during the outage. Rearrangement of the system may be necessary to reduce the loading of equipment and improve the voltage profile. Depending on the planning criteria for the system, single outage $(n-1)$-criteria or multiple outages$(n-k)$-criteria are allowed without loss of supply. Figure 5.13 indicates the structure of a meshed power system, which is typically planned to fulfill the $(n-1)$-criterion for voltage level above 110 kV.

The system is characterized by:

- Complicated system structure
- High expenditure for system planning
- Operation for normal operating conditions normally with all breakers closed
- Loading of lines under normal operating conditions according to the planning criteria

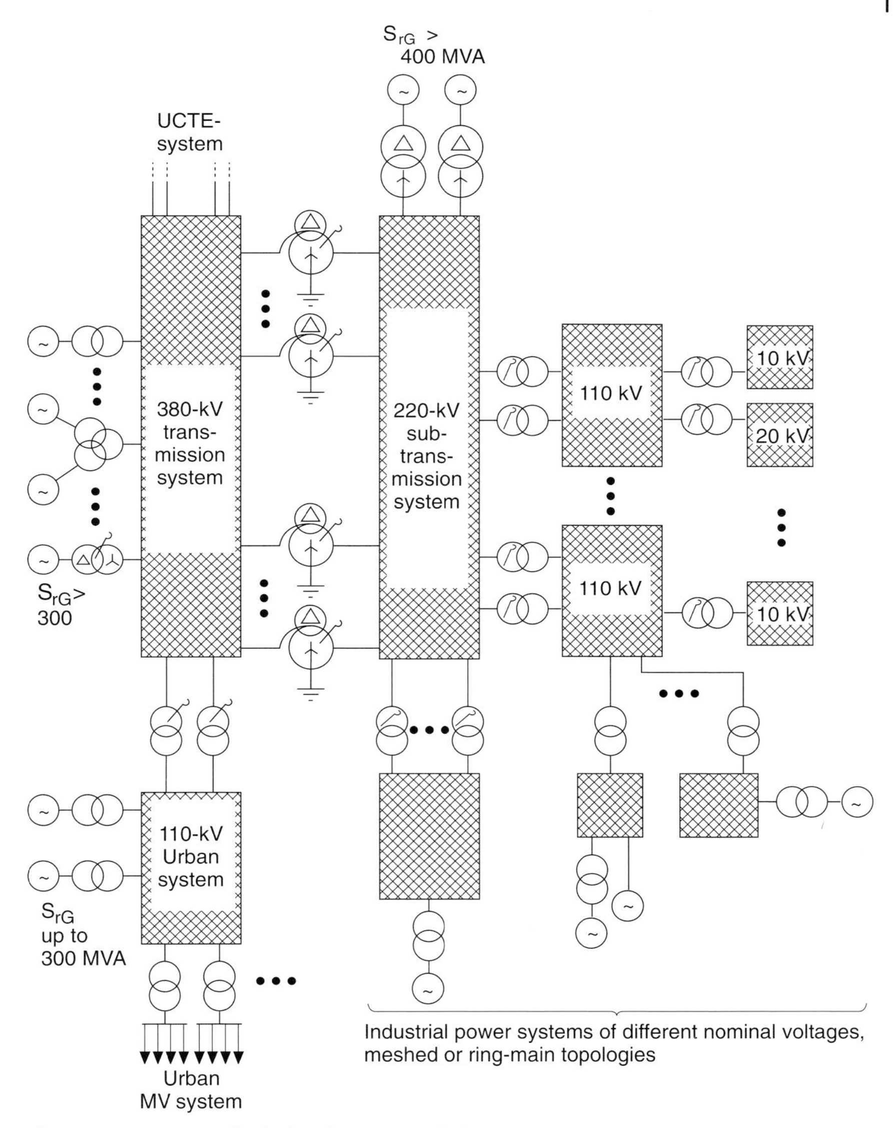

Figure 5.13 Diagram of a high-voltage transmission system with different voltage levels.

- Reserve for outages according to planning criteria
- High investment cost
- High cost of maintenance
- System losses minimal; in some cases the system topology needs to be changed
- Voltage profile for normal operating and for outage conditions according to planning criteria

- High flexibility for changed load conditions
- No interruption of supply according to planning criteria
- Standardization of cross-section and rating of equipment is possible
- Protection with distance protection relays or with differential protection
- Applicable for HV and EHV systems for high load density.

5.3.3.2 Meshed MV Systems

A typical meshed structure for MV systems is the ring-main system with all switches in closed position. Ring-main systems as described in Section 5.3.2 can also be operated as meshed systems if the protection is designed accordingly. As an example, a ring-main system with feeding remote station is indicated in Figure 5.14. All breakers and load-switches are closed; the loading of the lines depends on the load, the line impedance, and the r.m.s. value and the phase-angle of the voltages in the feeding substations. An uninterruptible supply for the LV load is guaranteed if the LV transformers are connected with circuit-breakers. because of this, the scheme is only justified in public MV systems supplying important loads. It can be found widely in industrial power systems.

The system is characterized by:

- Clear and simple structure
- Moderate to high planning expenditure
- Simple operation under normal operating conditions
- Loading of lines under normal operating conditions up to 70%
- Reserve for outages of more than one line
- Investment cost within the high range, especially if circuit-breakers are installed

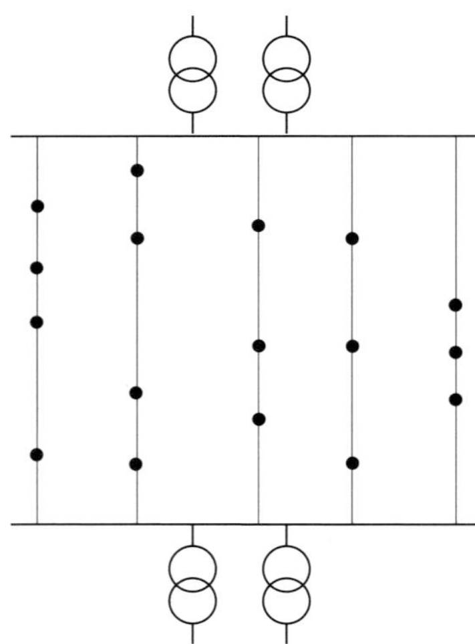

Figure 5.14 Meshed MV system based on a ring-main structure.

- Maintenance costs high
- System losses minimal
- Voltage profile in the system optimal; small differences between feeding station and remote station
- High flexibility for changed load conditions
- Reserve for outage of the feeding MV transformer and station available
- Standardization of cross-sections of all lines (usually one cross-section is sufficient)
- Feeder protection usually with distance or differential protection relays
- Application in medium-voltage systems up to 35 kV, also in case of high load density, for the supply of important consumers and in industrial power systems.

5.3.4 Meshed Systems at the LV Level

Meshed systems at the LV level are still found in older installations and in industrial power supply. The reliability is very high and multiple outage of equipment is covered with this scheme. The voltage profile is flat and the system losses are minimal. Depending on the connection of the feeding primaries, three different types of systems can be distinguished:

- Supply station-by-station
- Single-line supply
- Multiple-line supply.

5.3.4.1 Meshed System Supplied Station-by-Station

Figure 5.15 indicates the general structure of such a LV system. Each individual LV system is supplied by one LV transformer. The MV system can be designed as a ring-main system in urban areas as indicated or in rural areas as radial system. The system is safe against outages only at the LV level; outages of LV transformers and of the MV line will lead to a loss of supply.

5.3.4.2 Single-Line Supply

The single-line supply of a LV system is outlined in Figure 5.16. The meshed LV system is supplied by more than one LV transformer; however, the transformers are connected at MV level to one line only. The system covers outages in the LV system and of the LV transformers. Outages of the MV line will result in a loss of supply.

5.3.4.3 Multiple-Line Supply

A LV system with multiple-line supply is fed by several LV transformers which are connected on the MV side to different lines. The system is safe against outages in the LV system, outages of LV transformers and also outages in the MV system as well if the MV system is designed accordingly. Figure 5.17 indicates the general structure.

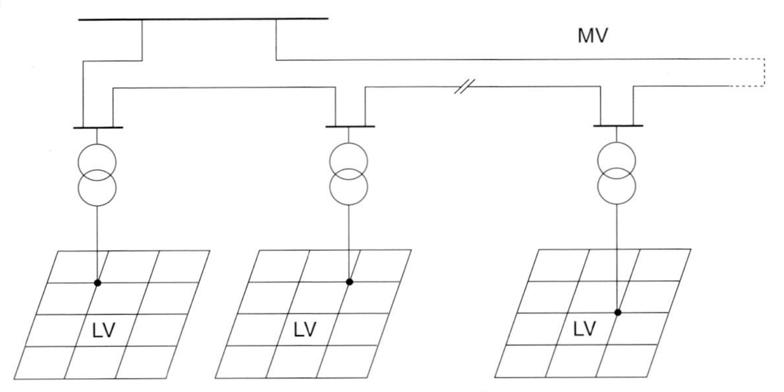

Figure 5.15 Structure of a meshed LV system supplied station-by-station.

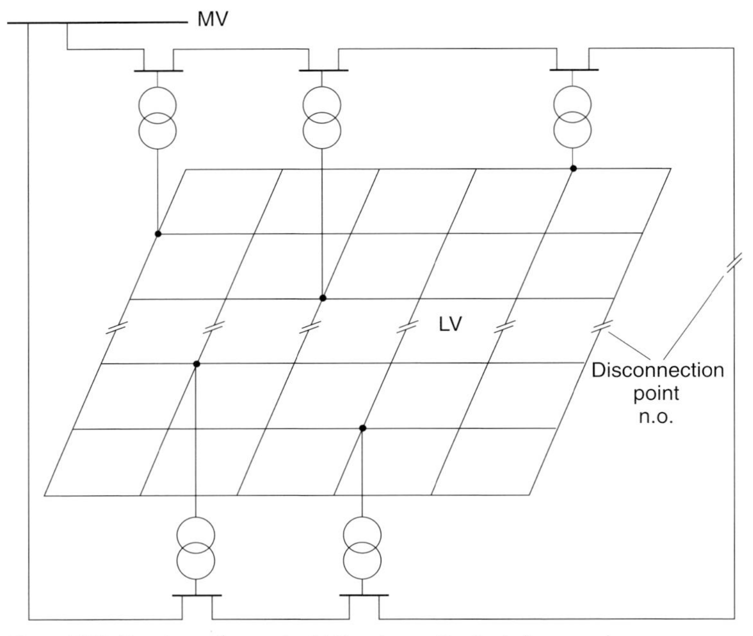

Figure 5.16 Structure of a meshed LV system with single-line supply.

The system is characterized by:

- Complicated topology
- High planning expenditure
- High maintenance cost
- Easy operation under normal operating conditions, but outage of LV lines due to fuse-break will not be recognized
- Loading of the lines for normal operating condition up to more than 70%

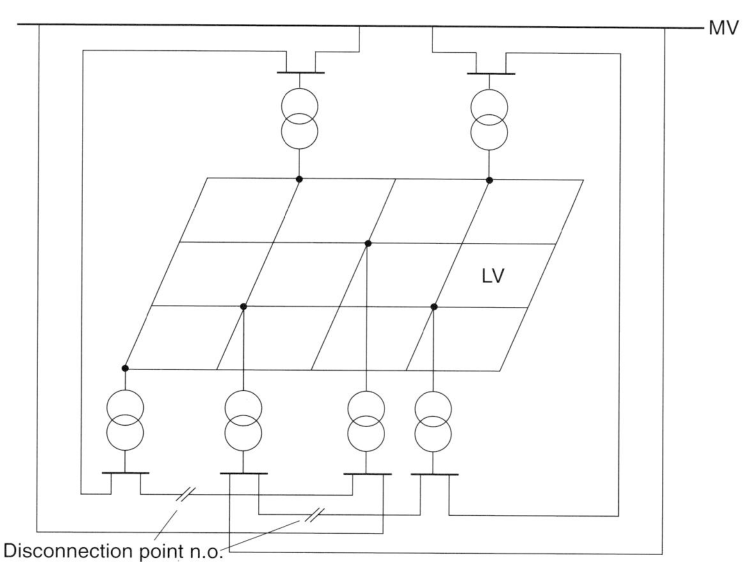

Figure 5.17 Structure of a meshed LV system with multiple-line supply.

- System losses minimal
- Voltage profile flat
- Flexibility for changed load conditions high
- Reserve for outage of the feeding MV transformer available
- Standardization of cross-sections of lines possible
- Protection with overcurrent protection, back-power relays necessary
- Application in low-voltage systems with high load densities and in systems with high reliability requirements.

5.4 Special Operating Considerations

Planning of power system must always take into account operational constraints and a long-term view of load development. Generally the system planner is free to select a suitable voltage level. It should further be considered that a general change of the system topology and of planning and operating criteria can only be introduced into the system over a longer time period that allows for gradual restructuring of existing systems.

Beyond that, the interaction of the different voltage levels has to be considered. Medium- and low-voltage systems have to be planned and operated in such a way that a suitable structure is maintained on both levels, for example, if the 10 kV system is operated as a meshed system, the 0.4 kV system can be operated as meshed or radial system.

Table 5.2 Aspects for combination of different system topologies on different voltage levels.

Feeding system HV/MV	Supplied system MV/LV	Loading	Reliability	Remarks
Meshed HV system	Meshed HV or MV system	Load-flow calculation	Very high according to planning criteria	Common combination
Meshed MV system	Meshed LV system	Simulation of loading	Very high in both MV and LV systems	Back-power relay necessary in LV-system
	Radial LV system	Simulation of loading	High in MV system, low in LV system	No special considerations for planning and operation
Ring-main system with open disconnection point in MV system	Meshed LV system	Simulation of loading	Fair in MV system, very high in LV system	Back-power relay necessary in LV-system
	Radial LV system	Simulation of loading	Fair in MV system, low in LV-system	Common combination, no special considerations for planning and operation

High-voltage transmission systems are operated with few exceptions as meshed systems. The loading of equipment for normal and emergency conditions as well as the losses and the voltage profile for normal operating and for emergency conditions must be determined by load-flow calculations. Regarding MV and LV systems that both are operated as meshed systems, the impedances of MV and LV systems normally do not allow any load-flow through the LV system. Loading of the LV transformers is determined only by the LV load, and cannot be changed except by changing the system topology. An exact determination of the loading of the LV transformers is difficult, as knowledge of the LV load in most cases is based on the energy correlated with typical load curves (see Section 2.6).

When the MV system is to be operated as a ring-main system with open disconnection point, the LV system should be operated in a similar way. This is of special importance in case of short-circuits in the MV system if branch short-circuit currents are flowing through the LV system. Special protection equipment (back-power relays) have to be installed in this case. If the LV system is supplied from one LV transformer only, any topology in the MV system can be chosen. Table 5.2 summarizes some special aspects of the combination of different system topologies at different voltage levels.

6
Arrangement in Gridstations and Substations

6.1
Busbar Arrangements

6.1.1
General

The arrangement and connection of incoming and outgoing feeders in gridstations and substations and the number of busbars have an important influence on the supply reliability of the power system. Gridstations and substations and the topology of the power system must be designed in a similar way and must therefore be included in the context of planning as a single task.

6.1.2
Single Busbar without Separation

The simplest arrangement of a substation is presented in Figure 6.1a. The outgoing feeders are connected to a single busbar and a single transformer is installed. Independently of the number of feeders supplied according to the topology of the system, no supply reserve exists for the outage of the transformer or of the busbar. The transformer can therefore be loaded up to 100% of its permissible (rated) load. This arrangement is found in MV and LV systems but also in 110/10 kV systems, where a three-winding transformer can be installed to feed two MV systems (see Figure 6.1c).

The arrangement with two transformers, as in Figure 6.1b, offers a supply reserve for the outage of one transformer if both transformers are loaded under normal operating conditions only to the extent that each one can take over the total load of the substation in case of outage of the other transformer, which is usually not substantially more than 50% of the rated load.

If circuit-breakers are installed in the outgoing feeders, short-circuits of the lines affect only the consumers attached to the faulted line, since the network protection disconnects the faulted line selectively. If load-break switches are installed in the outgoing feeders then one circuit-breaker is needed either on the MV or on the LV side of the transformer. In case of short-circuits on any feeder, the total load

Power System Engineering: Planning, Design and Operation of Power Systems and Equipment, Second Edition.
Jürgen Schlabbach and Karl-Heinz Rofalski.

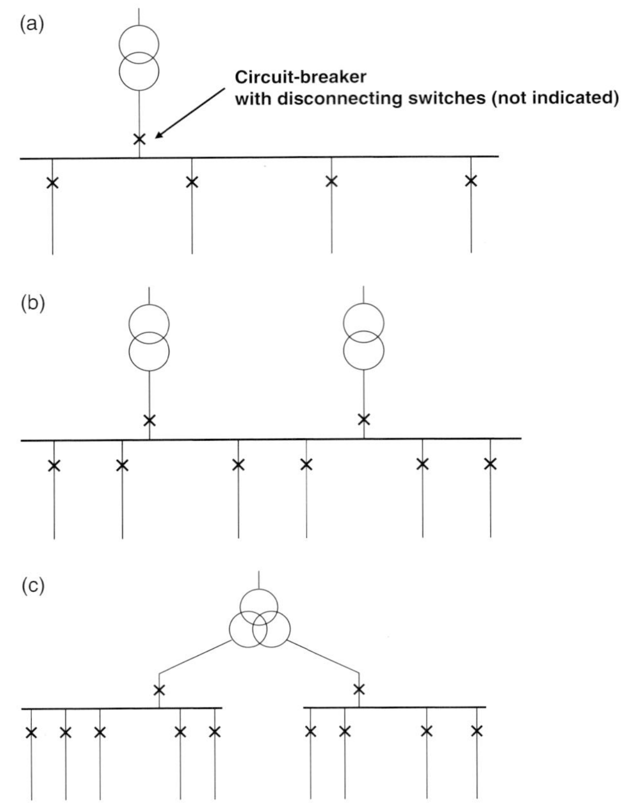

Figure 6.1 Substation single busbar on load-side. (a) Supply by one transformer; (b) supply by two transformers; (c) block arrangement to supply two MV systems.

is switched off and supplied again only after isolation of the faulted line by the associated load-break switch.

The arrangement is characterized by the following features:

- Supply reserve in the case of busbar faults not provided by the substation itself
- Supply reserve against transformer outage only given with second transformer
- Deenergizing the busbar requires interruption of supply
- Installed usually only in areas with small load density in the LV and MV voltage range
- Flexibility for operation is comparatively low
- Feeder arrangement in radial systems possible without circuit-breakers
- Supply of ring-main systems advisable only if a remote station is available.

6.1.3 Single Busbar with Sectionalizer

The disadvantages described above can be avoided by the arrangement of a busbar with sectionalizer. In the first case the sectionalizing function is realized by a load-break switch, in the second case by a circuit-breaker. It is generally not meaningful to construct substations having two transformers with single busbar without sectionalizer. In principle with use of two transformers further arrangements of the substations are possible. The block arrangement outlined in Figure 6.1c with sectionalizer or coupling is one of these possibilities. Different arrangements with single busbar and more than one transformer are outlined in Figure 6.2.

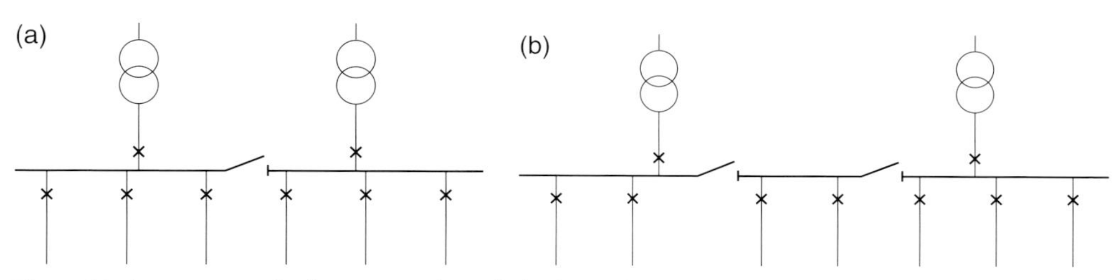

Figure 6.2 Arrangement of substations with single busbar and several transformers on the load-side. (a) Two transformers, two busbar sections; (b) two transformers, three busbar sections.

The arrangement is characterized by the following features:

- Supply reserve in the case of busbar faults available for ~50% of the load in the case of two busbar sections (66% in case of three busbar sections)
- Supply reserve in the case of transformer outages available, depending on the loading of the transformers
- Deenergizing of a busbar section requires supply interruption for ~50% or 33% of the load, depending on the number of busbar sections
- Installed usually only in areas with medium load density in LV and MV systems
- Used as an intermediate installation if three transformers are to be installed finally
- Flexibility in operation in the medium range
- Arrangement of feeders in radial systems possible without circuit-breakers
- Supply of ring-main systems with remote station enables higher loading of the transformers.

6.1.4 Special H-Arrangement

Substations with single busbar, longitudinal buscoupler and two transformers are also installed in the 110 kV systems in urban areas. The 110 kV cables are looped

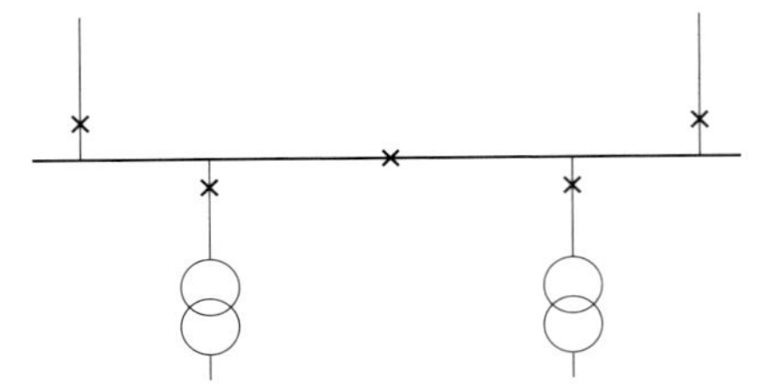

Figure 6.3 Arrangement of a substation in H-arrangement (fully equipped with circuit-breakers) on the feeding side.

in and out to the substation as outlined in Figure 6.3 [11] and [12]. If the parameters of the line protection are set in such a way that the longitudinal buscoupler is opened in case of short-circuits on the lines, circuit-breakers in the outgoing feeders can be avoided and only load-break switches are needed. A similar arrangement is applied for the transformer circuit-breakers, where in case of faults the feeding line is then possibly switched off also. A substation arrangement without any circuit-breakers is called load disconnecting substations, but this requires two load-break switches in the busbar in order to be able to deenergize each section of the busbar.

The characteristic features of the arrangement are:

- Supply reserve in the case of busbar faults available for ~50% of the load, depending on the arrangement of the lower voltage side also for 100% of the load
- In the case of busbar faults, no energy supply through the connected cables or overhead lines
- Supply reserve in the case of outage of transformer faults available
- Deenergizing of a busbar section possible without supply interruption with an appropriate arrangement on the lower voltage side
- Used in areas with medium load density, sometimes also with high load density in urban systems
- Limited flexibility in operation
- Reduced investment cost for 110 kV also possible.

6.1.5
Double Busbar Arrangement

Switchgear with double busbar is a typical arrangement for gridstations in MV, HV and EHV systems. All incoming and outgoing lines and transformers are connected with circuit-breakers and disconnecting switches to the busbars, as outlined in Figure 6.4.

A buscoupler, consisting of circuit-breaker and disconnecting switches, is necessary to separate the two busbars in case of busbar faults. The arrangement offers a high degree of supply reliability and operational flexibility, since each outgoing line and transformer can be switched without supply interruption from one busbar to the other if the busbars are operated in coupled mode. For separate operation of the busbars, separated network groups can be operated.

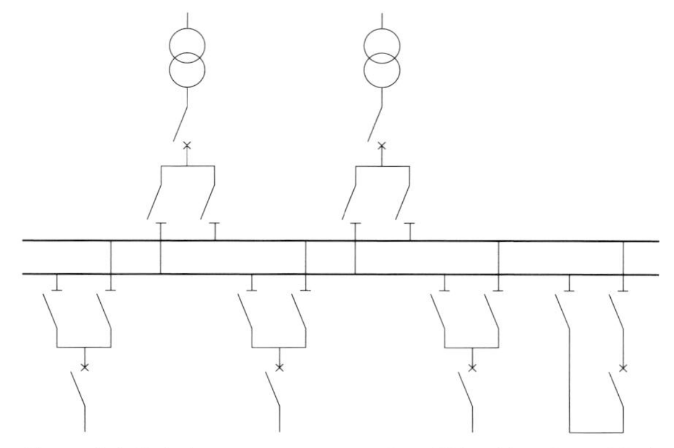

Figure 6.4 Switchgear arrangement in a HV gridstation with double busbar.

Typical characteristics of the arrangement with double busbar are:

- Supply reserve in the case of busbar faults available for the entire load
- Supply reserve for outage of transformers available, depending on the loading of the transformers
- Deenergizing of a busbar section possible without supply interruption
- Used even in urban areas with very high load density
- Used in HV and EHV transmission systems
- Used in important substations in MV systems
- Used for supply of industrial systems
- Very high operation flexibility.

6.1.6 Double Busbar with Reserve Busbar

The arrangement of having double busbar and one additional reserve busbar is very costly and is therefore advisable only for very important gridstations in the HV and EHV system, for example, in gridstations of national importance and for the connection of power stations of high importance. During operation all three busbars are energized; the outgoing transformers and lines are connected to two busbars only while the third one is operated with no-load and is available as a reserve busbar and for switching purposes. The arrangement has the advantage that a busbar can be completely deenergized without reducing the operational flexibility (two busbars remain in operation). Also, in case of loss of one busbar due to faults, the two other busbars remain available.

The characteristics of this arrangement are:

- Supply reserve in case of busbar faults available for the entire load
- Supply reserve for outage of transformers available, depending on the loading of the transformers
- Deenergizing of one busbar possible without supply interruption and without reduction of the operational flexibility

- Used rarely, usually in HV and EHV gridstations of very great importance
- Used in very important power stations
- Used in industrial power systems and sometimes for urban supply in the 110 kV system
- Operational flexibility very high
- Very high investment cost.

6.2 Arrangement in Switchyards

Switchyards need to be designed with respect to the foreseeable voltage stress, switching and breaking capability, short-circuit withstand capability, loading under normal and emergency conditions taking account of the requirements of load dispatch management, operational security and supply reliability. A switchyard consists of:

- Electrotechnical equipment such as switchgear, current and voltage transformers, surge arresters, insulation joints, armatures
- Mechanical structural parts such as conductors, bars and pipes for busbars and gantries, partly seen as electrotechnical equipment as well
- Secondary devices such as measuring and protection transformers and transducers, protective relays, coupling for remote control, batteries and so on
- Civil engineering structures such as buildings, foundations, fire-extinguishing equipment and fences.

The documentation of switchyards becomes extensive when account is taken of the multiplicity of items of equipment, their interdependency and their importance. Knowledge of the appropriate standardized symbols according to the different parts of IEC 60617 (DIN 40101) is therefore necessary.

A twelve months access to online database comprising parts 2 to 13 is available from IEC.

6.2.1 Breakers and Switches

The different types of breakers and switches to be used in switchgears are described in different parts of IEC 60947 and IEC 60890 (VDE 0660) for low-voltage installations as well as in EN 50052 and EN 50064 (VDE 0670) for high-voltage installations.

- Circuit-breakers have a switching capability for switching on and off any kind of current up to the rated current, that is, load current and short-circuit currents.
- Circuit-breakers installed in overhead systems should have the capability of operating sequences for successful and unsuccessful autoreclosing.
- Load-break switches are capable of switching load currents under normal operating conditions, but have no capability for switching short-circuit currents.

- Disconnecting switches can be operated only under no-load conditions. Currents of busbars without load and no-load currents of transformers with low rating can be switched on and off as well. Interlocking with the circuit-breaker is necessary.
- Earthing switches are used for earthing of equipment. The combination of earthing switch with disconnecting switch is common.
- Fuses are installed in LV and MV systems only. They interrupt currents of any kind by the melting of a specially designed conductor and must thereafter be replaced. Combination of fuses with disconnecting switches can be found especially in LV systems.

Circuit-breakers are named according to the method of arc quenching they use. Vacuum circuit-breakers are nowadays installed in MV systems; in the HV and EHV range outdoor circuit-breakers are operated with compressed air or sulfur hexafluoride (SF_6). Circuit-breakers in gas-insulated switchgear (SF_6-isolated) are of the SF_6-type.

6.2.2 Incoming and Outgoing Feeders

Incoming and outgoing feeders in switchgear are equipped with circuit-breakers and disconnection and earthing switches. Current and voltage transformers for the connection of protection and measurement devices are usually installed at each feeder in HV switchyards. The current transformer is placed at the busbar side of the voltage transformer in order to detect short-circuits of the voltage transformer by the protection device. Installations without voltage transformers in each feeder are also found: in this case the voltage transformer is placed at the busbar. In addition, the feeders are equipped with surge arresters and coupling devices for frequency carrier signals depending on the requirements of the switchgear. A typical arrangement of the individual devices of feeder arrangement in a HV switchyard is outlined in Figure 6.5.

6.2.3 Current Transformers

Current transformers are used for the connection of protection and measuring devices. In order to avoid damage by high voltages, current transformers are not allowed to be operated without load and must not be protected by fuses on the secondary side. One terminal of the current transformer on the secondary side is to be grounded for definition of ground potential. Current transformers are named according to the intended purpose (M: Measurement; P: Protection) and accuracy. For measuring purposes, accuracy classes 0.1, 0.2 and 0.5 are used. The total measuring error remains below 0.1%, 0.2% and 0.5% if the current transformer is operated in the range of 100–120% of the rated current. Current transformers with accuracy classes 1, 3 and 5 (accuracy 1%, 3% and 5%) are used for operational measurements.

Measuring (instrument) current transformers should be saturated in case of overcurrent in order to protect the attached measuring devices, whereas

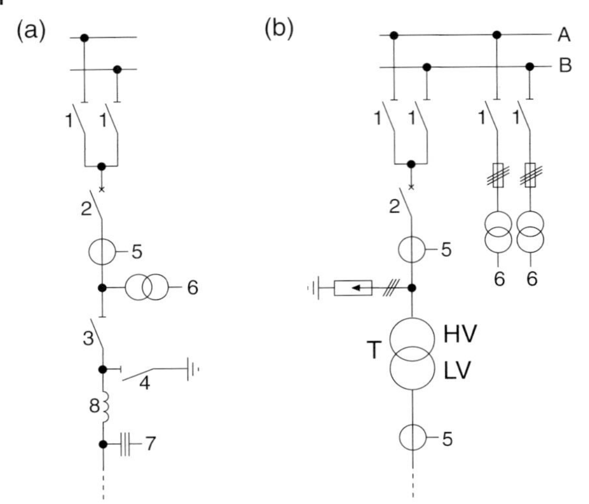

Figure 6.5 Typical feeder arrangement in a HV switchyard [12]. (a) Overhead line feeder with double busbar; (b) transformer feeder with double busbar. **1**, Busbar disconnecting switch; **2**, circuit-breaker; **3**, feeder disconnecting switch; **4**, earthing switch; **5**, current transformer; **6**, voltage transformer; **7**, capacitive voltage transformer with coupling for frequency carrier signal; **8**, blocking reactor against frequency carrier signals.

protection current transformers should have sufficiently small errors while operated with overcurrent. This is defined by the overcurrent factor n of the current transformer according to IEC 60044, see also IEC 61869. The rated overcurrent factor n_r defines multiples of the rated current at rated load S_r and power factor $\cos\varphi = 0.8$ of the load (burden), for which the error remains below the value defined in the accuracy class. For measuring purposes, overcurrent factors of $n_r <$ 5 are sufficient, while the overcurrent factor for protection purposes should be n_r > 5–10. The total error is indicated according to IEC 60044 as for the accuracy class, for example:

- 40 VA 5 P10 Rated load (burden) 40 VA: protection current transformer P; total error below 5% with the rated overcurrent factor $n_r = 10$
- 30 VA 0.5 M5 Rated load (burden) 30 VA: measuring current transformer M; total error below 0.5% with rated overcurrent factor $n_r = 5$.

The total error is given by the geometric addition of magnitude error F_i and the phase-angle error δ_i, if the instantaneous values of the primary and secondary current are taken into account for the determination of the total error. According to IEC 60044, the phase-angle error for a current transformer of the class 1M5 shall thereby remain below 60 minutes (one degree). The rated load (burden) of a current transformer is to be selected in such a way that the load of the connected devices and the losses of the connecting cables on the secondary side are covered. With reduced burden S_B, the overcurrent factor increases in accordance with Equation 6.1.

$$n_B \approx n_r \cdot \frac{S_r}{S_B} \tag{6.1}$$

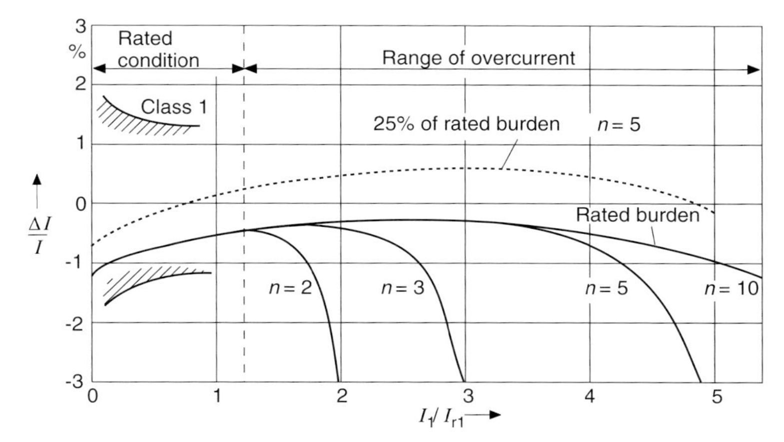

Figure 6.6 Error of a current transformer for different overcurrent factors and different burden.

Figure 6.6 represents the error of a current transformer for different rated overcurrent factors and different burden.

Current transformers have to withstand the expected short-circuit currents. The withstand capability is determined by the rated short-time current I_{thr} and the rated peak short-circuit current i_{pr}. The rated short-time current I_{thr} is the r.m.s.-value of that current in the primary winding, permitted for one second during short-circuit of the secondary winding. The rated short-time current must be higher than or equal to the thermally relevant short-time current I_{th}. The rated peak short-circuit current i_{pr} should be larger than 1.8 times the peak value of the rated short-time current I_{thr}.

Apart from the measuring error of the current transformer, the transient characteristic of short-circuit currents with maximal DC component and the magnetic remanence flux remaining in the core after disconnection of short-circuits are of significance. In automatic reclosing it is possible to switch into the still-existing short-circuit, with the consequence that the remaining remanence flux in the current transformer leads to a very high magnetizing current, driving the core into saturation and thus impairing the performance [13]. IEC 60044 additionally defines the current transformer class PR with special requirements on the remanence flux. The ratio of remanence flux to saturation flux (remanence factor K_{r}) may not exceed 10%.

Short-circuit currents with maximal DC component must be transferred by the current transformer with a low error. The maximum flux caused by the short-circuit current with maximal DC component must be lower than the saturation flux. This can be achieved by means of larger cross-section of the iron core and by smaller secondary burden, including larger cross-sections for the connection cables. An increased transformation ratio with reduced burden must be avoided with respect to difficulties with protective excitation. Requirements for the transient characteristic of current transformers are defined in IEC 60044. The requirements are realized by a linearization of the iron cores: four categories are defined, as specified in Table 6.1.

Table 6.1 Categories of current transformers and their parameter according to DIN EN 50482 and DIN EN 61869.

Category	Implementation	F_i	δ_i	K_r	T_s
P	• Iron closed core • Total error defined for symmetrical current on primary side • Remanence flux not limited	–	–	–	–
TPS	• Iron closed core • Low residual flux • Remanence flux not limited • To be used for differential protection	±0.25%	–	–	Some seconds
TPX	• Iron-closed core • Defined limits for error in magnitude and phase-angle • Remanence flux not limited • Suitable for automatic reclosure	±0.5%	±30′	~0.8	Some seconds
TPY	• Small air-gap to limit remanence flux • Remanence flux ≤10% of saturation flux • Suitable for automatic reclosure only if time-constant is smaller than reclosing time	±1.0%	±60′	~0.1	0.1 s up to 1 s
TPZ	• Large air-gap (linear core) • Remanence flux negligible • Protective devices remain in excitation for a long time	±1.0%	±18′	~0	60 ms

The transfer from primary to secondary of the fundamental current (50 Hz or 60 Hz) is achieved adequately with these current transformers; the transfer of the DC component is less with increasing linearization of the core.

6.2.4 Voltage Transformers

Voltage transformers are used for the measurement of line-to-line (phase-to-phase) voltages and line-to-earth (phase-to-neutral) voltages. Voltage transformers are constructed either as inductive transformers (U_n = 1–765 kV) or as capacitive transformers ($U_n \geq 60$ kV). Capacitive voltage transformers are suitable for the connection of frequency carrier signals. Voltage transformers are named according to their intended application (M: Measurement; P: Protection) and their accuracy class. One has to take into account error in magnitude F_U and phase-angle error δ_U. The indicated error margins are to be guaranteed with secondary burden

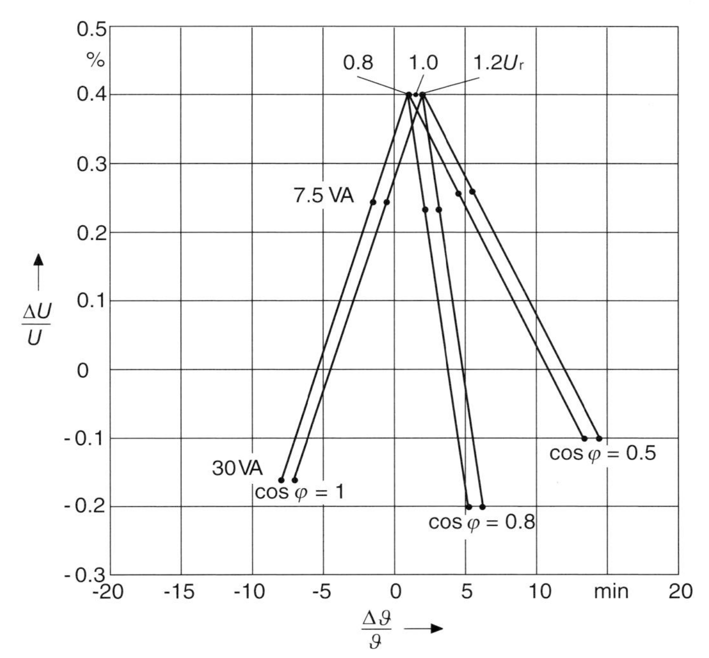

Figure 6.7 Error in phase-angle and magnitude of a voltage transformer $\frac{500\,kV/\sqrt{3}}{100\,V/\sqrt{3}}$; 30 VA 0.5 M for different burdens.

of 25% and 100% of the rated burden at power factor $\cos\varphi = 0.8$. The accuracy of measuring voltage transformers (class M) is to be guaranteed in the voltage range between $0.8 \times U_n$ and $1.2 \times U_n$, and for protective applications the voltage should be $1.0 \times U_n$.

Voltage transformers with classes 0.1, 0.2 and 0.5 (error less than 0.1%, 0.2% and 0.5% within the range $0.8–1.2 \times U_n$) are used for exact measurement purposes. For operational measurement, voltage transformers with accuracy class 1 and 3 (error less than 1% and 3%) are used. For protection purposes, voltage transformers 3P and 6P are used, having voltage error less than 3% and 6% and a phase-angle error less than 120 minutes and 240 minutes respectively in the voltage range $0.8–1.2 \times U_n$. Figure 6.7 indicates the dependence of the phase-angle error and the error in magnitude with different burden for a voltage transformer $\frac{500\,kV/\sqrt{3}}{100\,V/\sqrt{3}}$; 30 VA 0.5 M.

Single-pole isolated voltage transformers are exposed to increased voltages of varying duration depending upon the neutral earthing of the system. The rated voltage factor indicates multiples of the nominal system voltage. The voltage transformer is allowed to be operated for times of 30 s, 4 h or 8 h. Voltage transformers with a rated voltage factor 1.5/30 s may be used only in systems with low impedance earthing (earth-fault factor $\delta \leq 1.4$); those having a rated voltage factor of 1.9/30 s or 1.9/4 h or 1.9/8 h can be installed in systems with all kinds of neutral earthing, that is, isolated neutral, resonance earthing and low impedance earthing. The factor 1.9 corresponds approximately to the maximal permissible voltage of the system (power frequency).

Capacitive voltage transformers, which are used in high-voltage transmission systems in particular, show on the secondary side transient oscillations arising from short-circuits, which exhibit decaying amplitudes up to 10% of the system nominal voltage with frequencies up to 10 Hz in the case of short-circuits at zero voltage crossing. These transient oscillations downgrade the performance of distance protection relays, especially with respect to the direction decision. The problem can be solved, however, by small load of the voltage transformer and/or by installation of a filter circuit. Furthermore, ferro-resonances can occur with capacitive voltage transformers.

7
Transformers

7.1
General

The design and rating of transformers for utilization in power stations and industrial systems as well as in transmission and distribution systems are subject to the particular conditions of the respective application. The main data, besides the construction type, are the rated apparent power, the rated voltages of the windings, the impedance voltage and the no-load and short-circuit losses. The rated apparent power can in principle be determined individually for each specific installation, but it should be noted that, with respect to reserve capability of spare transformers and replacement and maintenance requirements, the number of different types and ratings should be limited. For transformers to be installed in medium- and low-voltage systems, standardized values of apparent power, impedance voltage and losses should preferably be used. Transformers used within the high-voltage range in general are designed individually according to the Owner's Specification. Transformers are usually provided with additional tap-windings, to be switched on and off by tap-changers to control the voltage drop within a small range defined by the additional voltage of the tap-winding. Low-voltage transformers are constructed with tap-windings, but these can be operated only under no-load and no-voltage condition. The weight and dimensions of power transformers are determining factors concerning transport and maneuverability, that is, the load-carrying capacity of bridges, road and railway constructions, the minimum clearance profile in tunnels, under bridges and at road embankments. Figure 7.1 indicates the specific weight and specific losses of transformers.

7.2
Utilization and Construction of Transformers

7.2.1
Utilization of Transformers

Power transformers in transmission and distribution systems are installed with rated apparent power from 250 to 2.5 MVA (voltage ratio 10–30 kV to

Power System Engineering: Planning, Design and Operation of Power Systems and Equipment, Second Edition.
Jürgen Schlabbach and Karl-Heinz Rofalski.

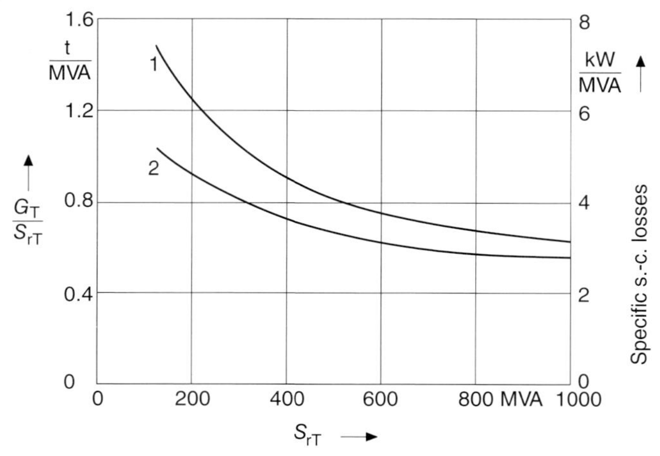

Figure 7.1 Specific weight (upper curve) and specific short-circuit losses (lower curve) of transformers with rated voltages above 115 kV.

low-voltage), and from 12 to 63 MVA (110–150 kV to 10–30 kV). High-voltage transformers (400 to 220–110 kV) have rated apparent power within the range 100–400 MVA. The transformers are constructed as three-phase transformers; transformers with higher rating are produced by connecting three single-phase transformers to a three-phase transformer bank, whereby rated apparent power up to 1600 MVA can be installed. The transformers in the European 400 kV system have rated power of 660 MVA (3 × 220 MVA) and 1000 MVA (3 × 333 MVA), respectively.

Block transformers in power stations are adapted to the rated power of the generator up to 1500 MVA (400 kV) in nuclear power stations. The vector group, as standard, is YNd5, tap-changers are normally available. Further transformers are needed in power stations for start-up operation, emergency generation and auxiliary supply. Depending on the type of power station, transformers are installed at the generator's main connection, directly at the high-voltage switchgear. Transformers are likewise needed for the supply of the low-voltage installations (0.66 or 0.4 kV) within the auxiliary power system of the power station.

In industrial power systems special transformers are used for electrolysis plants, medium-frequency heating devices, induction furnaces and heating systems, arc furnaces (rated transformation ratio, e.g. 110 kV/0.8–0.5 kV), rectifier-controlled drives (static inverter transformers) and welding equipment (three-phase AC-current and single-phase AC-current welding transformers).

Three-phase transformers are constructed with a three-leg or a five-leg core, with the respective windings for different voltage levels. The connections are designated by the primary voltage (HV, MV, LV) or according to the voltage levels (highest voltage No. 1, next highest voltage No. 2, etc.). The three phases are designated U, V and W (or in some cases X, Y, Z). Start and end of the winding are marked with 1 and 2. The windings of connection 1U1 to 1U2 (highest voltage, winding U) and the winding of 2U1 to 2U2 (second highest voltage, winding U) have the same direction of coil winding. Neutrals are designated N or n (also 1N, 2N for different voltage levels and/or windings). Transformers with two additional core legs (five-leg core) allow reduced construction height compared with transformers having a three-leg core.

Autotransformers have a galvanic connection of the high- and low-voltage windings in such a way that the low-voltage winding is part of the high-voltage one. The savings in construction material, volume and cost are significant. Autotransformers serve as network transformers at the highest voltage levels, with the two systems having direct, that is, low-impedance neutral grounding. Both three-phase units and single-phase units connected to three-phase transformer banks are in use in power systems. Normally a third winding is installed, acting as balance winding in case of unsymmetrical currents due to unsymmetrical short-circuits. If the third winding is constructed with connections outside the transformer tank through bushings, this winding can be used, for example, for the auxiliary supply of substations. This supply scheme can also be used in case of full winding transformers.

The use of three-winding transformers is common practice in power stations and in industry for the supply of two medium-voltage levels, sometimes with different nominal voltages. The connection of two generators through a three-winding transformer is applied only in small power stations, for example, with generator rating up to 70 MW. The three-winding type gains in importance in connection with the use of renewable energy, where comparatively small generation units are used. In rural supply systems, three-winding transformers are used likewise in order to supply, for example, a 10 kV system and a 30 kV system from the 110 kV network. The disadvantage is that outage of the transformer results in the loss of two generators or of the loss of supply of two voltage levels. Furthermore, the voltage control of one system depends on the loading of the other one. Figure 7.2 illustrates typical arrangements of transformers in power stations, industry and utilities.

7.2.2 Oil-Immersed Transformers and Dry-Type Transformers

Oil-immersed transformers are built for each voltage level. The winding insulation consists of vacuum-dried paper impregnated with mineral or synthetic oil. The insulating oil serves at the same time as cooling medium. Forced cooling by pumps is often installed to improve the heat dissipation. Transformers using silicone oil or ester are related to the oil-immersed transformers. Transformers with dry-type insulation of glass fiber-reinforced epoxy resin are air-cooled; the typical voltage range is up to 36 kV only with rated apparent power up to 2.5 MVA. The advantage of dry-type transformers is seen mainly in the reduced fire load and the smaller construction volume compared with oil-immersed transformers. Transformers with XLPE- insulated windings are applied likewise in the medium-voltage range, when the winding consists of a XLPE cable which is wound around the iron core. The decrease of the fire load is only marginal compared with oil-immersed transformers.

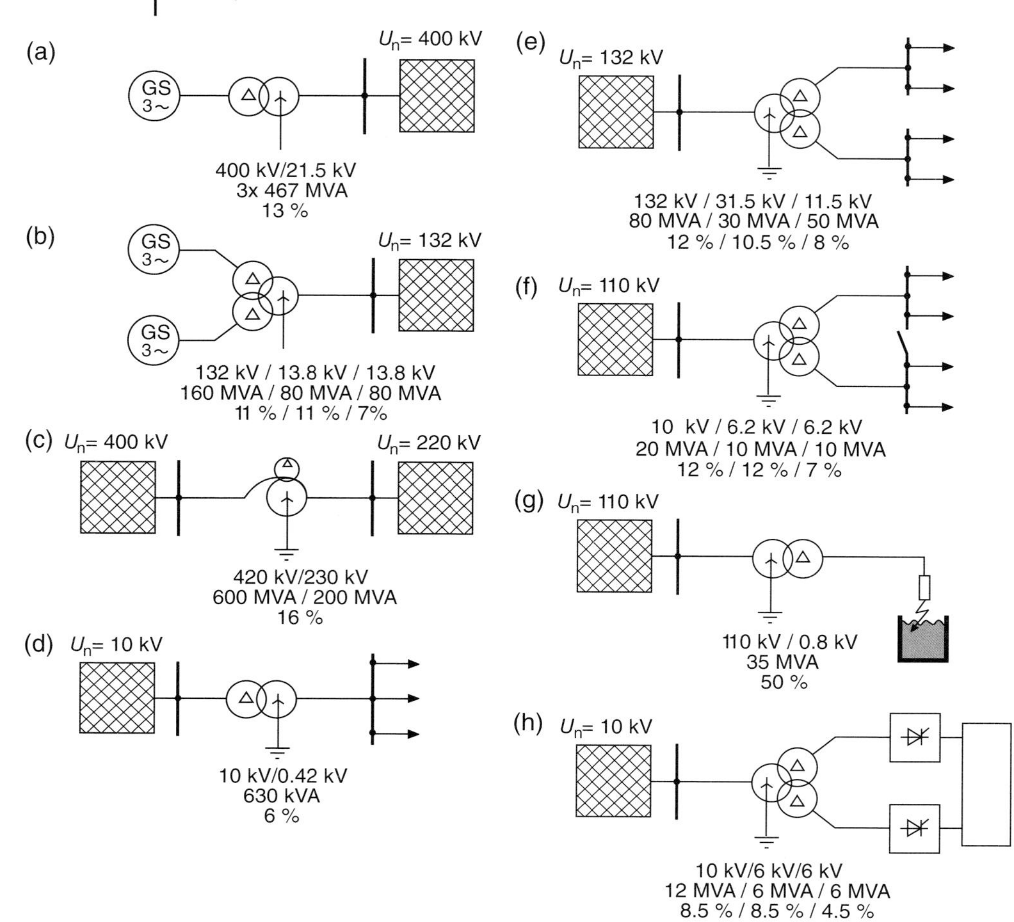

Figure 7.2 Arrangements of transformers in power systems. (a) Generator transformer (transformer bank of three single-phase transformers); (b) three-winding block transformer in a gas-turbine power station; (c) high-voltage autotransformer coupling two power systems; (d) low-voltage transformer; (e) three-winding transformer for the supply of two rural MV-systems; (f) three-winding transformer in an industrial power system; (g) transformer for the supply of a three-phase alternating-current arc furnace; (h) static inverter transformer for the connection of two static frequency inverters.

Standard operating conditions for air-cooled transformers are specified in IEC 60076-2 (VDE 0532-76-2). Cooling air temperatures (ambient temperatures) range between 40 °C and −5 °C (indoor applications) and as low as −25 °C (for outdoor installation) with mean daily temperatures below 30 °C and mean yearly temperatures below 20 °C. Additionally, contamination of the surfaces has to be considered, since deposition of layers can impair the heat dissipation and condensation of moisture can reduce the insulation strength. The environmental and climatic classes are defined as follows:

E0 No condensation at transformer surfaces, pollution negligible (can be disregarded in dust-free, dry areas)
E1 Occasional condensation on transformer surfaces (only without load or in no-load condition), minor pollution possible
E2 Frequent condensation and/or pollution at transformer surfaces
C1 Operation above −5 °C, temperature during storage and transport down to −25 °C
C2 Operation, transport and storage at temperatures down to −25 °C.

For classification of fire loads, the following are defined:

F0 No measures for reduction of fire risk need be planned
F1 Measures to reduce fire risk necessary, delivery of toxic materials and smoke possible.

The type of cooling and the type of insulation can be taken from the nameplate as outlined in Table 7.1.

Table 7.1 Types of cooling and insulation of transformers.

	Internal cooling agent		**External cooling agent**	
	Type	**Circulation**	**Type**	**Circulation**
Oil-immersed transformers	O Mineral oil or synthetic fluid with flame temperature ≤300 °C	N Natural circulation of cooling agent	A Air	N Natural convection
	K Insulation fluid with flame temperature >300 °C	F Forced circulation by cooler	W Water	F Forced movement by ventilation
	L Insulation fluids without measurable flash point	D Forced circulation by cooler and through windings		
	G Gas			
Dry-type transformers	G Gas	N Natural circulation	A Air	F Forced circulation
	A Air	F Forced circulation		

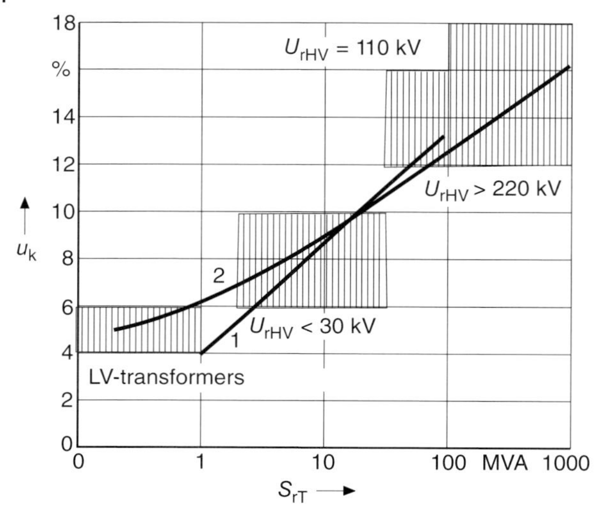

Figure 7.3 Characteristic value of impedance voltage of two-winding transformers. 1, Average values; 2, average values of full-winding transformers.

7.2.3
Characteristic Data of Transformers

The characteristic data of transformers cannot be determined independently, as they depend strongly on the voltage level and the rated apparent power. With increasing rated apparent power and/or rated voltage, the magnetic leakage flux of the transformer will increase, the volume of the insulation increases and the impedance voltage will increase as well. As a secondary effect the determination of the hot-spot temperature becomes more difficult. This also affects the thermal maximal permissible loading beyond the rated apparent power, which generally decreases with rising rated apparent power. Recommended values of the rated apparent power, which are common in medium-voltage systems, must be considered.

Figures 7.3 and 7.4 indicate reference values for the impedance voltage u_{kr}, the short-circuit losses P_k, the relative no-load current I_l and the no-load losses P_l of two-winding three-phase transformers in relation to the rated apparent power S_{rT}. The standard DIN 42504 is withdrawn and currently under revison.

7.3
Operation of Transformers

7.3.1
Voltage Drop

For the simplified representation and calculation of the voltage drop of transformers during operation the so-called "Kapp's triangle" is often used, based on the

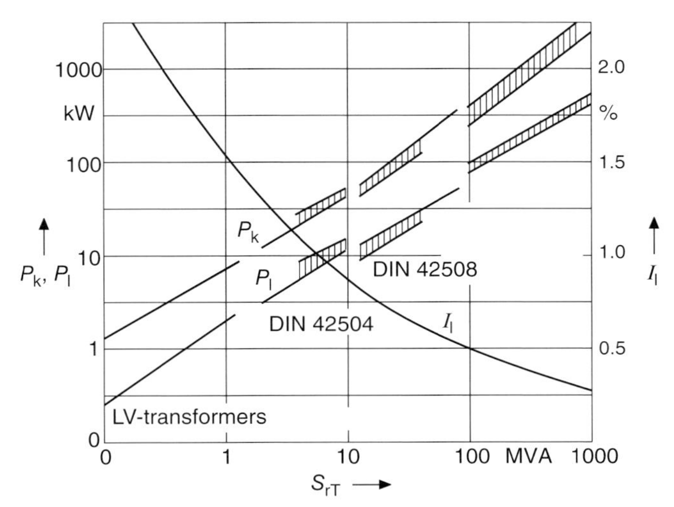

Figure 7.4 Characteristic values for I_l and P_l versus S_{rT} of two-winding transformers according to DIN 42504 and DIN 42508 (standard values in Germany).

voltage drop at the short-circuit reactance $X_k = j\omega L$ and winding resistance R_{Cu} only. The voltage drop Δu or ΔU is calculated using Equation 7.1.

$$\Delta U \approx (R_{Cu} \cdot \cos\varphi + X_k \cdot \sin\varphi) \cdot \frac{I}{I_{rT}} \tag{7.1a}$$

$$\Delta u \approx (u_R \cdot \cos\varphi + u_k \cdot \sin\varphi) \cdot \frac{I}{I_{rT}} \tag{7.1b}$$

with

R_{Cu} = ohmic resistance of windings
X_k = leakage reactance (short-circuit reactance)
u_R = ohmic part of impedance voltage
u_k = inductive part of impedance voltage
φ = phase angle of load current
I_r = rated current
I = actual load current.

For the reduction of the voltage drop, tap-changers operated under load are used. Tap-changers operate either according to the reactor principle with slow-switching reactors (e.g. in the USA) or according to the resistance principle with fast-switching resistance, as in European practice. The tap-changer with resistance switching has advantages regarding the transformer's weight, its transportability and economical design, especially in the high-voltage and extra-high-voltage systems, and with respect to the low switching time [14].

7.3.2 Permissible Loading of Transformer Neutral

The three phases of transformer windings can be arranged in star-, delta- or zigzag-connection. The vector group (star: Y or y; delta: D or d; zigzag; Z or z) is indicated on the transformer's nameplate, with a capital letter for the highest voltage side and small letters for the windings on the other voltage levels. The neutral connection, if brought out of the transformer tank, is indicated by the letters N and n. The respective number of the vector group indicates the phase-shift as multiples of 30° of the phase angle of the line-to-earth voltages of the lower voltage side relative to the related line-to-earth voltages of the higher voltage side (mathematically positive sense). Table 7.2 shows common vector groups of two-winding transformers.

The arrangement of the windings, whether in star-, delta- or zigzag-connection, has a significant influence on the impedance of the zero-sequence component and on the permissible loading of the transformer neutral. Star- or zigzag-connected windings have a defined zero-sequence impedance if the neutral is earthed. Delta-connected windings and ungrounded star- or zigzag-connections have an infinite zero-sequence impedance.

The maximal permissible loading of transformer neutral is defined as the capability to carry the rated current of the earthed winding through the neutral, expressed as the ratio of the current through earth I_E to the transformer rated current I_{rT}. A high maximal permissible loading of the neutral is necessary in case of low-impedance earthing of the power system or if the system is operated with resonance earthing through a Petersen coil.

Transformers with vector groups Yd, Dy and Yyd (with additional compensation winding in delta-connection) as well as low-voltage transformers Yz have 100% neutral loading capability. The relation of zero-sequence impedance to positive-sequence impedance of these transformer types is in the range of 0.1 (zero-sequence impedance of z-winding) to 2.4 (Yy with compensation winding in delta-connection). The ratio X_0/X_1 of a transformer depends on the detail of its construction. For transformers of vector group Yy, Yz and with three-leg core, the ratio is X_0/X_1 = 3–10; with a five-leg core or for three single-phase transformers the ratio is X_0/X_1 = 10–100. Low-voltage transformers in general have a ratio of zero-sequence impedance to positive-sequence impedance of $X_0/X_1 = 1$. Table 7.3 gives an overview of the maximal permissible loading of transformer neutrals for different vector groups and construction types.

7.4 Thermal Permissible Loading

7.4.1 Temperature Models

The maximal permissible loading of transformers is determined by the maximal permissible temperature of the materials used, especially of insulation materials.

Table 7.2 Vector group and circuit arrangement of three-phase transformers.

Vektor group[a]	Vector diagramm[b] HV LV	Connection diagramm[c] HV LV
Dd0		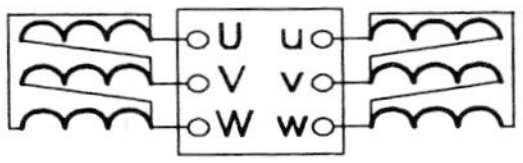
Yy0		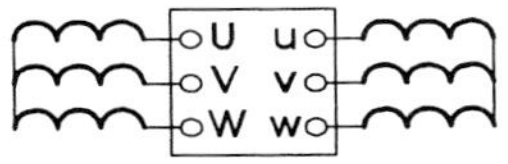
Dz0		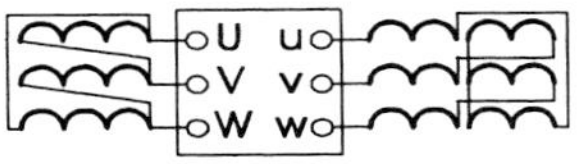
Dy5		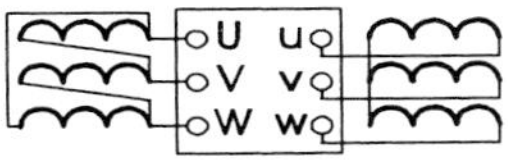
Yd5		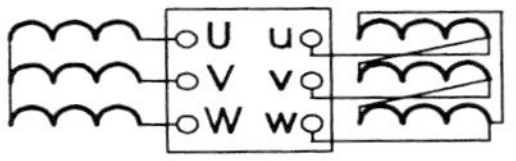
Yz5		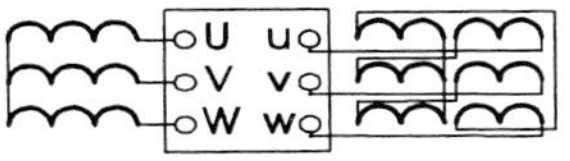
Dd6		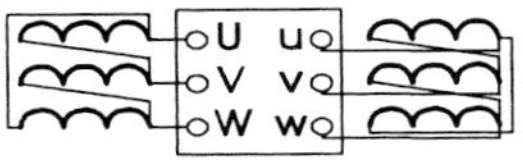
Yy6		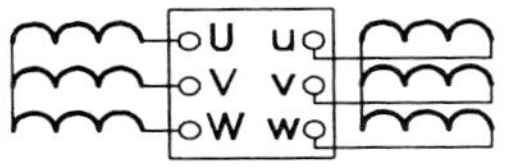
Dz6		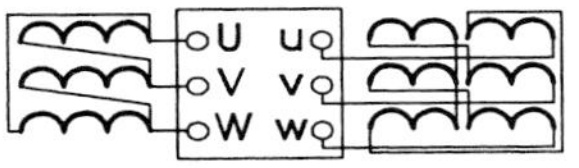
Dy11		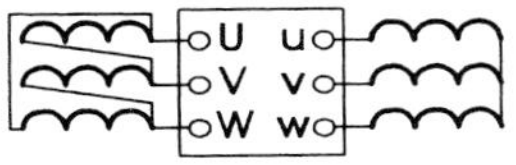
Yd11		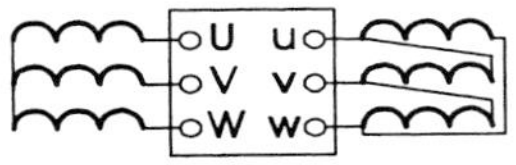
Yz11		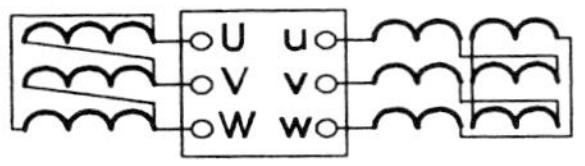

a In case the neutral is brought out of the tank, also include N and/or n.
b 1U, 1V, 1W for the highest voltage; 2U, 2V, 2W and/or 3U, 3V, 3W for lower voltage connections.
c Same direction of coil winding direction starting from connection point.

Table 7.3 Maximal permissible loading of transformer neutrals (N or n means that in this case the winding is actually grounded) [15] and [16].

Vector group	Construction of leg	Permissible loading of neutral I_E/I_{rT}
Yyn	Without compensation winding	Up to 0.1
	Jacket-type transformer	
	Core-type transformer	
	Single-phase transformer	
	3-limb-core transformer	<0.25 for up to 1.5 h
		<0.2 for up to 3 h
		<0.1 permanent
Yyn	Without compensation winding	<1.0; in case U_0 is small
Yyn+d	With compensation winding	<0.33; in case $S_r = 0.33 S_{rT}$
Dyn		1.0
Zyn	Without compensation winding	Up to 0.1
	Jacket-type transformer	
	Core-type transformer	
	Single-phase transformer	
	3-limb-core transformer	<0.25 for up to 1.5 h
		<0.2 for up to 3 h
		<0.1 permanent
ZNy		1.0

If this temperature is exceeded occasionally or permanently, the aging of the insulation material results in a change of the chemical and physical characteristics and consequently in reduced insulation strength and reduced thermal conductance, thus in a reduction of the life time of the transformer. The materials used in the construction of transformers are classified into different classes in accordance with IEC 60085 with associated temperature limits as per Table 7.4.

The maximal temperature in the transformer is only close to the winding at the so-called hot-spot. At greater distance from the winding the temperature is lower and insulation materials of a lower insulating class can be used without reducing the thermal class of the transformer itself. Operational experience of many years with oil-immersed transformers indicates that the aging process can essentially be neglected if the maximum temperature is kept below 90 °C. First outages due to aging of the insulation are expected at the earliest after 50 years. An increase in the temperature leads to a doubling of the rates of aging processes for each additional 8 °C above the specified 90 °C (Montsinger's law) and thus to an average lifetime of 25 years if the temperature is sustained at 98 °C. There is as yet no similar experience for dry-type transformers. The relative aging rate V is given for normal paper insulation in IEC 60354:1991-09 according to Equation 7.2a [17]. The document is currently under revision.

$$V = \frac{V_{HS}}{V_{98}} = 2^{\left(\frac{(\vartheta_{HS}-98)}{6}\right)} \tag{7.2a}$$

Table 7.4 Thermal classes of insulation materials of transformers acc. to IEC 60085 [17].

Thermal class	Temperature (°C)	Example of insulation material
Y	90	Organic fibers (paper, molded wood, wood, cotton) not impregnated PVC
A	105	Organic fibers (paper, molded wood, wood, cotton) impregnated Nitrile-rubber
E	120	Foils and shaped parts of polyester Varnish made from polyvinylformal, polyvinyl acetate or epoxy resin
B	130	Shaped parts of epoxy resin and polyester resin with inorganic filler Foils from polyethersulfon Varnish with higher temperature-resistance
F	155	Molded shaped parts of epoxide-isocyanate resin Varnish made from polyterephthalate Polyamide foils Glass fiber-reinforced insulation made from modified silicone resin
H	180	Glass-fabric with silicon or silicone-rubber Varnish made from polyamide, silicone-rubber or aramid fibers

The need to include the higher temperatures in the standards was discussed in the IEC (International Electrotechnical Commission) and they are included in the revision of the standards under IEC 60076-7. For thermally stabilized paper insulation with higher permissible temperature, the relative aging rate is given by Equation 7.2b.

$$V = e^{\left(\frac{15000}{110+273}+\frac{15000}{\vartheta_{HS}+273}\right)} \tag{7.2b}$$

with

V_{HS} = aging rate at actual hot-spot temperature
V_{98} = aging rate at hot-spot temperature of 98 °C
ϑ_{HS} = hot-spot temperature in degrees.

The relative aging rate of $V = 1$ for oil-immersed transformers actually refers to a hot-spot temperature of 98 °C, which leads to a lifetime of 25 years, according to the standard IEC 60076-7. The use of thermally stabilized isolating papers allows for higher hot-spot temperatures up to 110 °C [14].

The reduction of lifetime L is determined by addition of all relative aging rates V_i related to the total number of the events N in the time interval ($t_i = t_2 - t_1$) according to Equation 7.3.

$$L = \int_{t_1}^{t_2} V \, dt \tag{7.3a}$$

$$L = \sum_{i=1}^{N} V_i \cdot t_i \tag{7.3b}$$

with

V_i = relative aging rate according to Equation 7.2
t_i = time interval.

All determinations (see Figure 7.5) take into account that the oil temperature increases linearly along the winding from bottom to top (from ϑ_{OL} to ϑ_{OU}) and that the winding temperature is always higher by a constant temperature difference ϑ_{WO} than the oil temperature. The hot-spot temperature ϑ_{HS} is higher by ϑ_{HSW} than the temperature at the upper end of the winding in order to allow additional heating by increased leakage flux at the end of the winding and to take into account any temperature increase due to nonlinearity. The coolant temperature ϑ_C is lower than the oil temperature at the lower end of the winding. The mean winding

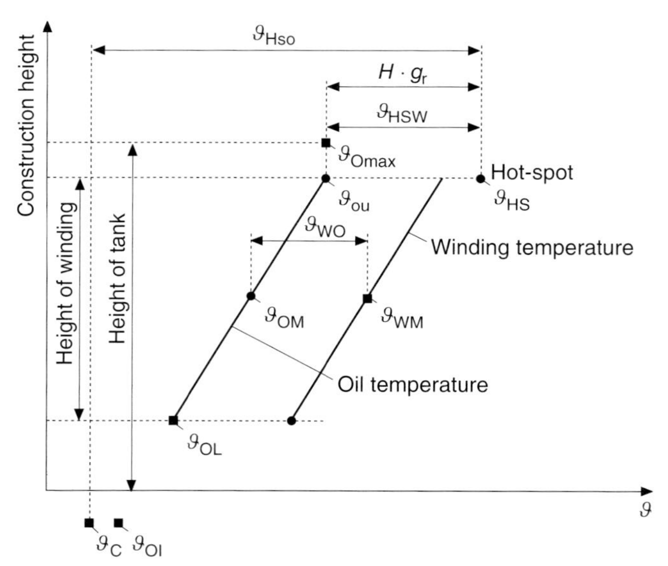

Figure 7.5 Temperature rise within the transformer for consideration of the maximum thermal loading; see text for explanation.

temperature ϑ_{WM}, to be determined by impedance measurement, exceeds the mean oil temperature ϑ_{OM} by approximately $\Delta\vartheta_{WO}$. The difference in temperature between the mean oil temperature and the hot-spot temperature is calculated by multiplying the temperature difference $\Delta\vartheta_{WO}$ by the hot-spot factor H. The maximal oil temperature may exceed the oil temperature at the upper end of the winding, but it is not seen as critical with respect to the reduction in insulation strength.

The following temperatures can be obtained by measurement:

- Maximal oil temperature ϑ_{Omax}
- Mean winding temperature ϑ_{WM}
- Oil temperature ϑ_{OI}
- Coolant temperature ϑ_C

The temperature of the hot-spot can also be measured by sensors, if they are placed near the hot-spot during manufacturing. The temperature difference between the hot-spot and the coolant is called the hot-spot temperature rise ϑ_{HSO}.

Depending on variations of ambient temperature and the varying load of the transformer, the temperature of the hot-spot will vary. All considerations related to the maximal thermal loading have therefore to be accomplished either using average values (annual average values) or accepting a higher permissible hot-spot temperature. For the determination of the ambient and/or cooling agent temperature, the following considerations apply according to IEC 60076-7.

For air-cooled transformers in outdoor installation the actual ambient temperature is to be used. If the outside temperature is not constant during the period of peak load, then the actual temperature profile (1) or a simulated temperature profile course (double sinusoidal) ϑ_a (2) or a weighted ambient temperature ϑ_{aw} (3) can be used together with the monthly average temperature for the thermal aging processes as well as for the determination of the maximal hot-spot temperature.

1. The actual temperature profile can be used if the peak load period is limited to a few days in the year.
2. The simulated temperature profile (double sinusoidal) according to Equation 7.4 considers the seasonal and daily variation of the temperature:

$$\vartheta_a = \bar{\vartheta}_{ay} + A\cdot\cos\left(\frac{2\pi}{365}\cdot(d - d_{max})\right) + B\cdot\cos\left(\frac{2\pi}{24}\cdot(h - h_{max})\right) \tag{7.4}$$

with

ϑ_{ay} = yearly average temperature
A = amplitude of the daily average temperature for one year
B = maximal amplitude of the daily temperature for one year (to be used for the aging rate) or amplitude of the maximal daily temperature for one year (to be used for hot-spot temperature)
d = day of the year (d = 1 for 01 January)
d_{max} = hottest day of the year
h = hour of each day
h_{max} = hour of maximal daily temperature.

3. The weighted ambient temperature ϑ_{aw} is calculated according to Equation 7.5.

$$\vartheta_{aw} = \overline{\vartheta} + 0.01 \cdot (2 \cdot \Delta\overline{\vartheta})^{1.85} \qquad (7.5)$$

with

ϑ = average temperature in the time period under consideration
$\Delta\vartheta$ = difference between mean maximal and minimal temperatures in the period under consideration.

The weighted ambient temperature ϑ_{aw} is higher than the average (arithmetic mean value) ambient temperature. For the calculation of the hot-spot temperature it is recommended to use the average value of the maximum temperatures during the period under consideration.

For distribution transformers in indoor installations, temperature correction is made for the ambient temperature. This temperature correction is to be added to the difference of hot-spot temperature ϑ_{HS} and maximum oil temperature ϑ_{Omax}. It is recommended to measure this temperature correction during a test. If this is not possible, the figures in Table 7.5 can be used.

Table 7.5 Temperature correction to be used for distribution transformers in indoor installations according to IEC 60076-7.

Type of indoor installation	Number of transformers	Rated apparent power (kVA)			
		250	500	750	1000
		Temperature correction in °C			
Basement with natural convection	1	11	12	13	14
	2	12	13	14	16
	3	14	17	19	22
Ground floor and building with bad ventilation	1	7	8	9	10
	2	8	9	10	12
	3	10	13	15	17
Building with good natural ventilation, basement and ground floor with forced ventilation	1	3	4	5	6
	2	4	5	6	7
	3	6	9	10	13
Kiosk housing	1	10	15	20	–

For water-cooled transformers the temperature of the cooling water intake is to be used.

Operation periods with high hot-spot temperatures (increased aging) are compensated by operation periods with low hot-spot temperatures. In addition, the hot-spot temperature must be determined and monitored as exactly as possible by means of measurement or thermal imaging.

7.4.2
Maximum Permissible Loading of Oil-Immersed Transformers

7.4.2.1 General

The maximum permissible loading of oil-immersed transformers with rated apparent power up to 100 MVA can be determined according to IEC 60076-7 [18]. For higher rated apparent power, the recommendations of the manufacturer are to be used. The VDE standard 0536 is replaced at international level by the revised version of IEC 60354 "Loading guide for oil-immersed transformers" dated 1991 [19]. A revised version is published as IEC 60076-7 (Power Transformers–part 7: Loading guide for oil-immersed power transformers) [20]. Differences between IEC 60354:1991 and VDE 0532:1977 are not completely dealt with in the context of this book, however; details can be found in [17]. In this book the determination of the permissible loading is explained on the basis of IEC 60076-7.

Different types of transformers, such as distribution transformers, medium-sized transformers and large power transformers have to dealt with separately.

Distribution Transformers The maximal permissible loading of distribution transformers with rated apparent power ≤2.5 MVA and with natural oil-cooling, without on-load tap-changer, is determined only by the hot-spot temperature and the thermal aging.

Medium-sized Transformers Medium-sized transformers have rated apparent power ≤100 MVA and an impedance voltage u_k according to Equation 7.6. The influence of the leakage flux is comparatively low.

$$u_k \leq \frac{\left(25 - 0.1 \cdot \frac{3 \cdot S_r}{W}\right)}{100} \tag{7.6}$$

S_r = rated apparent power in MVA
W = number of legs with windings.

Different methods of cooling are considered. For autotransformers an equivalent apparent power S_t and/or an equivalent impedance voltage u_{kt} must be determined according to Equations 7.7 and 7.8.

$$S_t = S_r \cdot \left(\frac{U_{rHV} - U_{rLV}}{U_{rOS}}\right) \leq 100 MVA \tag{7.7a}$$

$$u_{kt} = u_{kr} \cdot \left(\frac{U_{rHV}}{U_{rLV} - U_{rHV}}\right) \leq 0.25 - \frac{S_t}{1000} \tag{7.7b}$$

For other types of autotransformers the rated power per phase (per leg) is important. Equivalent apparent power and equivalent impedance voltage are calculated according to Equation 7.8.

$$S_{\mathrm{t}}=\frac{S_{\mathrm{r}}}{W}\cdot\left(\frac{U_{\mathrm{rHV}}-U_{\mathrm{rLV}}}{U_{\mathrm{rHV}}}\right)\leq 33.3MVA \tag{7.8a}$$

$$u_{\mathrm{kt}}=u_{\mathrm{kr}}\cdot\left(\frac{U_{\mathrm{rHV}}}{U_{\mathrm{rHV}}-U_{\mathrm{rLV}}}\right)\leq 0.25-\frac{S_{\mathrm{t}}}{1000\cdot W} \tag{7.8b}$$

with

U_{rHV}; U_{rLV} = rated voltages of the transformer
S_{r} = rated apparent power (transferable power from HV to LV)
W = number of legs with windings
u_{kr} = rated impedance voltage.

The international standard IEC 60076-7 does not include the restriction regarding the impedance voltage. As for three-phase autotransformers, Equation 7.7 applies.

Large Power Transformers Large power transformers have rated apparent power >100 MVA and an impedance voltage higher than defined by Equation 7.6, with a comparatively high influence of the leakage flux. Regarding the maximum permissible loading of the transformers, IEC differentiates between four different loading conditions:

1. Continuous loading with constant load and constant coolant temperature.
2. Normal cyclic load, with which a basic load and an increased load alternate during a daily load cycle and during the time interval *t*. The load cycle does not lead to so much reduction of the lifetime as does the continuous loading.
3. Long-time emergency operation, which is likewise defined as a cyclic load, the duration of which can amount to weeks or months. Long-time emergency operation will result in a reduction of lifetime, but not to a reduction of the insulation strength.
4. Short-time emergency operation, which can lead to high temperatures in the transformer and to a temporary reduction of the insulation strength. Short-time emergency operation by definition should be accepted only for short time and not regularly, and only if no alternative means are available for a secured supply by the power system. The emergency operation time should be shorter than the thermal time constant of the transformer, typically less than 30 minutes.

In general the temperatures are to be kept below those given in Table 7.6.

The calculation of the maximal permissible loading of the transformers is based on the calculation of the hot-spot temperature and/or the hot-spot temperature rise. Different methods are possible, and in principle one should differentiate between the determination of the temperature from the results of the temperature rise test, suitable for quasi-constant conditions, and the application of thermal models with variable load and variable ambient temperatures (coolant temperatures). The mathematical computation equations are included in IEC 60076-7. Results of these calculations are given in tabulated and graphic form.

Table 7.6 Current and temperature for thermal loading of transformers according to IEC 60067-7.

Type of loading	Unit	Distribution transformer	Medium-size transformer[a]	Large power transformer[b]
Normal cyclic load				
Current related to rated current	%	150	150	130
Temperature of hot-spot or metallic parts having contact with insulation material	°C	120 (140)	120 (140)	120
Temperature of other parts having contact with oil, glass-fiber, and so on	°C	140	140	140
Maximal oil temperature	°C	105	105	105
Long-time emergency operation				
Current related to rated current	%	180	150	130
Temperature of hot-spot or metallic parts having contact with insulation material	°C	140 (150)	140	140 (130)
Temperature of other parts having contact with oil, glass-fiber, and so on	°C	160	160	160
Maximal oil temperature	°C	115	115	115
Short-time emergency operation				
Current in relation to rated current	%	200	180	150
Temperature of hot-spot or metallic parts having contact to insulation material	°C	[c]	160	160
Temperature of other parts having contact with oil, glass-fiber, and so on	°C	[c]	180	180
Maximal oil temperature	°C	[c]	115	115

a The voltage at each winding should not exceed 105% of the rated voltage and/or of the voltage of any step of the tap-changer.
b The restrictive handling of the maximum permissible loading is based on the increase of the heating effect by higher leakage fluxes (in particular with loads above the rated apparent power), on the uncertainties with the measurement and/or computation of the hot-spot temperature from the temperature test and on the high importance of the large power transformers for a reliable electrical power supply.
c Since the duration of long-time and short-time emergency operation of distribution transformers can hardly be controlled, no limit values for hot-spot and oil temperature are indicated in IEC 60067-7. For installation in buildings the increased ambient temperature should be considered.
The limit values for temperature and current shall not be applied at the same time.
Values in parentheses refer to IEC 60354.

7.4.2.2 Continuous Loading

The maximal permissible loading S_{dT} of oil-immersed transformers with constant load and changing cooling temperature and/or ambient temperature, is given by Equation 7.9.

$$S_{dT} = K_{24} \cdot S_{rT} \tag{7.9}$$

The factor K_{24} (one-day-factor) for different kinds of cooling is given in Table 7.7.

Table 7.7 Relative permissible loading (factor K_{24}) of oil-immersed transformers with constant load according to IEC 60067-7.

Transformer cooling		Cooling or ambient temperature (°C) / hot-spot temperature rise(°C)							
		−25/123	−20/118	−10/108	0/98	10/88	20/78	30/68	40/58
Distribution transformer	ONAN	1.37	1.33	1.25	1.17	1.09	1.00	0.91	0.81
Medium-size and large power transformer	ON	1.33	1.30	1.22	1.15	1.08	1.00	0.92	0.82
	OF	1.31	1.28	1.21	1.14	1.08	1.00	0.92	0.83
	OD	1.24	1.22	1.17	1.11	1.06	1.00	0.94	0.87

7.4.2.3 Normal Cyclic Load

Temperatures and aging as described are basic parameters, whereas the data refer in each case to "the normal aging" of the insulation at a hot-spot temperature of 98 °C, according to a hot-spot temperature rise of 78 °C with a coolant temperature of 20 °C. Hot-spot temperatures exceeding 140 °C are not permitted. The typical daily load cycle of the transformer is represented in Figure 7.6. After an initial loading (relative to the rated power) K_1, relative loading K_2 for a period t follows, continued by a load in the range of the initial load for the remaining time as in Figure 7.6a. With two or more high loading periods (Figure 7.6b) with intermediate periods of lower load, the duration of the load t can be set equal to the sum of the durations of the individual high loads. With varying load cycle, an equivalent area of the load line with same peak load and average basic load can be described as in Figure 7.6a. Higher values than 150% of the rated apparent power should be allowed for emergency operation only, while the oil temperature should remain below 115 °C.

Tables and diagrams included in IEC 60067-7 specify separately the different types of cooling, details see table 7.1, ONAN and ONAF as well as OFAF and OFWF, for coolant temperatures between 0 and 40 °C. The loading curves according to IEC 60067-7 can be used also for the determination of the transformer rated apparent power with given load cycle as in Figure 7.6. The procedure is described by two examples as shown in Figure 7.7.

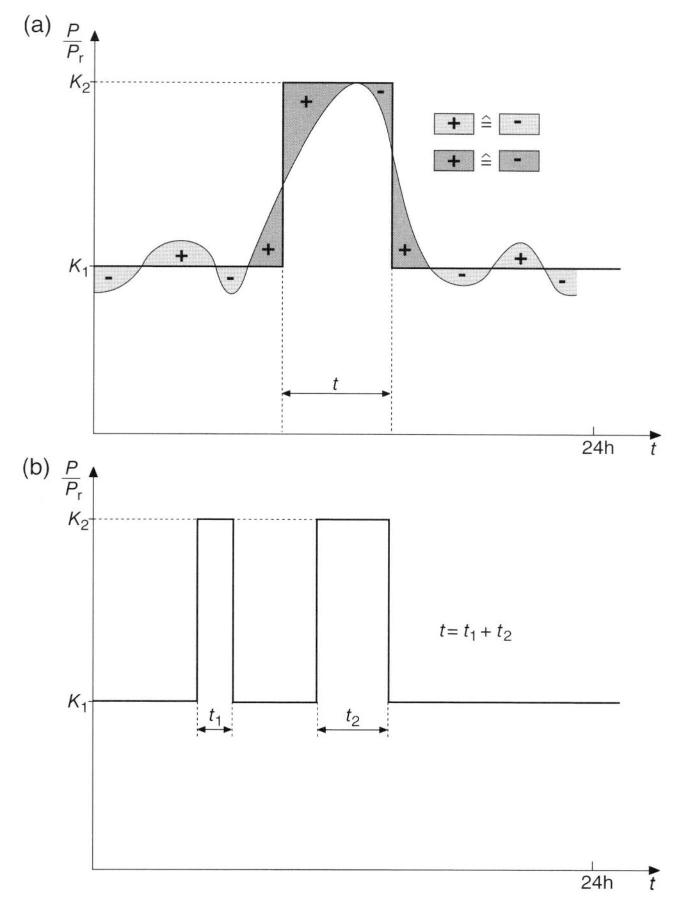

Figure 7.6 Simplified diagram to determine the daily load cycle of oil-immersed transformers. (a) Cyclic load with pronounced peak load and approximation by defined peak load of duration t; (b) series of high loads with same total time duration t as in (a).

Example A The maximal permissible loading of a transformer, ONAF cooling, rated apparent power 40 MVA is to be determined for normal cyclic load. Preloading condition (basic load) is 24 MVA, coolant temperature of 30 °C. The relative preloading condition is $K_1 = 0.6$, thus the maximal permissible loading for 2 hours is ~56 MVA ($K_2 = 1.39$), for 4 hours ~48 MVA ($K_2 = 1.19$) and for 6 hours ~43 MVA ($K_2 = 1.08$).

Example B The rated apparent power of a transformer, ONAF cooling, is to be determined. The transformer supplies a power system with a daily load cycle of 40 MVA/20 h and 68 MVA/4 h, coolant temperature of 30 °C. The ratio of basic to peak load is $k = 68/40 = 1.7$. The rated power is determined as outlined in

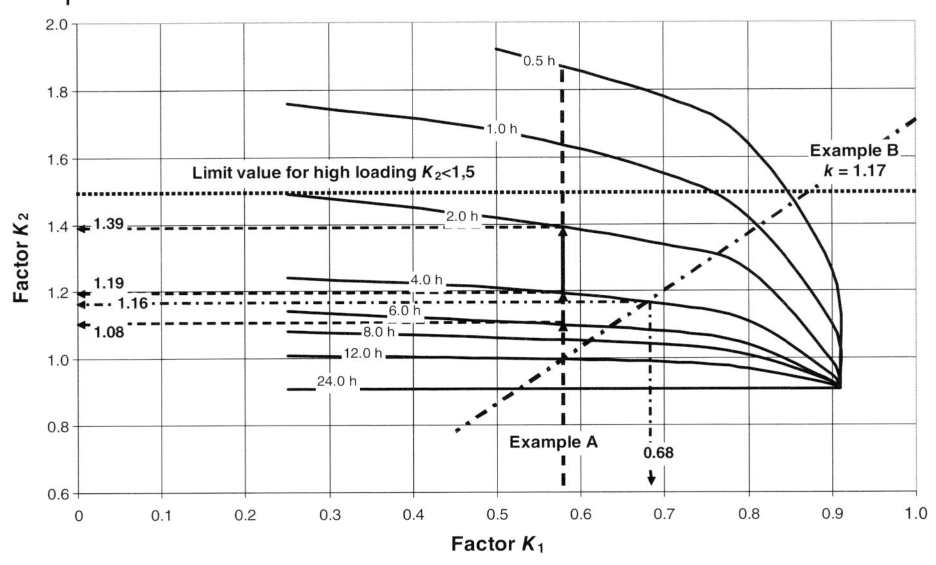

Figure 7.7 Determination of the maximal permissible loading of oil-immersed transformers according to IEC 60067-7 with sample application. Example A: determination of the maximal permissible loading (normal cyclic load) as a function of the loading period, Example B: determination of the rated apparent power with given load cycle [20].

Figure 7.6. Curve B with an upward gradient $k = 1.7$ cuts the 4 h load-curve at $K_2 = 1.16$ and/or $K_1 = 0.68$. Thus the required rated apparent power is $S_{rT} = 68\,\text{MVA}/1.16 = 40\,\text{MVA}/0.68 = 59\,\text{MVA}$.

7.4.2.4 Long-Time and Short-Time Emergency Operation

In the determination of the maximal permissible loading for long-time and short-time emergency operation, the reduction in lifetime is important. The reduction in lifetime can be calculated based on mathematical models. Characteristic values can be obtained in tabulated form from IEC 60067-7. For other data, information from the manufacturer should be taken into account for the determination of maximal permissible loading. For a given load cycle, as explained for the normal cyclic operation, the hot-spot temperature can be determined for defined ambient temperatures as well as the reduction in lifetime resulting from the emergency operation.

The procedures are explained for a medium-size transformer (63 MVA; 14%; ON) in accordance with Table 7.8 (excerpted from IEC 60067-7). The hot-spot temperature is calculated using Equation 7.10a.

$$\vartheta_{HS} = \vartheta_a + \vartheta_{HSO} \tag{7.10a}$$

For long-time emergency operation the hot-spot temperature should not exceed 140 °C, during short-time emergency operation the hot-spot temperature should

Table 7.8 Factors for calculation of hot-spot temperature rise and lifetime reduction of oil-immersed transformers (medium size, ON) according to IEC 60067-7.

	Cooling temperature or ambient temperature ϑ_a (°C)							
	40	**30**	20	10	0	−10	−20	−25
Factor L_1 for the calculation of lifetime reduction	10	**3.2**	1.0	0.32	0.1	0.032	0.01	0.0055

Top value: Factor L_2 for calculation of lifetime reduction
Bottom value: Hot-spot temperature rise ϑ_{HSO} (°C)

K_2	K_1 0.5	0.7	**0.8**	0.9	1.0	1.1
1.0	0.042	0.085	0.154	0.347	1.0	
	72	74	75	76	78	
1.1	0.118	0.205	0.316	0.585	1.39	4.3
	83	85	86	87	89	91
1.2	0.386	0.608	0.844	1.32	2.48	6.15
	94	96	98	99	101	102
1.3	1.41	2.11	**2.76**	3.88	6.12	11.7
	107	109	**110**	112	113	115
1.4	8.27	10.5	**14.0**	19.9	31.7	61.6
	122	123	**125**	126	128	130
1.5	36.0	44.9	58.2	78.7	113	182
	136	137	139	140	142	144
1.6	123	172	213	271	356	490
	148	151	152	153	155	156
1.7	649	901	1100	1390	1800	2410
	164	166	167	168	170	172

K_1 and K_2 are the relative loadings, see Figure 7.7 and text.
Bold type indicates the examples given in the text.

remain below 160 °C. The reduction of lifetime in "normal" days is calculated with Equation 7.11a.

$$L_V = L_1 \cdot L_2 \tag{7.11a}$$

with factors L_1 and L_2, the ambient temperature ϑ_a and the hot-spot temperature ϑ_{HSO} as in Table 7.8.

The transformer is assumed to be operated at a load cycle of 82 MVA during 4 hours daily with a basic load of $K_1 \times S_{rT} \approx 50$ MVA. The ambient temperature is $\vartheta_a = 30$ °C. The load factors are $K_1 \approx 0.8$ und $K_2 \approx 1.3$. In accordance with Table 7.8 (upper part) the factor is $L_1 = 3.2$ and for the lower part the factor is $L_2 = 2.76$, with a hot-spot temperature rise of $\vartheta_{HSO} = 110$ °C. Thus the reduction in lifetime is about nine "normal" days and the hot-spot temperature is 140 °C according to Equation 7.10.

$$\vartheta_{HS} = \vartheta_a + \vartheta_{HSO} = 30\,°C + 110\,°C = 140\,°C \qquad (7.10b)$$

$$L_V = L_1 \times L_2 = 2.76 \times 3.2 = 8.83 \qquad (7.11b)$$

The load cycle assumed is permissible for long-time emergency operation, since the hot-spot temperature remains below 140 °C. This is also valid for short-time emergency operation. For a high load of $K_2 \times S_{rT} = 88$ MVA with otherwise the same conditions, the factor is $L_2 = 14.0$ and the hot-spot temperature rise is $\vartheta_{HSO} = 125$ °C. The hot-spot temperature is therefore 155 °C; long-time emergency operation is no longer permissible, only the short-time emergency operation would be permitted. The reduction of lifetime for short-time emergency operation cannot be defined, as the short-time emergency operation by definition arises only for a short time and not regularly, that is, not every day. It should be noted that a duration of 4 h with high loading is hardly to be defined as short-time emergency operation.

7.4.3 Maximal Permissible Loading of Dry-Type Transformers

The determination of maximal permissible loading of dry-type transformers is defined in IEC 60076-12 "Loading guide for dry-type power transformers" [21] and [22]. Since long-term experiences regarding the aging of insulation of dry-type transformers is so far lacking, and the development of new insulation materials is a continuing process, all standardization attempts must accordingly be only short-dated.

Investigations on insulation of dry-type transformers have indicated that the aging rate is doubled with a rise in temperature of 10 °C. Thus, a basis is found similar to considerations for oil-immersed transformers for the determination of the maximal permissible loading.

The maximal permissible loading depends also on the maximal permissible winding temperature, which is given for different classes of insulation materials (see Table 7.4) in Table 7.9. The base value of the hot-spot temperature serves as reference value for the normal aging rate with an ambient temperature of 20 °C. The maximum value of the hot-spot temperature must not be exceeded under any circumstances because of the danger of insulation failure. Dry-type transformers have air as cooling agent. In contrast to oil-immersed transformers, with the insulating oil serving at the same time as cooling agent, a thermal time constant for air as the cooling agent is lacking.

If the temperatures given in Table 7.9 are exceeded by less than 5 °C, the permissible temperature of the winding is to be reduced by 5 °C. When the excess over ambient temperature is between 5 °C and 10 °C, the permissible temperature of the windings is to be reduced by 10 °C. In installations located higher than 1000 m above sea level the permissible temperature rise is to be reduced by 2.5% in case of natural air cooling and by 5% in case of forced air cooling for each 500 m above 1000 m.

Table 7.9 Maximal permissible temperature rise of dry-type transformers according to IEC 60076-12.

	Temperature of insulation (°C)	Thermal class	Maximal permissible winding temperature rise (°C)	Hot-spot temperature (°C)	
				Base value	Maximal value
Winding temperature rise measured through impedance measurement	105	A	60	95	140
	120	E	75	110	155
	130	B	80	120	165
	155	F	100	145	190
	180	H	125	175	220
	220	C	150	210	250

For the determination of the maximal permissible loading of dry-type transformers the hot-spot temperature measured during the heat-run test is taken as the basic value, while taking into account different load cycles. In accordance with IEC 60076-12 this procedure is preferable to calculation.

The calculation method proceeds from a standardized daily load cycle as described for oil-immersed transformers. Starting from an initial load K_1 (ratio of the rated apparent power) for a period of t_1 hours a higher load K_2 for a period of t_2 hours follows, and then a load at the magnitude of the initial load for the remaining time, see Figure 7.6a. The hot-spot temperature ϑ_{HS} is increased by the load increase from ϑ_{HS1} to ϑ_{HS2} during the time t_2 of the higher load. The increase of the hot-spot temperature is calculated from Equation 7.12,

$$\vartheta_{HS} = \vartheta_{HS1} + (\vartheta_{HS2} + \vartheta_{HS1}) \times (1 - e^{t/\tau}) \tag{7.12}$$

with the hot-spot temperature ϑ_{HS2} at the end of the higher loading K_2 as in Equation 7.13.

$$\vartheta_{HS2} = Z \times \vartheta_{HS} \times K_2^q \tag{7.13}$$

where

ϑ_{HS} = basic value (rated value) of hot-spot temperature as in Table 7.9
K_2 = high loading value (Figure 7.6a)
Z = relation of hot-spot temperature rise to average winding temperature ($Z \approx 1.25$)
q = factor taking account of air circulation; $q = 1.6$ for natural convection
τ = thermal time-constant of winding (typical $\tau \approx 0.5$–$1.0\,h$).

Typical figures indicating the loading of dry-type transformers are given in IEC 60076-12 for the different thermal classes according to Table 7.9, thermal time-constants $\tau = 0.5\,h$ and $\tau = 1.0\,h$ and ambient temperatures 10, 20 and 30 °C. It is assumed that the ambient temperature during the daily load cycle (24 h) is con-

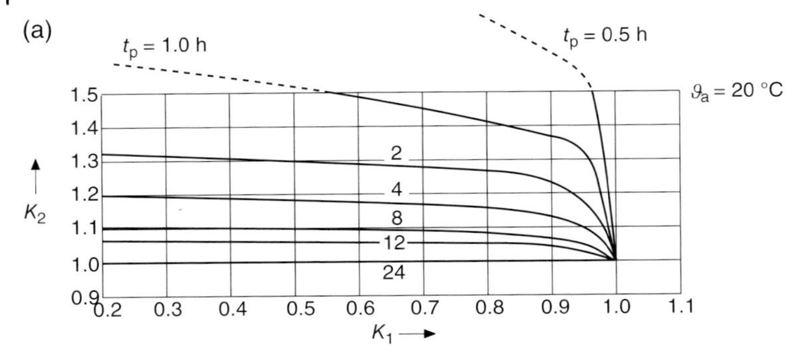

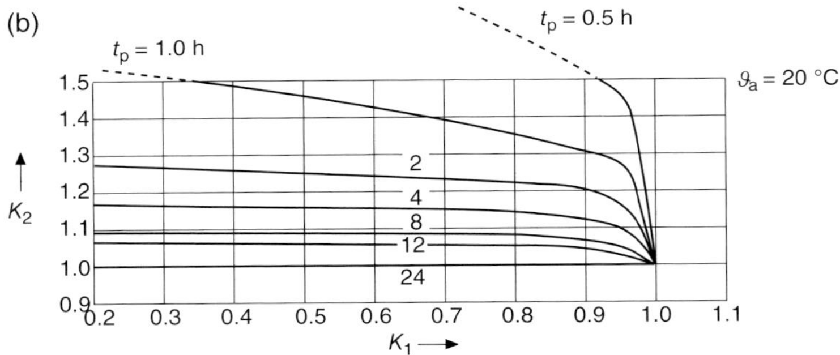

Figure 7.8 Diagram for determination of maximal permissible loading of dry-type transformers according to IEC 60076-12, τ = 1.0 h; ambient temperature ϑ_U = 20 °C. (a) Thermal class F (155 °C); (b) thermal class H (180 °C).

stant. The maximum current is limited to 150% of the rated current. Two examples for transformers of thermal classes F and H are given according to Figure 7.8.

The rated apparent power of a transformer for a given load cycle, 1.0 MVA for 20 h and 1.4 MVA for $t_2 = 4$ h, is to be determined. From Figure 7.8a (thermal class F, 155 °C), the values $K_1 = 0.83$ and $K_2 = 1.16$ are obtained. Thus the required rated apparent power is given by Equation 7.14a.

$$S_r \geq \frac{1.0\,\text{MVA}}{0.83} = \frac{1.4\,\text{MVA}}{1.16} = 1.20\,\text{MVA} \tag{7.14a}$$

For a transformer of thermal class H (180 °C), Figure 7.8b indicates $K_1 = 0.81$ and $K_2 = 1.13$. Thus the rated apparent power is given by Equation 7.14b.

$$S_r \geq \frac{1.0\,\text{MVA}}{0.81} = \frac{1.4\,\text{MVA}}{1.13} = 1.235\,\text{MVA} \tag{7.14b}$$

The transformer rated power can be chosen to be 1.25 MVA of the thermal class F or H. The loading reserves differ somewhat depending on the class selected.

7.5 Economical Operation of Transformers

The costs of transformers are determined by the investment cost, the cost of ohmic and no-load losses and the cost for reserve in case of outages. Additionally, transformer costs are influenced by the load factor, the loss factor (see Chapter 4) and the relation of yearly peak load to the rated apparent power.

The investment cost rises nonlinearly with the rated apparent power. The doubling of the rated apparent power will result in a cost increase of only around 30 to 40% if all other parameters are kept constant as far as possible. The tendency to install only a few transformers with high rated apparent power is challenged by the necessity for reserve for outages. The reserve for outage in case of two transformers operated in parallel must be 50%; with three transformers the reserve is 33% of the rated apparent power.

The short-circuit losses depend on the current, whereas the no-load losses depend on the voltage. Due to construction constraints, the no-load losses increase with decreasing impedance voltage, whereas the short-circuit losses increase with increasing impedance voltage. In particular, this dependence is seen with low-voltage transformers, leading to the efficiency η_T according to Equation 7.15.

$$\eta_T = 1 - \frac{(P_o - P_k)}{S_{rT}} \tag{7.15}$$

with

P_o = no-load losses
P_k = short-circuit losses
S_{rT} = rated apparent power.

The related graph is called the efficiency curve, as outlined in Figure 7.9. As can be seen from the figure, the maximal efficiency is with a loading of 55–70% of the rated apparent power. The total losses then amount to less than 1% of the rated power.

The present value K_0 of the total costs for a transformer is calculated with Equation 7.16.

$$K_0 = K_A + (k_p + K_W \cdot T_m) \cdot \frac{P_l}{r_n} + (k_P + k_W \cdot T_m \cdot \vartheta) \cdot \left(\frac{S_{max}}{S_{rT}}\right)^2 \cdot \frac{P_k}{r_n} \tag{7.16}$$

with

K_A = investment cost
k_P = cost per unit power
k_W = cost per unit energy
T_m = yearly operating hours
r_n = annuity factor (see Eq. 4.3)

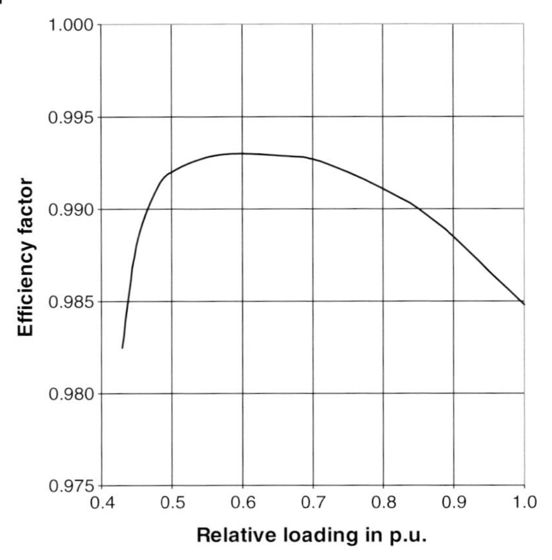

Figure 7.9 Efficiency curve of a LV transformer (S_{rT} = 630 kVA; u_k = 4%).

ϑ = loss factor (see Section 4.2, Chapter 4)
P_l = no-load losses
P_k = short-circuit losses
S_{max} = yearly peak load
S_r = rated apparent power.

For low-voltage transformers the most economical operation, determined by the minimal relative cost per kWh, is given for approximately 5000 yearly full-load hours. For fewer yearly full-load hours, economical operation is possible only with a loading exceeding the rated apparent power. LV transformers in the public energy supply generally exhibit less than 4500 yearly full-load hours. Hence it follows that the transformer will be operated temporarily with higher load than the rated power. If this period coincides with times of lower outside temperatures, this will not result in a reduction of the lifetime. Generally it can be said that for yearly full-load hours below 3500, the maximal permissible loading according to IEC 60076-7 and/or IEC 60076-12 is applicable; for full-load operating hours exceeding 3500, the economic aspects of operation are dominant.

7.6 Short-Circuit Strength

Calculation methods for the electromagnetic and thermal effects of short-circuit currents are defined in IEC 60865-1 (VDE 0103), which applies to single-phase AC current and three-phase AC installations with rated voltages up to 420 kV. Here, the calculation of the thermal withstand capability to short-circuit currents is

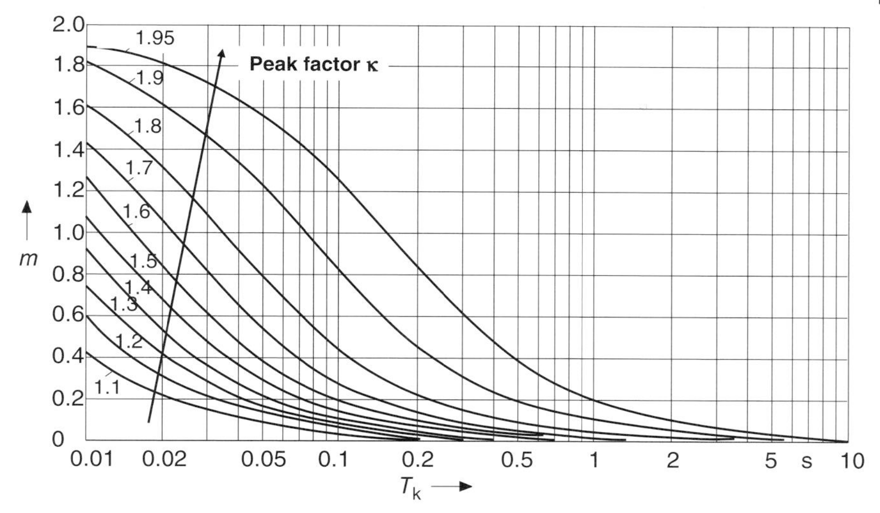

Figure 7.10 Factor *m* for calculation of short-duration short-circuit (heat dissipation of DC component) according to figure 21 of IEC 60909-0.

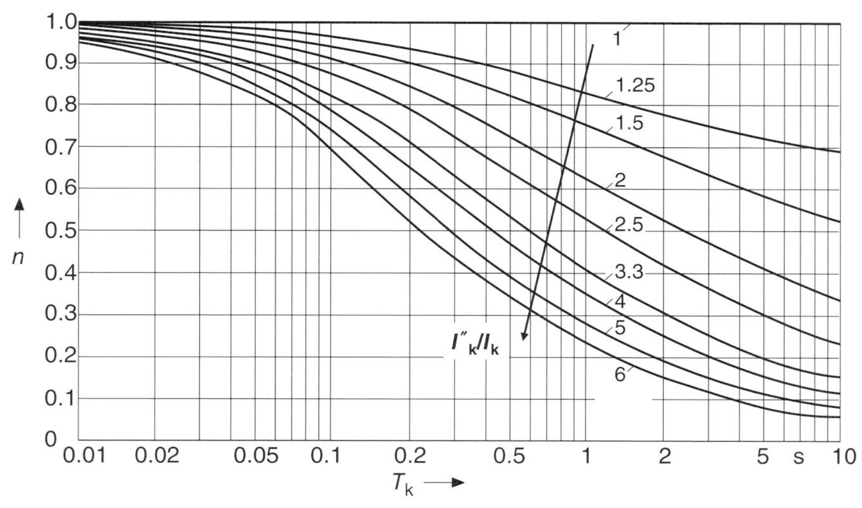

Figure 7.11 Factor *n* for calculation of short-duration short-circuit (heat dissipation of AC component) according to figure 22 of IEC 60909-0.

described. More details are outlined in [23]. According to IEC 60865-1, separate conditions are to be established for equipment (transformers, transducer, etc.) and conductors (bus ducts, bars, cables, etc.).

The thermally equivalent short-time current I_{th} is calculated according to Equation 7.16a, based on the amount of heat Q generated in a conductor of resistance R during the short-circuit duration T_k.

$$I_{th} = \sqrt{\frac{Q}{R \cdot T_k}} = \sqrt{\frac{\int_0^{T_k} i_k^2(t)\mathrm{d}t}{R \cdot T_k}} \tag{7.16a}$$

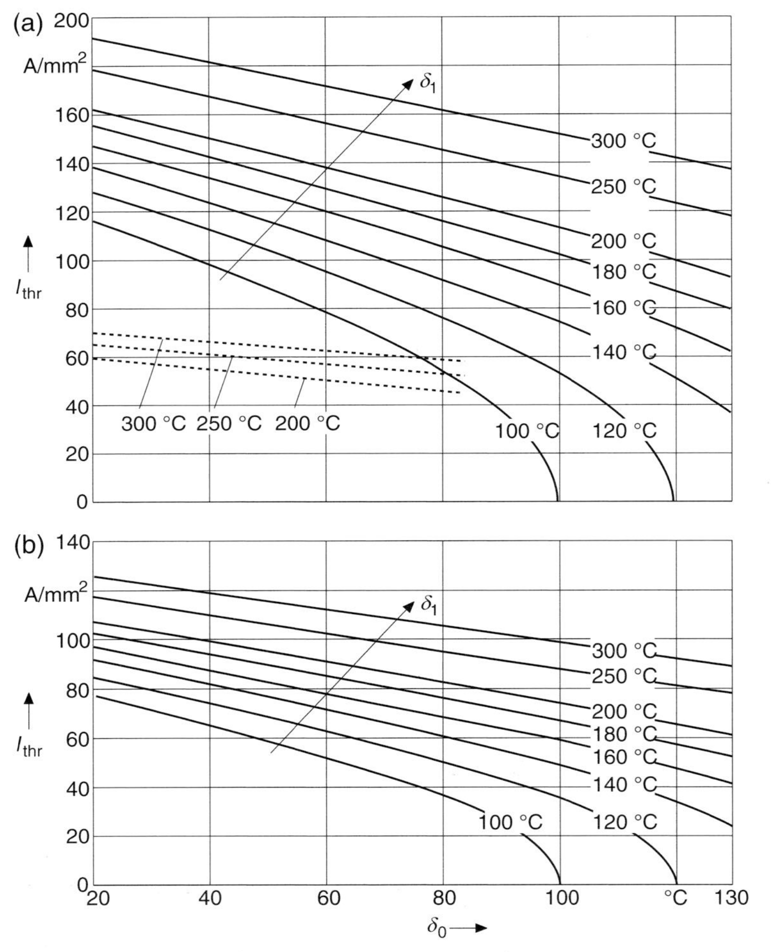

Figure 7.12 Rated short-duration short-circuit current density of conductors. δ_0, Temperature before short-circuit; δ_1, temperature at end of short-circuit. (a) Copper; (b) aluminum.

The thermally equivalent short-time current can also be determined taking account of the heat dissipation factors m and n according to Figures 7.10 and 7.11, considering the thermal effect of the DC and AC components of the short-circuit current with Equation 7.16b,

$$I_{th} = I_k'' \cdot \sqrt{m+n} \tag{7.16b}$$

with the initial short-circuit current I_k''. IEC 60909-0 includes calculation equations for the heat dissipation factors m and n as in Figure 7.10 and 7.11. The thermally equivalent short-circuit current I_{th} in the case of several short-circuits with different time durations T_{ki} and currents I_{thi} is calculated as in Equation 7.17a.

$$I_{th} = \sqrt{\frac{1}{T_k} \sum_{i=1}^{n} I_{thi}^2 \cdot T_{ki}} \tag{7.17a}$$

$$\text{with} \quad T_k = \sum_{i=1}^{n} T_{ki} \tag{7.17b}$$

The thermal withstand capability is achieved if the rated short-duration short-circuit current I_{thr}, given by the manufacturer is lower than the value calculated from Equation 7.18a in case the short-circuit duration is less than 1 s. If the short-circuit duration is longer, Equation 7.18b applies. The thermally permissible values for different conductor materials are to be obtained from Figure 7.12.

$$I_{th} \leq I_{thr} \quad \text{for} \quad T_k \leq T_{kr} \tag{7.18a}$$

$$I_{th} \leq I_{thr} \cdot \sqrt{\frac{T_{kr}}{T_k}} \quad \text{for} \quad T_k \geq T_{kr} \tag{7.18b}$$

8
Cable Systems

8.1
General

Cables are used for the transmission and distribution of electrical energy in public and industrial power systems. The permissible loading of the cables is determined by different parameters such as environmental conditions, type of laying (in ground or in air), cable design and type of insulation, operating conditions and so on. Conductors are made of aluminum or copper. The insulation is of various materials: PVC (polyvinyl chloride) and PE (polyethylene) are used as standard used in LV and MV cables; oil-insulation and gas-pressure cables can still be found in HV systems ($U_n > 110\,kV$), whereas XLPE (cross-linked polyethylene)-insulated cables are today standard in power systems with nominal voltages of 110 kV and above. Mass-impregnated paper-insulated cables are still in use in the medium-voltage range, but are found only on older cable routes; this type will no longer be installed.

Cable abbreviation codes are used that indicate the material of the cable from the inner layer to outer layers. Copper conductors, mass-impregnated paper-insulated cables, and internal protection shields are not specially indicated. In addition to the coding of the inner construction, the number of conductors, the cross-section and the shape of the conduct as well as the nominal voltage (line-to-ground / line-to-line) is indicated. Special coding is defined in the specific cable standards or can be found in [24]. Abbreviation codes for impregnated paper-insulated cables and cables with PVC or XLPE insulation are listed in Tables 8.1 and 8.2.

Conductor shape and type are identified as

RE Solid round conductor
RM Stranded round conductor
SE Solid sector-shape conductor
SM Stranded sector-shape conductor
RF Flexible stranded round conductor.

Power System Engineering: Planning, Design and Operation of Power Systems and Equipment, Second Edition.
Jürgen Schlabbach and Karl-Heinz Rofalski.

Table 8.1 Alphanumeric abbreviations for cables with impregnated paper insulation.

Abbreviation	Meaning
A	Aluminum conductor
H	Screening for "Hochstadter" cable
E	Individual wires wrapped with metal screen and corrosion protection
K	Lead alloy screen
KL	Pressed or smooth extruded aluminum jacket
KLD	Pressed or smooth extruded aluminum jacket, extension elements
u	Nonmagnetic
D	Pressure bandage
E	Protective covering with embedded layer of elastic bands or foils
D	Nonmagnetic pressure bandage
v	Twisted conductors
F	Armoring of galvanized steel strips with retaining metallic spiral of metal strip
Gl	Nonmagnetic gliding wires
u	Nontwisted conductors
St	Steel tube
B	Armoring of galvanized steel strips
F	Armoring of galvanized flat steel wires
FO	Armoring of galvanized flat steel wires, open
R	Armoring of galvanized round steel wires
RO	Armoring of galvanized round steel wires, open
Gb	Retaining spiral of galvanized steel strip (anti-twist tape)
A	Protective cover of fibers
AA	Double protective cover of fibers or glass-fiber tapes
Y	Outer PVC-sheath
2Y	Outer sheath of thermoplastic material (PE)
YV	Reinforced PVC sheath

8.2 Construction Details

Within the low-voltage range, PVC or PE insulated cables with an outer protective covering or sheath of PVC are used. For increased protection against touch voltages, concentric copper screens of round wires are applied to the insulation. If waveform screens are applied, the screen must not be cut for connecting coupling T-joints (cable type NYCWY). For higher tensile strength, cables with longitudinal galvanized steel wires as outer armoring are used (cable type NYFGY). To protect cables against penetration of aromatic hydrocarbons, a lead alloy sheath is used covered by the outer sheath of PVC (cable type NYKY).

Within the medium-voltage range (3.6/6 kV to 18/30 kV), cables without electric field limiting screens can still be used, as the inhomogeneous field causes

Table 8.2 Alphanumeric abbreviations for cables with plastic insulation (PVC, PE, XLPE).

Abbreviation	Meaning
A	Aluminum conductor
I	House wiring cables
Y	PVC thermoplastic insulation
2Y	PE thermoplastic insulation
2X	Polymerized (cross-linked) PE insulation (XLPE)
H	Conductive, electric field limiting layer, covering conductor and insulation
HX	Insulation of interlaced halogen-free polymer mixture
C	Concentric conductive layer of copper
CW	Concentric corrugated conductive layer of copper
CE	Concentric conductive layer of copper, applied to each core of multiple-core cables
S	Copper screen or screen
SE	Conductive electric field limiting layer, covering conductor, insulation, and copper screen, applied to each core of multiple-core cables
K	Metal screen of lead alloy
F	Metal jacket screen of galvanized flat steel wires
R	Armoring of galvanized round steel wires
Gb	Retaining spiral of galvanized steel strip (anti-twist tape)
HX	Coating of cross-linked halogen-free polymer mixture
Y	Protective cover between screen or concentric layer and armoring made of PVC
Y	PVC outer covering
2Y	Outer covering of thermoplastic material (PE)
FE	Flame resistance of insulation, flash point to be specified

relatively low electric field strength in the tangential direction in the insulation. For higher voltages, only cables with homogeneous electric field are used. An electric field-limiting screen or foil ensures a homogeneous electric field in the dielectricum; the electric field strength is only effective in the radial direction. Each cable core is covered with its own sheath of lead alloy. To protect the sheath against mechanical damage, additional armoring by steel tapes or wires is used (cable type NEKEBA). Today cables with plastic insulation (PVC, PE or XLPE) are used in the medium-voltage range, whereas PVC insulations are used only for voltages up to 10 kV due to the high dissipation factor (cable type NYSEY). XLPE insulation (cable type N2XS2Y) is used for higher voltages (110 kV and above). The main electrical parameters of cable insulation materials are shown in Table 8.3.

Table 8.3 Electrical parameters of insulation materials for cables.

	Oil	PVC	XLPE	Impregnated paper
Dissipation factor $\tan\delta$	3	100	0.55	10
Max. temperature ϑ_{max} (°C)	85	70	90	65–85
Specific resistance ρ_I (Ω m)	5×10^{12}	7×10^{11}	10^{14}	5×10^{14}
Maximum electric field strength E_D (kV mm^{-1})	15–25	40	95	15–40
Permittivity ε_r at 20 °C	2.2–2.8	3–4	2.4	3.3–4.2

Cables with oil insulation and XLPE insulation are installed in MV and HV systems ($U_n > 30$ kV). A typical cable used in 110 kV systems is the gas-pressure type cable. The cable is placed in a hermetically sealed steel pipe filled with nitrogen (pressure 1.6 MPa). The steel pipes are either constructed as seamless pipe or as lengthwise-welded pipe according to DIN EN 10220. Gas-pressure cables are suitable because of their mechanical withstand capability, the low inductive interference and the possibility of laying the steel pipe in short segments, which is an outstanding advantage in crowded urban areas. In practice, the importance of oil-insulated cables and gas-pressure cables has decreased and XLPE-insulated cables are installed even in EHV systems.

8.3 Electrical Parameters of Cables

Insulation materials of cables have a high permittivity ε_r compared with air or insulation gases. As the distance between the conductors under high voltage and the outer screen at ground potential is small, the capacitances of cables are much larger than those of overhead transmission lines. Values for the capacitances in the positive-sequence system (symmetrical operation) and the capacitive no-load current of cables are given in Figures 8.1 and 8.2 for different types of HV cables.

The reactances of cables in the positive-sequence and zero-sequence component depend strongly on the cable construction and the type of laying (triangle or flat formation in the case of single-core cables). Exact data can only be obtained through measurements and/or with detailed analysis of the magnetic field. Reference values for the inductive reactances of different types of cables in the positive-sequence component are presented in Figure 8.3.

The reactance in the zero-sequence component depends additionally on the presence of other conducting material in the ground or in the vicinity, such as pipelines, earthed shields, screens and armoring of other cables, earthing conductors, and so on, as the current in the zero-sequence component flows through all earthed and conducting materials and not only through the screen and sheath

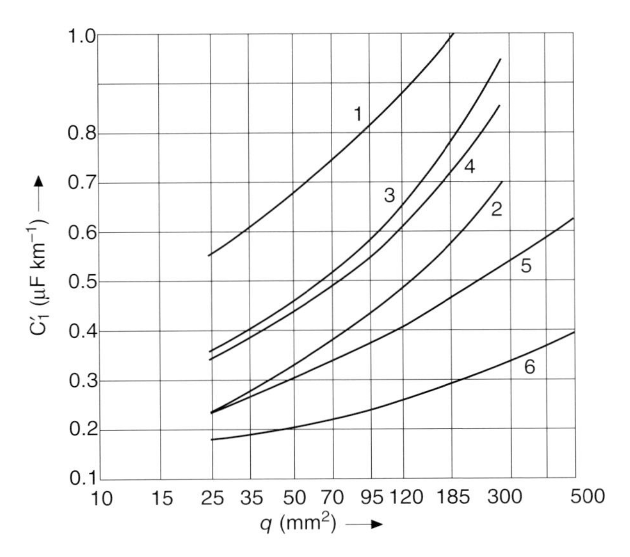

Figure 8.1 Capacities of high-voltage cables ($U/U_0 <$ 12/20 kV). **1**, Single-core cable NKBA 0.6/1 kV; **2**, as 1, 6/10 kV; **3**, three-core cable NEKEBY 6/10 kV; **4**, PVC-insulated cable NYSEY 6/10 kV; **5**, XLPE-insulated cable N2XSEY 6/10 kV; **6**, XLPE-insulated cable N2XSEYBY 12/20 kV.

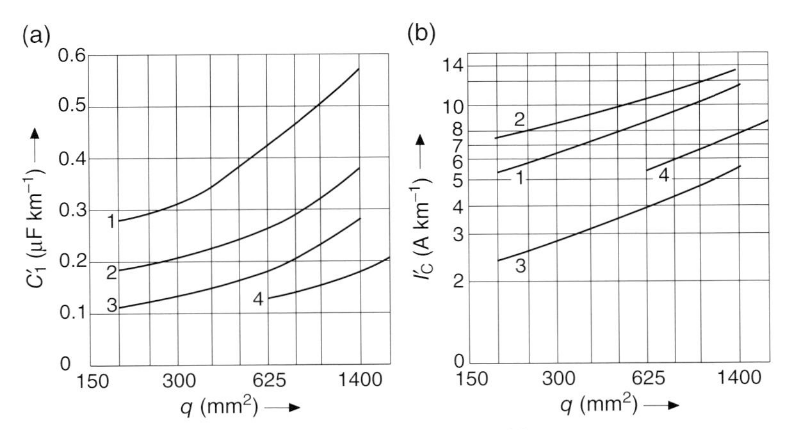

Figure 8.2 Characteristic parameters of high-voltage cables. **1**, Single-core oil-insulated (oil-filled) cable 110 kV; **2**, as 1, 220 kV; **3**, XLPE-insulated cable 110 kV; **4**, XLPE-insulated cable 220 kV. (a) Capacities C_1' (identical to C_0'); (b) charging currents (no-load) I_C'.

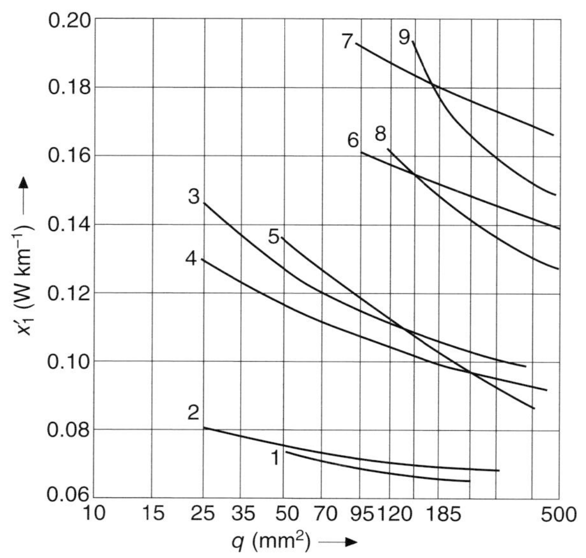

Figure 8.3 Inductive reactances (positive-sequence component) of three-phase cables. **1**, 0.6/1 kV, four-core cable, NKBA; **2**, 0.6/1 kV, four-core cable, NA2XY; **3** three-core cable with armoring 10 kV; **4**, PVC-insulated cable 6/10 kV, NYFGbY; **5**, VPE-insulated cable 6/10 kV, NA2XSEY; **6**, single-core oil-filled cable 110 kV, triangle formation; **7**, single-core oil-filled cable 110 kV, flat formation; **8**, XLPE-insulated 110 kV triangle formation; **9**, XLPE-insulated cable 110 kV flat formation, distance 15 cm.

of the cable concerned. For this reason, measurements of the impedances in the zero-sequence component should be carried out during commissioning and in high-voltage systems should be repeated at intervals of at most some years.

8.4
Losses and Permissible Current

8.4.1
General

The permissible current of cables is determined by maximal permissible temperature of the cable insulation, which should not be exceeded at any time so as to protect the insulation against deterioration. Table 8.4 indicates the maximal permissible temperatures of insulating materials.

Cable losses originate from current-dependent losses in conductor, metallic sheath, coatings, screens and armoring and possibly in outer metal tubes as well as from current-independent losses in the dielectricum of the cable insulation. These losses heat the individual parts of the cable construction and must be trans-

Table 8.4 Maximal permissible temperature of cable insulation.

Cable type	Maximal permissible temperature (°C)	
	Operating temperature	At end of short-circuit
	Copper conductor	
Cable with soft-soldered joints	—	160
VPE and XLPE insulation	90	250
PVC insulation		
≤300 mm²	70	160
>300 mm²	70	140
Oil-impregnated insulation		
0.6/1 kV	80	250
3.6/6 kV	80	170
6/10 kV	70	170
12/20 kV	65	170
18/30 kV	60	150
Electric field-limited insulation 12/20 kV	65	170
	Aluminum conductor	
VPE and XLPE insulation	90	250
PVC insulation		
≤300 mm²	70	160
>300 mm²	70	140
Oil-impregnated insulation		
0.6/1 kV	80	250
3.6/6 kV	80	170
6/10 kV	70	170
12/20 kV	65	170
18/30 kV	60	150
Electric field-limited insulation 12/20 kV	65	170

ferred through the different layers to the surrounding soil, or to the surrounding air if laid in air. Generally one starts from the thermal equivalent diagram as shown in Figure 8.4. The maximal permissible current is calculated by Equation 8.1.

$$\Delta\vartheta = \sum T \cdot P_{\text{tot}} \tag{8.1}$$

with

$\Delta\vartheta$ = temperature difference
T = thermal resistance
P_{tot} = losses.

The different types of losses are to be determined in detail. Based on the knowledge of the thermal resistances of the cable construction and the surrounding soil

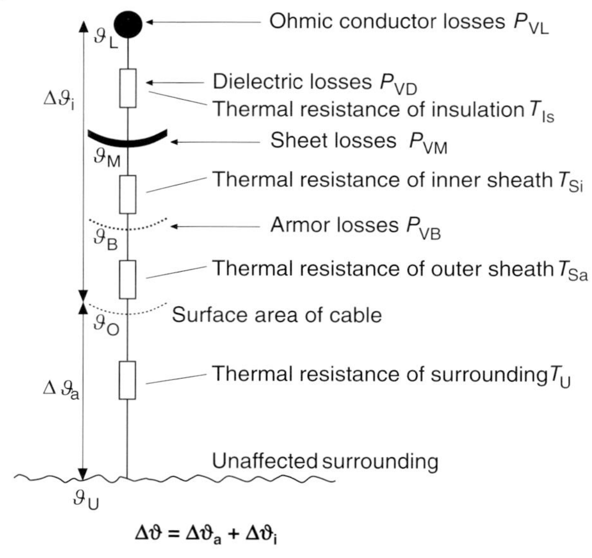

Figure 8.4 Thermal equivalent diagram for cables.

or air, either the temperatures at individuals layers in the cable for given load current or the maximal permissible current for given temperatures can be calculated.

8.4.2 Calculation of Losses

Cable losses consist of two parts, current-dependent and current-independent losses. Losses of the conductor are given by Equation 8.2.

$$P_{VL} = I^2 \cdot R'_{20} \cdot [1 + \alpha \cdot (\vartheta_L - 20\,°C)] \cdot F_S \cdot F_P \tag{8.2}$$

with

R'_{20} = DC resistance per unit length at 20 °C
I = conductor current
α = temperature coefficient
ϑ_L = conductor temperature in °C
F_S = factor of skin effect (to be considered with cross-sections above 185 mm^2)
F_P = factor of proximity effect (to be considered with cross-sections above 185 mm^2).

Methods for calculation of the factors of skin effect and proximity effect are outlined in [23, 24], denoted there by y_S and y_P. Losses in a steel pipe considered as outside casing are taken into account by adjusted factors for skin effect and proximity effect. Depending upon the arrangement of the cable cores in the pipe and the presence of armoring, the losses in the steel pipe can amount up 70% of the losses by skin and proximity effects [25]. The calculation of the resistance per unit length considering the skin and proximity effect is valid up to cable cross-sections of 1500 mm^2 (copper conductors) and up to 2000 mm^2 (aluminum conductors). For larger cross-sections and for large hollow conductors, additional considerations are needed [25].

Losses occur also in screens, sheaths and armoring due to eddy-current and longitudinal currents and are likewise current-dependent. In the case of magnetic materials, as used in pressure bandages and in metallic casings and pipes, magnetization losses arise. The losses in sheaths and screens depend strongly on the kind of earthing of sheaths and screens. With earthing at one end, no longitudinal current can flow and the losses are accordingly small; with earthing at both ends, the losses in sheaths and screens can be up to 40% of the total losses of the cable depending on the conductor cross-section [25].

If sheaths are cross-bonded, that is, the sheath is grounded at the beginning, at the end and with one-third and two-thirds of the length with cyclic exchange of the sheath sections, the sheath losses are reduced to an insignificant part of the total losses. In practice, an exact cross-bonding of the sheaths (at each third of the cable length) usually cannot be realized; the sheath currents of the three length sections do not compensate themselves completely and the sheath losses are higher in this case. The sheath losses by longitudinal currents are taken into account by a factor λ_1' and the eddy-current losses by a factor λ_1'' as multiples of the conductor losses according to Equation 8.3a.

$$P_{VM}' = P_{VL}' \cdot (\lambda_1' + \lambda_1'') \tag{8.3a}$$

The detailed procedure for the calculation of losses by Equation 8.3a is outlined in [23, 24]. If the cross-bonding locations of the cable system are not exactly known in the planning and design stage, the respective factor λ_1' should be assumed to be $\lambda_1' = 0.03$ for cables buried directly in the ground and $\lambda_1' = 0.05$ for cables buried in pipes and ducts.

Longitudinal current losses and eddy-current losses in armoring and steel tubes are considered in a similar way by the factors λ_2' for the losses by longitudinal current and λ_2'' for the eddy-current losses according to Equation 8.3b.

$$P_{VB}' = P_{VL}' \cdot (\lambda_2' + \lambda_2'') \tag{8.3b}$$

The detailed procedure for the calculation of losses by Equation 8.3b can be found in [25, 26].

The current-independent dielectric losses depend on the dissipation factor $\tan\delta$ and on the capacitance of the cable C_1'' (positive-sequence component) and can be calculated with Equations 8.4 and 8.5.

$$P'_{VD} = \omega \cdot C'_1 \cdot \left(\frac{U_n}{\sqrt{3}}\right)^2 \cdot \tan\delta \tag{8.4}$$

$$C'_1 = \frac{2 \cdot \pi \cdot \varepsilon_0 \cdot \varepsilon_r}{\ln\left(\frac{d_I}{d_L}\right)} \tag{8.5}$$

with

ε_0 = permittivity ($\varepsilon_0 = 8.8542 \times 10^{-12}\,\mathrm{A\,s\,V^{-1}m^{-1}}$)
ε_r = relative permittivity (see Table 8.3)
d_I = outer diameter of insulation
d_L = conductor diameter.

For detailed analysis, the thermal dependence of the dissipation factor $\tan\delta$ has also to be considered.

8.4.3
Soil Characteristics

The determination of the heat dissipation to the surrounding soil requires knowledge of the thermal characteristics (temperature and thermal conductivity) of the soil. The temperature of the soil changes with laying depth, yearly seasonal changes, the mean air temperature and the type of surface (grass, bitumen, concrete). The mean monthly soil temperature for Central European conditions is outlined in Figure 8.5. The temperature of a grass-covered earth surface corresponds approximately to the ambient air in the respective period. If the surface is sealed with bitumen, the temperature of the earth surface varies in a wide range, summer season having far higher and winter season lower temperatures. As can be seen in Figure. 8.6, the soil temperature at depths larger than 3–4 m is nearly constant at ~10 °C; typically no cables are laid in this depth.

Figure 8.6 indicates the seasonal variation of soil temperature under grass-covered surfaces in Central Europe in different depths [27]. It can be observed that the soil temperature follows a sine function similar to that of the solar radiation with a time delay of several weeks depending on the depths. At larger depth the maximum soil temperature is lower and at ~10 m depth is roughly constant at 8.5–10 °C.

The thermal conductivity of the soil depends on the water content of the soil, expressed by the humidity content *m*, and on the porosity *G* (the ratio of cavity volume to the total volume). For conditions surrounding cables those types of soil are desirable which consist of consolidated sand and sand–gravel mixtures with flat grain distributions, as the fine sand portions fill up the cavities between the coarser constituents. A loam proportion of ~10% leads to a lower drainage of the soil, improves the compressibility of the soil and increases the heat conductivity. Figure 8.7 indicates the thermal conductivity of sandy soil as function of the temperature and of the humidity content *m*. It is assumed that the humidity content

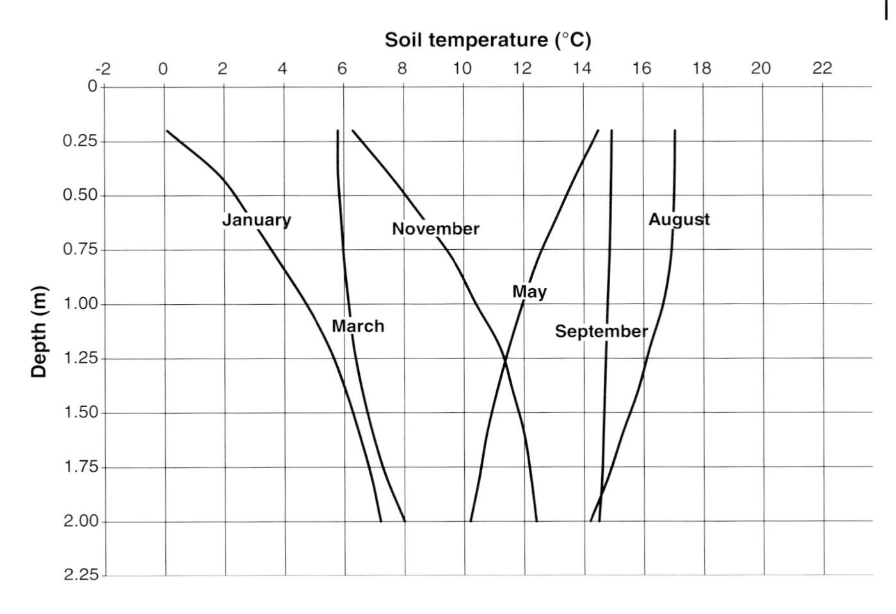

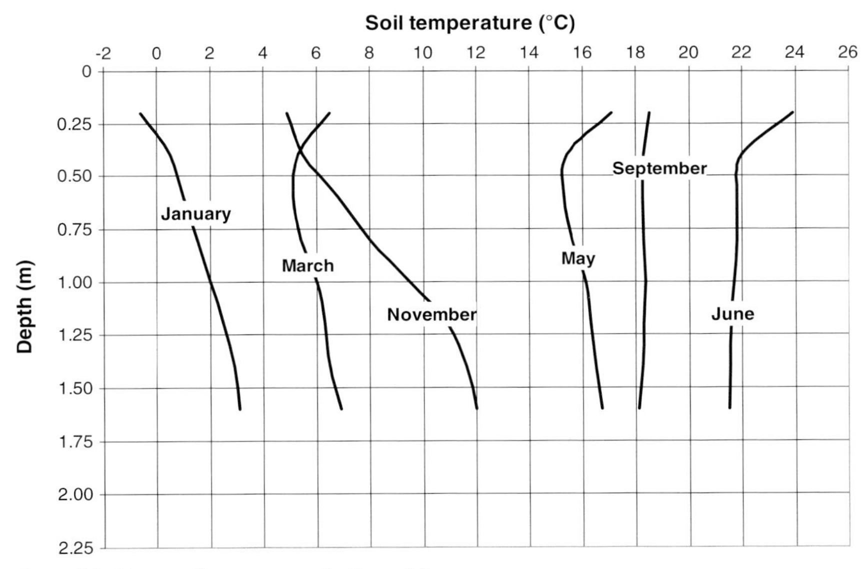

Figure 8.5 Mean soil temperature in Central Europe as a function of the soil depth [24]. (a) Grass covered; (b) sealed with bitumen.

m of sandy soils at depth greater 1 m remains above 25% even after long periods of suppressed rehumidification, for example, when the surface has been sealed. The thermal conductivity of different soils as function of the saturation factor h is outlined in Figure 8.8.

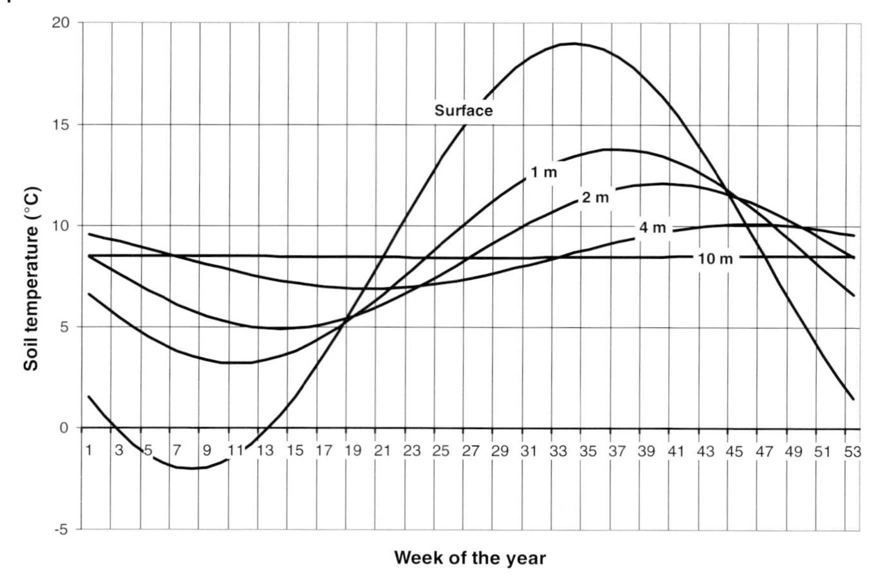

Figure 8.6 Seasonal changes of the soil temperature under grass-covered surfaces in Central Europe at different laying depths.

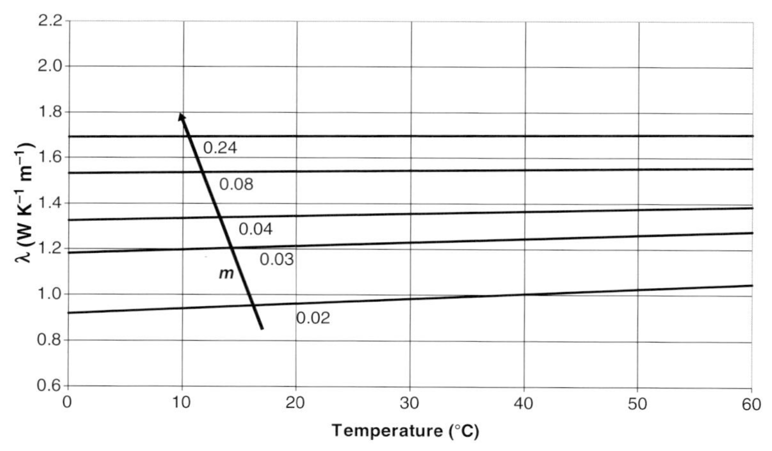

Figure 8.7 Thermal conductivity of sandy soil as a function of temperature and humidity content m [28].

The thermal conductivity remains constant within the range up to $h < 0.25$ for thermally stabilized sandy soils and $h < 0.5$ for loam soils. These figures characterize the range of the dried soil. Soil drying is to be attributed to the temperature field in the ground with a defined temperature gradient, which is related to the humidity gradient depending on the type of soil. Each isotherm corresponds to a

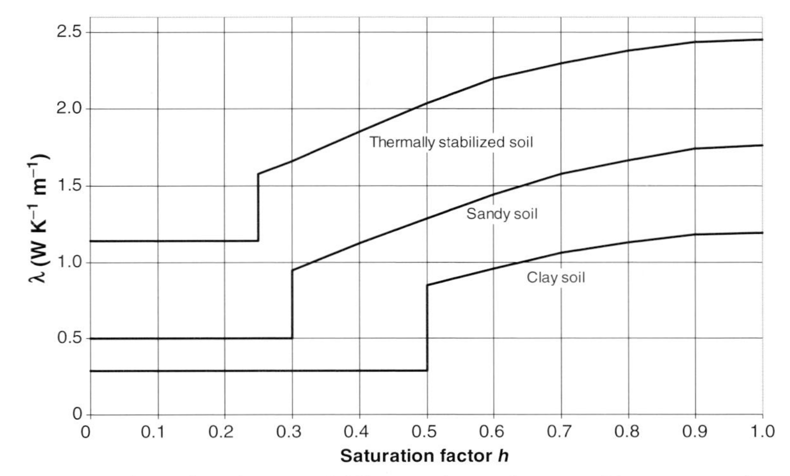

Figure 8.8 Thermal conductivity of different soils as a function of the saturation factor *h* [29].

defined humidity content. If the temperature exceeds a limit value, characteristic for each type of ground, the rehumidification of the soil is insufficient to compensate for the drying by evaporation, drainage and water transportation; consequently the soil dries out completely. A critical temperature for soil drying can be assigned to this humidity content, which is dependent on the soil type, the soil temperature, the humidity content of the unimpaired soil and the load and/or the load cycles of the buried cable. The thermal conductivity of the soil remains constant below the critical humidity content.

Based on Figures 8.7 and 8.8, thermal conductances of the ground can be assumed as outlined below:

- Thermally stabilized sandy soil: $\lambda = 1.8–2.2\,\mathrm{W\,K^{-1}m^{-1}}$
  during soil drying: $\lambda = 1.1\,\mathrm{W\,K^{-1}m^{-1}}$
- Sandy soil: $\lambda = 1.0–1.5\,\mathrm{W\,K^{-1}m^{-1}}$
  during soil drying: $\lambda = 0.5\,\mathrm{W\,K^{-1}m^{-1}}$
- Loamy soil: $\lambda = 0.9–1.1\,\mathrm{W\,K^{-1}m^{-1}}$
  during soil drying: $\lambda = 0.25\,\mathrm{W\,K^{-1}m^{-1}}$.

8.4.4
Thermal Resistances of Cables

For the calculation of the maximal permissible loading (current), the thermal resistances of the cable and/or the individual layers of the cable such as conductor, insulation, screens, sheaths, coating and so on need to be known. Reliable data can only be given by the manufacturer and should be required in detail in each specification. Values for typical materials used in the cable construction and of given laying conditions are shown in Table 8.5.

Table 8.5 Typical thermal parameters of materials used in cable construction and laying material [25].

Material	Thermal resistance $(\mathrm{K\,m\,W^{-1}})$	Thermal capacity $10^6\ \mathrm{W\,s\,K^{-1}m^{-3}}$
Paper insulation	5.0–6.5	2.0
VPE, XLPE and PE insulation	3.5	2.4
PVC insulation (depending on voltage)	5.0–6.0	1.7
Butyl-rubber and rubber insulation	5.0	2.0
Fiber and textile coatings	6.0	2.0
Rubber protective layers	6.0	2.0
Polychloroprene layers and coatings	5.5	2.0
PVC protective layers	5.0–6.0	1.7
PVC/bitumen on Al-coatings	6.0	2.0
PE protective coatings	3.5	2.4
Concrete	1.0	2.0
Glass fiber	4.8	2.2
Stone materials	1.2	2.0
PVC covers	7.0	1.7
PE covers	3.5	2.4

8.4.5 Calculation according to VDE 0276-1000

Calculation of maximal permissible loading is described in VDE 0276-1000 for cables with rated voltages up to 30 kV. One proceeds from standardized laying arrangement, soil surrounding conditions and operating cycles that are typical in power systems of public utilities and in industry. The application of VDE 0276-1000 is subject to the following conditions:

- Cables of the same type are assumed in one trench.
- Cables of same rated voltage are assumed in the trench.
- Cables of the same loading conditions and load cycles are assumed.
- Only standardized cable types are included.
- Only standard laying arrangements are considered.
- Electrically parallel operated cable systems are not considered.
- Sheaths are assumed to be earthed on both sides.
- No external heat sources are assumed.
- Temporary overloading of the cables is not considered.

Operating and laying conditions deviating from standard conditions as in Table 8.6 are considered in VDE 0276-1000 with factors f_1 and f_2.

The maximal permissible current is found from Equation 8.6.

$$I_{\mathrm{zul}} = I_{\mathrm{r}} \times f_1 \cdot f_2 \times \prod f \tag{8.6}$$

Table 8.6 Standard conditions for the determination of the maximal permissible loading of cables in earth according to VDE 0276-1000.

Standard operating and trench conditions for the calculation of rated current	Deviation in operating conditions
Operation mode (Section 5.3.1.1 of VDE 0276-1000) load factor $m = 0.7$; peak load for laying in earth	Reduction factors according to Tables 4 to 9 of VDE 0276-1000
Laying conditions (Section 5.3.1.2 of VDE 0276-1000) Laying depth 0.7 m	See Section 5.3.1.2.1 of VDE 0276-1000
Formation • 1 multiple-core cable • 1 single-core cable (DC system) • 3 single-core cables (AC system), triangle formation • 3 single-core cables (AC system) flat formation, distance between cables 7 cm	Reduction factors • Multiple cable trench as per Tables 4, 5, 9 and 13 of VDE 0276-1000 • — • Multiple cable trench as per Tables 4 to 7 of VDE 0276-1000 • Multiple cable trench as per Tables 4, 5 and 8 of VDE 0276-1000
Bedding in sand or refilled soil Protective covering from concrete tiles or flat plastic plates	Reduction factors • For protective covering with air inclusions see Section 5.3.1.2.3 of VDE 0276-1000 • For protective pipes see Section 5.3.1.2.4 of VDE 0276-1000
Trench surrounding (Section 5.3.1.3 of VDE 0276-1000) • Soil temperature in laying depth: 20 °C	Reduction factors • As per Tables 4 to 9 of VDE 0276-1000
Specific thermal resistivity of soil • Wet areas: 1.0 $\mathrm{Km\,W^{-1}}$ • Dry areas: 2.5 $\mathrm{Km\,W^{-1}}$	• As per Tables 4 to 9 of VDE 0276-1000
Earthing of sheaths, screens and armoring at both ends	See Section 5.2 of VDE 0276-1000

with

I_r = permissible current for standard conditions
Πf = Product of other conversion factors, for example, for laying in pipes and ducts.

The calculation method according to VDE 0276-1000 is based on the "characteristic diameter," which depends on the energy loss factor ϑ as per Equation 8.7, see also Section 4.2.

$$\vartheta \approx 0.17 \cdot m + 0.38 \cdot m^2 \tag{8.7}$$

with load factor m according to Equation 8.8.

$$m = \frac{\int_0^{T_Z} P(t)\mathrm{d}t}{P_{\max} \cdot T_Z} \tag{8.8}$$

Furthermore, the load is assumed to consist of a constant base load and higher maximal load for several hours. The model "characteristic diameter" assumes the losses from the variation of the peak load to be inside the cylindrical area of the characteristic diameter and the losses due to the constant base load to be in the outside area of the "characteristic diameter."

For the load diagram (base load–peak load) a standardized load pattern is defined such that, after a daily period of maximum 10 hours' duration with full load, a period of at least equal length will follow with maximum 60% of the full load. From this a load factor m = 0.67–0.77 is determined depending upon the load for the remaining 4 hours of the daily load profile. Comparison with actual load factors in power systems (Central Europe) reveals

- m = 0.75–0.82 in 110 kV systems
- m = 0.61–0.76 in 10 kV systems and other MV systems
- m = 0.50–0.77 in LV systems.

Investigations described in [25, 26] indicated that the maximal permissible loading determined with this method is in accordance with detailed models if the load shape is similar to that assumed. For other load patterns, detailed investigations with other methods have to be undertaken.

8.4.6
Determination of Maximal Permissible Loading by Computer Programs

As the VDE-method, as mentioned above, is to be used only under certain conditions, the calculation of the maximal permissible loading of cables and cable routes by means of a computer program is required, taking account of any load pattern, external heat sources and sinks, different laying arrangements and cable types [29]. Based on the method of IEC 60287 concerning the continuous current rating and taking account of IEC 60853-2 [32], the maximal permissible loading during cyclic load (cyclic rating) can be determined for each cable type and cable arrangement, taking account of the preloading (emergency rating). As an example, the results of the calculation of the maximal permissible loading of an oil-filled cable in flat formation, installed in concrete pipes and with cross-bonded sheaths, are given as follows. The maximal continuous loading of the cable (100% load) is $I_{\mathrm{thmax}} = 1536\,\mathrm{A}$. The cable can be loaded with higher current, depending on the preload conditions and the load cycle, as outlined in Figure 8.9. Other methods for the calculation of the current rating of electric cables, such as by finite element method, are discussed by standards committees and published for example, in IEC/TR 62095.

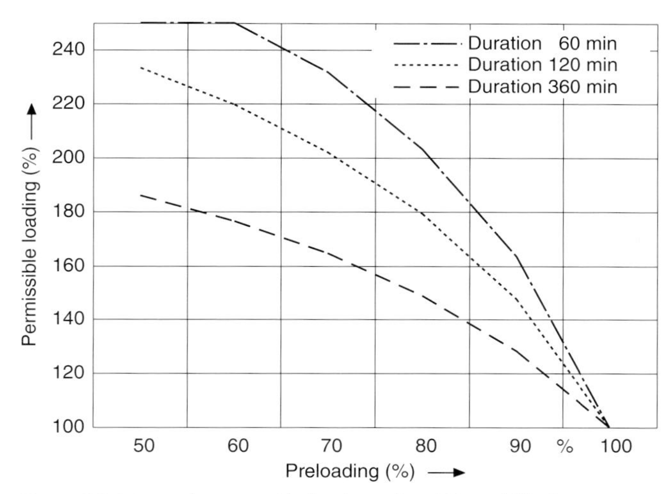

Figure 8.9 Maximal permissible loading of a 380 kV oil-filled cable, 1600 mm^2, as a function of the preload conditions. Laying depth 3900 mm; ground temperature 30 °C; thermal resistance of the soil 1 Km W^{-1}; distance of the conductors 500 mm.

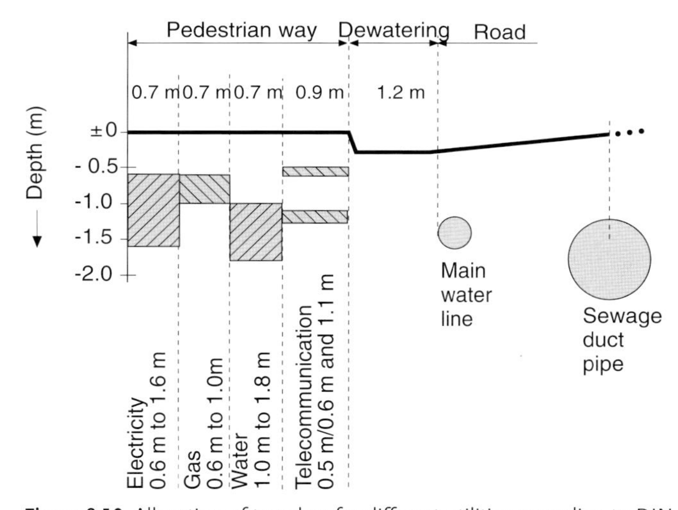

Figure 8.10 Allocation of trenches for different utilities according to DIN 1998 [33].

8.5
Planning and Design of Cable Routes and Trenches

8.5.1
Coordination with Other Cables and Pipes

The planning and design of cable routes and trenches cannot be treated as an isolated task since the cables and pipes of other utilities, such as gas, water, telecommunications and sewage water, have to be taken into account. According to DIN 1998, sidewalks, cycle tracks and greeneries without trees and vegetation and so on are to be used for cable and other supply routes. The suggested rights-of-way for different suppliers within the area outside of the roadways are usually limited; the rights-of-way suggested in DIN 1998 for separate routing according to Figure 8.10 sum to a total width of more than 3 m. If one considers that in urban areas, especially in city centers, several MV and LV cables must usually be laid, it is obvious that the laying arrangements recommended by DIN 1998 cannot be realized. It should further be noted that at road crossings and in the vicinity of consumer connections, crossings and connection joints are unavoidable. Additional space requirements are necessary for assembly reasons.

Separate horizontal laying of the different supply cables and pipes according to Figure 8.10 requires substantial costs for earthworks and for the reinstatement of the surface (interlocking tiles, bitumen, grass, etc.). Potential saving are offered by common laying and vertical arrangement of the cables and pipes in the trench.

The rules of the DVGW (German association of gas and water suppliers) [34–36] as well as DIN EN 805 [37] have to be considered. Minimum distances against other pipes and cables are defined (minimal requirements). DIN EN 805 stipulates a distance of 40 cm for lateral approximations and parallel routing; in case of bottlenecks and crossings, 20 cm is the minimal value. If even this is not possible, direct contact is to be avoided by suitable measures, such as separation layers or plates. According to the DVGW rules, in the case of crossings with lines and LV cables (up to 1 kV) without special preventive measures distances are to be kept at least 10 cm; in case of parallel running cables and pipes at least 20 cm or the half outside diameter of the cable or pipe having the larger cross-section. Distances to MV and HV lines with voltages above 1 kV must be more than 40 cm in case of parallel running pipes and cables and at least 20 cm in case of crossing. If this is not possible, the DVGW rules recommend suitable measures against direct touching, for example, by heat adjusting plates. The permissible rise in temperature of drinking water pipes due to electrical cables is not defined in norms; a maximum drinking water temperature of 18 °C as a typical limit value was established. In Figure 8.11 a trench is shown with different supply cables and pipes (gas, water, telecommunication) designed on the basis described.

The maximal permissible loadings of the electrical cables in the trench as shown in Figure 8.11 were calculated with computer program [31]. The temperature of the water pipe was limited to 17.5 °C.

Cable 1, NAYY 150 1 kV:	196.5 A	$\vartheta = 44.5\,°C$
Cable 2, NAYY 150 1 kV:	171.1 A	$\vartheta = 50.2\,°C$
Cable 3, NAYY 150 1 kV:	148.3 A	$\vartheta = 54.7\,°C$
Cable 4, NAXSY 240 10 kV:	346.2 A	$\vartheta = 49.5\,°C$

The safety regulations according to DIN 4124, such as sloping, safety device with planks and edge pressing and so on must to be considered.

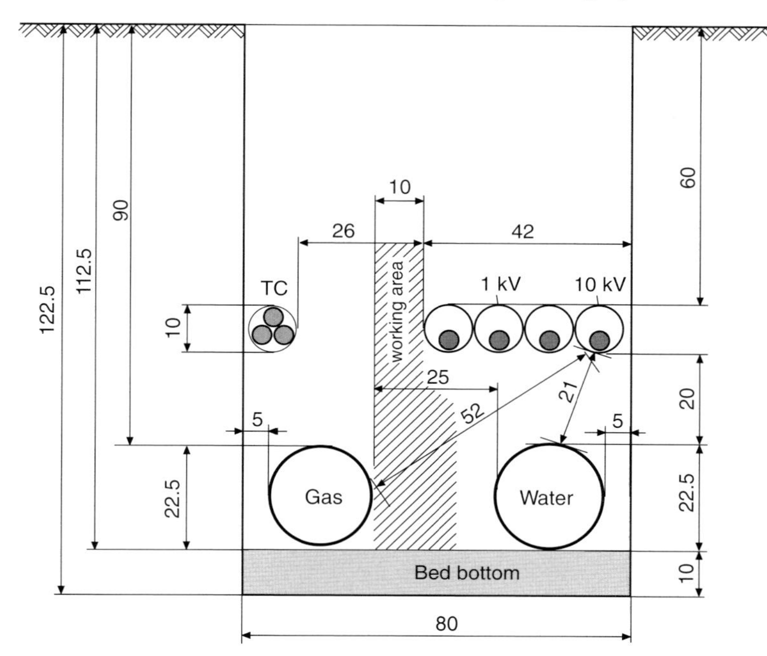

Figure 8.11 Supply trench for gas, water and electricity with minimum space requirements [30].

8.5.2 Effect of Thermally Unfavorable Areas

The influence of thermally unfavorable areas in the cable surroundings within the trench on the maximal permissible loading can be determined only with accurate knowledge of the parameters of the cable route. The results of parameter studies are presented in [25]. The following have to be considered:

- Crossing of cables with heating pipes at specified distance reduces the maximal permissible loading less than parallel routing at the same distance.
- Crossing of cables with other cables at a specified distance has less effect on the maximal permissible loading than parallel routing at same distance. The difference in general is smaller for the upper cable than for the lower cable in the trench.
- Having thermal insulation for distances longer than 5 m affects the permissible loading in such a way as if the entire route length were thermally insulated.
- Thermally unfavorable areas affect the permissible loading, even if the cable trench is filled with material of low thermal resistance (thermally favorable material).
- Cables crossing road layers for distances of more than 5 m, which have high thermal resistance, determine the permissible loading in such a way that the adjacent areas with low thermal resistance have no influence.

- If cables are laid in parallel to heat-insulated layers, for example, road surfaces, the maximal permissible loading is reduced only within distances less than 1 m.

8.5.3 Influence of Other Parameters

Apart from the thermal characteristic of the soil and the presence of external heat sources and sinks, the distances of the cables from each other and from other cable systems, the type and depth of laying, the kind of grounding of sheaths and armoring as well as the conductor cross-section and the outside diameter of the cable have an influence on the maximal permissible loading. For exact determination, knowledge of all parameters and a powerful computer program are necessary. Here, qualitative statements are made about these influences, based on the results of a parameter study as cited [25].

An increase of the spacing of the cables (in the case of single-core cables) has different effects depending upon voltage level, cross-section and sheath grounding. In the medium-voltage range as well as with cross-sections up to ~300 mm^2, sheaths are grounded at both ends. Increasing the distance nearly always leads to an increase of the permissible loading. In the high-voltage range, the sheaths should be cross-bonded for cross-sections exceeding 300 mm^2. The increase of spacing between the cables leads to a significant increase of the permissible loading and the effect is larger with larger cross-section. Without cross-bonding an increase of the cable cross-section beyond 500 mm^2 results in only a negligible increase of the permissible loading. Flat formation of single-core cables generally leads to higher maximal permissible loading than in case of triangle formation, if the spacing is identical. The attainable increase of the maximal loading is higher in the case of flat formation. Both formations (flat or triangle) have approximately the same permissible loading in case of small spacing.

The maximal permissible loading decreases if cables are laid at greater depth with the same soil temperature. Assuming laying depths between 1.2 and 1.5 m, the decrease in permissible loading is only less than 10%. The reduction with increased laying depth is smaller with increasing distance between the cables. In real cases, soil temperature decreases with increasing depth, thus the increase of laying depth normally has only negligible influence on the permissible loading.

The most significant influence on the maximal permissible loading is seen by cross-bonding of the sheaths. This effect is more significant with large conductor cross-sections. For a 132 kV XLPE cable of 500 mm^2 with cross-bonded sheaths, the increase of maximal permissible loading was up to 46% depending on the cable distance; for a conductor cross-section of 1200 mm^2 the increase amounted almost to 100% compared with double-side earthing.

Increasing the conductor cross-section in the case of double-sided earthing of the sheaths with view to increasing the maximal permissible loading is only recommended up to a cross-section of 500 mm^2. If cross-bonding is applied, the increase of cross-section always leads to an increase of the permissible loading. Increasing the outer cable diameter by use of thicker cable insulation can lead with some cable types to a clear increase of the maximal permissible loading. As the increase of cable cross-section, increase of outer diameter and other measures increase the cost of the cable installation, a detailed cost and economic analysis is required when designing cable trenches. Table 8.7 indicates some trends in cost and permissible loading for different cable parameters.

Table 8.7 Influence of cable parameters on the permissible loading of cables and on the cost of trenches and installations.

Parameter	Cable cost	Cost of transport	Cost of trench	Assembly cost at site	Permissible loading	Remarks
Cross-bonding of sheaths	—	—	—	+	++	Necessary in case of large cross-section
Flat or triangle formation	—	—	+	(+)	+	No influence with low spacing
Cable distance	—	—	+	(+)	+	With MV cables up to 300 mm^2
Laying depth	—	—	++	—	(–)	Standardized laying depth
Thermally stabilized soil	—	—	+	(+)	++	Recommended in all case
Cross-section	+	(+)	—	—	+	Together with cross-bonding
Outside diameter	+	(+)	—	—	(+)	Only in combination with increased dielectric strength

—, no influence; (+), low influence; (–) small reduction; +, increase; ++, significant increase.

The application of several of the named variations of parameters leads in certain cases to a significant increase of the maximal permissible loading. Significant increase of the permissible loading can be achieved by cross-bonding of the sheaths, by the use of thermally stabilized materials to refill the cable trench and by avoidance of thermally unfavorable areas in the cable route.

8.6 Short-Circuit Withstand Capability

8.6.1 General

The general considerations on short-circuit current calculation and short-circuit withstand capability were outlined in Section 7.6; more details are included in [38].

The short-circuit withstand capability of cables is determined by the maximal permissible temperature of the insulation, expressed by the maximal permissible temperature of the conductor in case of short-circuit, as defined in VDE 0276-1000; see Table 8.8. The permissible short-circuit temperature applies to a maximum

Table 8.8 Permissible short-circuit temperatures and rated short-time current density of cables.

Cable type	**Maximal permissible temperature (°C)**		**Temperature before short-circuit (°C)**							
			90	**80**	**70**	**60**	**50**	**40**	**30**	**20**
	Normal operation	**Short-circuit**	**Rated short-time current density A mm^{-2}; t_{kr} = 1 s**							
Copper conductor										
Cable with soft-soldered joints	—	160	100	108	115	122	129	136	143	150
VPE and XLPE insulation	90	250	143	149	154	159	165	170	176	181
PVC insulation										
≤300 mm	70	160	—	—	115	122	129	136	143	150
>300 mm^2	70	140	—	—	103	111	118	126	133	140
Oil-impregnated insulation										
0.6/1 kV	80	250	—	149	154	159	165	170	176	181
3.6/6 kV	80	170	—	113	120	127	134	141	147	154
6/10 kV	70	170	—	—	120	127	134	141	147	154
12/20 kV	65	170	—	—	—	127	134	141	147	154
18/30 kV	60	150	—	—	—	117	124	131	138	145
Electric field-limited insulation 12/20 kV	65	170	—	—	—	127	134	141	147	154
Aluminum conductor										
VPE and XLPE insulation	90	250	94	98	102	105	109	113	116	120
PVC insulation										
≤300 mm^2	70	160	—	—	76	81	85	90	95	99
>300 mm^2	70	140	—	—	68	73	78	83	88	93
Oil-impregnated insulation										
0.6/1 kV	80	250	—	98	102	105	109	113	116	120
3.6/6 kV	80	170	—	75	80	84	89	93	97	102
6/10 kV	70	170	—	—	80	84	89	93	97	102
12/20 kV	65	170	—	—	—	84	89	93	97	102
18/30 kV	60	150	—	—	—	77	82	87	91	96
Electric field-limited insulation 12/20 kV	65	170	—	—	—	84	89	93	97	102

short-circuit duration of $t_k = 5\,\text{s}$. An adiabatic heating of the cable (no heat dissipation) during the short-circuit duration is assumed. Skin and proximity effects are generally neglected, the specific heat is assumed constant and the change of resistance with temperature is assumed linear. Separate considerations are applicable with cross-sections exceeding $600\,\text{mm}^2$, as the skin effect has to be considered; however, this is not included in IEC 60865. Additional considerations according to IEC 60986 and IEC 60949 are to be taken into account.

The thermally equivalent short-time current I_{th} is calculated according to Equation 8.9a based on the amount of heat Q generated in a conductor with resistance R during the short-circuit duration T_k.

$$I_{th} = \sqrt{\frac{Q}{R \cdot T_k}} = \sqrt{\frac{\int_0^{T_k} i_k^2(t)\,dt}{R \cdot T_k}} \tag{8.9a}$$

The thermally equivalent short-time current can also be determined taking account of the heat dissipation factors m and n according to Figures 7.10 and 7.11, considering the thermal effect of the DC and the AC component of the short-circuit current as in Equation 8.9b,

$$I_{th} = I_k'' \cdot \sqrt{m + n} \tag{8.9b}$$

with the initial short-circuit current I_k''. The thermally equivalent short-circuit current in case of several short-circuits with different time durations T_{ki} and currents I_{thi} are calculated with Equation 8.10a.

$$I_{th} = \sqrt{\frac{1}{T_k} \sum_{i=1}^{n} I_{thi}^2 \cdot T_{ki}} \tag{8.10a}$$

$$\text{with} \quad T_k = \sum_{i=1}^{n} T_{ki} \tag{8.10b}$$

Based on the current density given by Equation 8.11,

$$J_{th} = \frac{I_{th}}{q_n} \tag{8.11}$$

with cross-section q_n and the thermally equivalent current I_{th}, a sufficient thermal short-circuit withstand capability is achieved if the calculated current density J_{th} is lower than the rated current density given by the manufacturer according to Equation 8.12.

$$J_{th} \le J_{thr} \cdot \sqrt{\frac{T_{kr}}{T_k}} \tag{8.12}$$

Data on the rated short-time current density of cables are contained in Table 8.8. Manufacturers' cable lists often include data on the thermally permissible

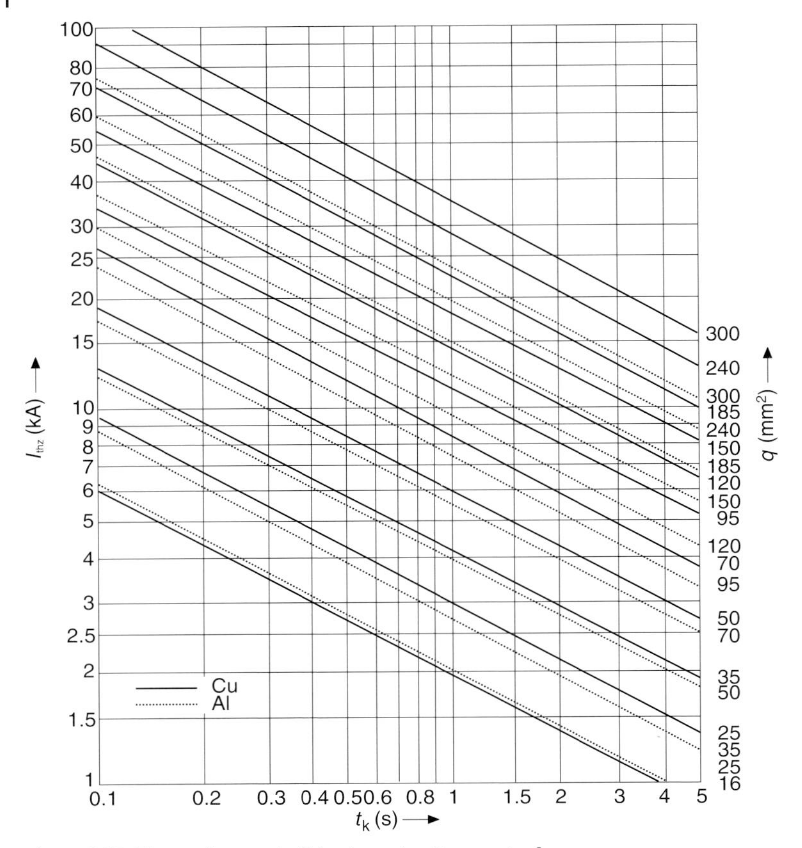

Figure 8.12 Thermally permissible short-circuit current of paper-insulated cables, voltages up to 10 kV. Data source: KABELRHEYDT [27].

short-circuit current I_{thz} as outlined in Figure 8.12. The rated short-time current is the value that results from a short-circuit duration of 1 s.

8.6.2 Rating of Cable Screens

Short-circuit currents flowing in cable sheaths, screens and armoring in unsymmetrical short-circuits are in the range of the short-circuit current of the cable, depending on the type of short-circuit and the type of laying. Generally, however, the current through the sheath, screen and armoring is lower, as a part of the short-circuit current is flowing through earth. For cables laid in air and/or on racks, the total short-circuit current flows over the sheaths, screens and armoring. Taking account of the high resistance of the outer protective sheath (lead alloy) and the armoring (steel) in comparison with the cable screens and sheaths of

Table 8.9 Constants of materials for sheaths, screens and armoring.

Material	Specific heat Q_c ($\mathrm{J\,K\,mm^{-3}}$)	Specific resistance ρ_{20} ($\Omega\,\mathrm{mm^2\,m^{-1}}$)	Temperature coefficient α_0 (K^{-1})
Copper	3.48×10^{-3}	17.28×10^{-6}	3.8×10^{-3}
Aluminum	2.39×10^{-3}	28.6×10^{-6}	4.0×10^{-3}
Lead alloy	1.45×10^{-3}	214×10^{-6}	4.35×10^{-3}
Steel	3.56×10^{-3}	143×10^{-6}	4.95×10^{-3}

copper or aluminum, it can be assumed that almost 100% of the current flows through the copper and aluminum screens and sheaths. Typical data of materials for sheaths, screens and armoring are given in Table 8.9. The data may differ from those for other sources because the type of materials and/or their manufacture can affect the parameters.

The maximal permissible short-circuit temperatures δ_e for the cable screens and sheaths are to be specified. If adiabatic heating of the cable is assumed for the duration of the short-circuit and the heat is dissipated only after disconnection of the short-circuit, the maximal permissible short-circuit current density J_{thz} and/or the maximal permissible short-circuit current I_{thz} for a given cross-section q_n are given by Equation 8.13.

$$J_{\mathrm{thz}} = \sqrt{\frac{Q_c \cdot (\beta + 20\,^\circ\mathrm{C})}{\rho_{20}} \cdot \ln\left(\frac{\delta_e + \beta}{\delta_b + \beta}\right) \cdot \frac{1}{T_{\mathrm{kr}}}} \tag{8.13a}$$

$$I_{\mathrm{thz}} = J_{\mathrm{thz}} \cdot q_n \tag{8.13b}$$

with

Q_c = specific heat in $\mathrm{J\,K\,mm^{-2}}$
α_0 = temperature coefficient in $\mathrm{K^{-1}}$
β = parameter $\beta = 1/\alpha_0 - 20\,^\circ\mathrm{C}$
δ_e = permissible short-circuit temperature
δ_b = permissible temperature before short-circuit
ρ_{20} = specific resistance at 20 °C in $\Omega\,\mathrm{mm^2 m^{-1}}$
q_n = cross-section of sheaths or screen in $\mathrm{mm^2}$
T_{kr} = rated short-circuit duration.

Other short-circuit durations and several short-circuits with different duration T_{ki} can considered using Equation 8.10.

9 Overhead Lines

9.1 General

Various aspects have to be considered during the design and project engineering of overhead lines.

- Determination of the permissible thermal loading (permissible current) in the context of the project planning period
- Design and determination of the tower arrangement and in special cases the placement of the towers along the route of the line route for project realization (and possibly in the tendering period)
- Mechanical design of the overhead line towers, again in the context of project realization.

9.2 Permissible Loading (Thermal) Current

9.2.1 Design Limits

The tensile strength and the modulus of elasticity of conductors for overhead lines are reduced by thermal stresses originating from the load current and from external heating by solar radiation. Additionally, the wind speed has an influence on the required mechanical parameters of the conductor. The reduction of the tensile strength and the modulus of elasticity result in an irreversible increase of the sag and finally in rupture of the line conductor. This effect depends on the temperature and the time of exposure. As an example, the tensile strength of an Aldrey conductor decreases non reversible by approximately 7% when heated to 75 °C over 12 months; when it is heated to 100 °C for the same period, the reduction is 21%. The tensile strength of pure aluminum conductors (99.5% Al) is reduced by approximately 8% for heating to 75 °C over 12 months; when it is heated to 100 °C during for same period, the reduction is approximately 9% and remains at this value even for longer loading periods. Copper conductors are less favorable in terms of the

Power System Engineering: Planning, Design and Operation of Power Systems and Equipment, Second Edition.
Jürgen Schlabbach and Karl-Heinz Rofalski.

reduction of tensile strength than are Al conductors. Operation of a Cu conductor at 75 °C for 12 months reduces the tensile strength by 25%; operation at 100 °C reduces it by nearly 40%. The permissible loading current of overhead line conductors therefore is limited with respect to the conductor temperature and the duration of exposure.

Overhead line conductors have to be designed in such a way that, for the total lifetime of the conductor, the reduction of tensile strength is less than 10%. The operating temperature should be less than 75 °C for an Aldrey conductor over a total period of 12 months summed over the total lifetime [39]. Figures 9.1–9.3 indicate the reduction of initial tensile strength (non reversible) as a function of temperature and loading period for typical overhead line conductors.

Several methods are existing to calculate the permissible thermal loading of overhead conductors, see CIGRE SC22 WG22-12 (1992), which are ased on the thermal balance or the conductor.

In determining the permissible load current of overhead line conductors, the heat balance of the conductor has to be considered. The heat balance is determined by the ohmic losses P_V, the heating by the sun exposure P_S, and the heat dissipation by radiation P_R and by convection P_K according to Equation 9.1.

$$P_V + P_S - P_R - P_K = 0 \tag{9.1}$$

9.2.2 Losses

The ohmic losses P_V of the conductor are determined by Equation 9.2a

$$P_V = I^2 \cdot R_{\approx} \tag{9.2a}$$

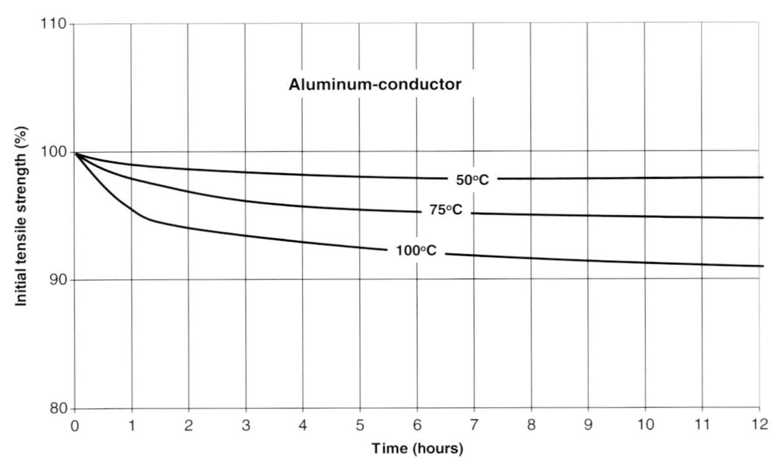

Figure 9.1 Reduction of initial tensile strength as a function of temperature and loading period for aluminum conductors.

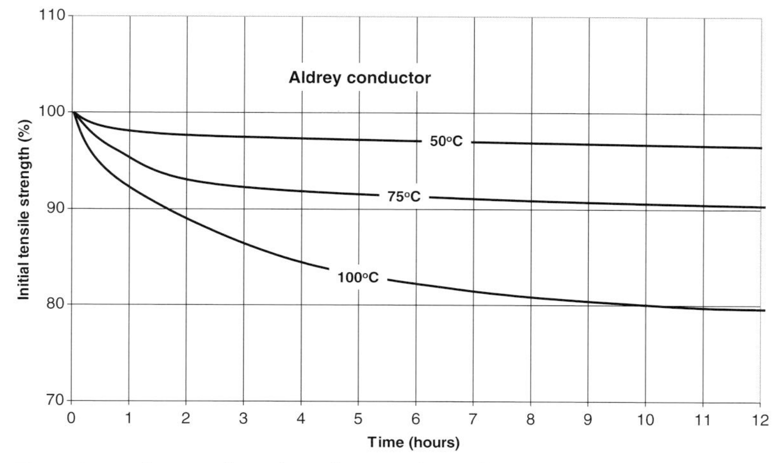

Figure 9.2 Reduction of initial tensile strength as a function of temperature and loading period for Aldrey conductors.

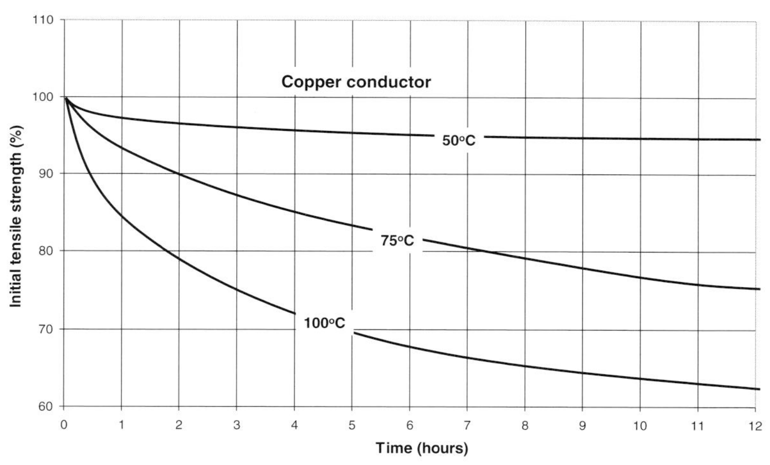

Figure 9.3 Reduction of initial tensile strength as a function of temperature and loading period for copper conductors.

The AC resistance $R_{\approx}$ is calculated as in Equation 9.2b.

$$R_{\approx} = \frac{1}{\kappa} \cdot \frac{1}{A_{\text{act}}} \cdot k_{\text{cond}} \cdot F_{\text{S}} \cdot F_{\text{P}} \cdot F_{\text{V}} \cdot [1 + \alpha \cdot (\delta_{\text{L}} - 20\,°\text{C})] \tag{9.2b}$$

with

I = r.m.s-value of the load current
κ = specific conductivity in $\text{m}\,\Omega^{-1}\,\text{mm}^{-2}$
A_{act} = actual conductor cross-section in mm^2

Table 9.1 Correction factors for the effective conductor length [37].

Conductor type	Correction factor k_{cond}
Single layer	1.01
Double- and triple layer	1.015
Four layers	1.02
Five layers	1.025

k_{cond} = factor for the consideration of the effective conductor length according to Table 9.1
F_S = skin-effect factor
F_P = proximity-effect factor
F_V = spirality-effect factor
α = resistance temperature coefficient
δ_L = conductor temperature.

Data for the AC resistance of typical overhead line conductors are outlined in Table 9.2.

The AC-resistance for other temperatures are calculated with the resistance temperature coefficient for the respective materials:

- Aluminum $\alpha = 4.0 \times 10^{-3}\,K^{-1}$ $\delta_{max} = 80\,°C$
- Aldrey $\alpha = 3.6 \times 10^{-3}\,K^{-1}$ $\delta_{max} = 90\,°C$
- Copper $\alpha = 3.92 \times 10^{-3}$ to $4.0 \times 10^{-3}\,K^{-1}$ $\delta_{max} = 70\,°C$

As well as conductor materials mentioned above (copper, aluminum, aluminum/steel and Aldrey) with standardized cross-sections, there are other cross-sections and materials and/or conductor structures. Conductors of aluminum-encased steel wires (ACSR or Alumoweld) and conductors made from temperature-stabilized aluminum alloys (TAL or STALUM) can be used for conductor temperatures up to 150 °C [49].

9.2.3 Heating by Solar Radiation

The heating of the line conductor by the sun exposure P_S according to Equation 9.3a depends on the solar radiation, the portion of direct P_D and indirect P_I radiation, the surface and the color of the conductor, expressed by the absorption coefficient α, and on the altitude of the sun, determined by the irradiation angle δ, which varies during the day and the year.

$$P_S = \alpha \cdot (P_D \cdot \sin\delta + P_I) \tag{9.3a}$$

The sum of direct and indirect radiation according to Equation 9.4 is termed the global radiation P_G:

$$P_G \approx P_D + P_I \tag{9.4}$$

Table 9.2 AC resistances of typical overhead line conductors, 50 Hz [40].

Nominal cross-section (mm^2)	Actual cross-section (mm^2)	Outside diameter (mm)	Specific AC resistance at conductor temperature ($\Omega\,km^{-1}$)				
			0 °C	20 °C	40 °C	60 °C	80 °C
Copper (δ_{max} = 70 °C)							
10	10.0	4.1	1.659	1.804	1.948	2.092	2.237
25	24.2	6.3	0.686	0.745	0.805	0.865	0.924
50	49.5	9.0	0.335	0.364	0.314	0.423	0.452
70	65.8	10.5	0.254	0.276	0.298	0.320	0.312
120	117.0	14.0	0.143	0.155	0.168	0.180	0.193
185	182.0	17.5	0.093	0.1000	0.109	0.116	0.124
240	243.0	20.3	0.071	0.076	0.098	0.088	0.094
Aluminum (δ_{max} = 80 °C)							
25	24.2	6.3	1.085	1.180	1.274	1.368	1.463
50	49.5	9.0	0.531	0.577	0.623	0.669	0.715
70	65.8	10.5	0.401	0.436	0.471	0.506	0.541
120	117.0	14.0	0.226	0.246	0.265	0.285	0.304
185	182.0	17.5	0.146	0.158	0.171	0.183	0.196
240	243.0	20.3	0.110	0.119	0.129	0.138	0.148
300	299.0	22.5	0.090	0.097	0.105	0.113	0.120
Aldrey (δ_{max} = 90 °C)							
25	24.2	6.3	1.270	1.368	1.467	1.566	1.664
50	49.5	9.0	0.621	0.669	0.717	0.765	0.814
70	65.8	10.5	0.469	0.506	0.542	0.579	0.615
120	117.0	14.0	0.264	0.285	0.305	0.325	0.346
185	182.0	17.5	0.170	0.183	0.195	0.210	0.223
240	243.0	20.3	0.129	0.138	0.148	0.158	0.168
300	299.0	22.5	0.105	0.112	0.121	0.133	0.137

It can be assumed that the global radiation P_G with irradiation angle δ causes the main external heating effect to the line conductor. In general the time of maximal solar exposure as well as the time duration of the maximal load current is of significant interest.

The solar radiation on the conductor surface is expressed as in Equation 9.3b.

$$P_S \approx \alpha \cdot (P_G \cdot \sin\delta) \tag{9.3b}$$

The irradiation angle δ can be determined from the maximal radiation condition according to Equation 9.5.

$$\delta = \arccos[\cos(h_S) \cdot \cos(180^\circ - \psi)] \tag{9.5}$$

The angle ψ is the angle between the geographical direction of the overhead line against north (azimuth) and the solar altitude h_S according to Equation 9.6, φ being the geographical degree of latitude:

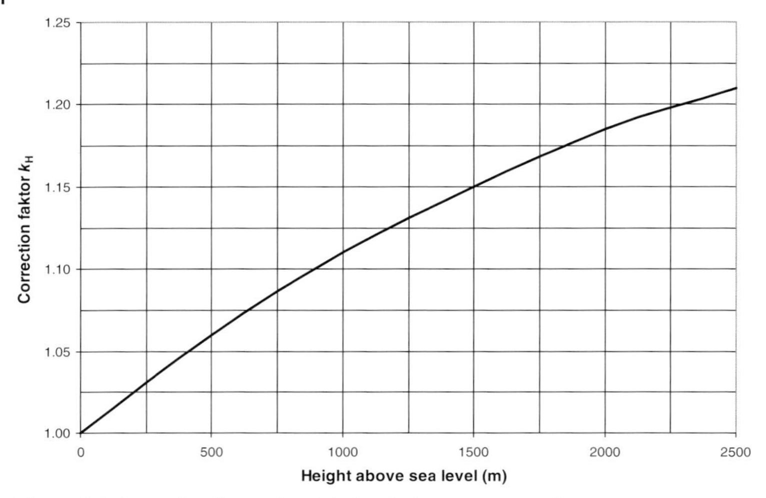

Figure 9.4 Correction factor for global radiation on a normal (horizontal) surface: influence of geographical height above sea level.

$$h_S = 113.5° - \varphi \tag{9.6}$$

The global radiation increases with increasing height above sea level as represented in Figure 9.4 by the factor k_H; the heat absorption per unit of length of the conductor with the diameter d is calculated according to Equation 9.7.

$$P_S = d \cdot \alpha \cdot k_H \cdot P_G \cdot \sin\delta \tag{9.7}$$

The parameters of Equation 9.7 are as described in the text.

Data for global radiation P_G can be taken from Figure 9.5. To calculate the heat absorption for any time of the day and the year, the respective data for the solar radiation can be taken from radiation tables or using suitable programs available through the internet. Other procedures for the computation of the irradiation on the conductor are possible, in particular with consideration of the direct and the indirect radiation, for example, as described in [40].

9.2.4 Heat Dissipation by Radiation and Convection

The heat emission of the line conductor is determined by radiation and convection. Radiation P_R in principle follows the Stefan–Boltzmann law according to Equation 9.8.

$$P_R = \sigma \cdot (T_1^4 - T_2^4) \tag{9.8a}$$

$$P_R = \frac{C}{10^8} \cdot (T_1^4 - T_2^4) \tag{9.8b}$$

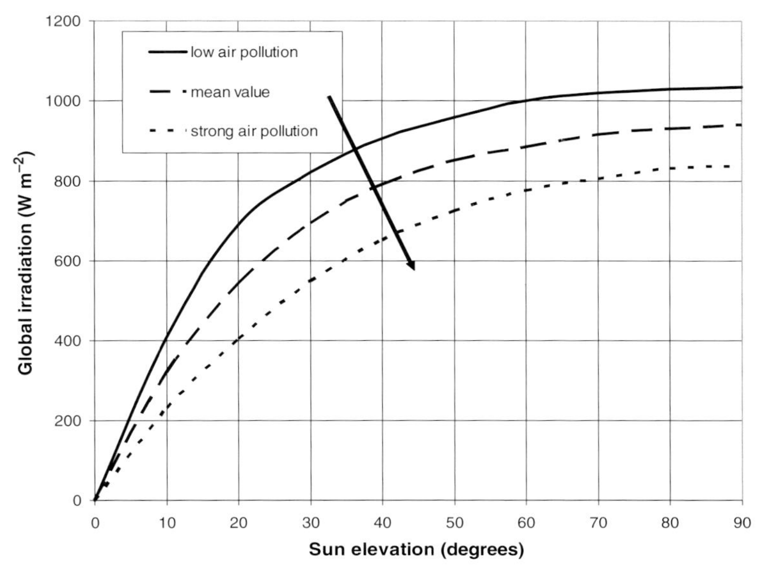

Figure 9.5 Global radiation on a normal surface: influence of air pollution.

Table 9.3 Emission coefficients for overhead line conductors ($T < 250\,°C$) [39].

Surface finish	Conductor material	
	Copper	Aldrey, Aluminum
Half-polished, new	0.15	0.08
Dull blank	0.24	0.23
Oxidized	0.5	0.35
Oxidized and less polluted	0.6	0.5
Heavily oxidized	0.75	0.7
Heavily oxidized and heavily polluted	0.87–0.93	0.87–0.93

where

T_1 = temperature of the radiating surface (conductor surface)
T_2 = ambient temperature
σ = Stefan–Boltzmann constant ($\sigma = 5.67 \times 10^{-8}\,\text{W m}^{-2}\,\text{K}^{-4}$)
C = radiation factor.

Since neither the line conductor nor the surrounding medium behaves like a physical black body, the radiation factor C is determined by the product of the radiation factor C_S of a black body and the emission coefficient ε of the line conductor. The emission coefficient ε varies between 0.08 and 0.93 depending on the conductor material and the surface finish, as can be seen from Table 9.3.

It should be noted that the radiation is dependent both on the temperature T_{21} in the immediate surroundings of the conductor and on the temperature T_{22} of the outside atmosphere, for which generally a value $T_{22} = 217\,\text{K}$ is taken (European practice). The portion of the radiation determined by the ambient temperature T_{21} amounts to ~75% and that determined by the radiation into the outside atmosphere amounts to ~25% [40]. The heat emission by radiation per unit of length of a conductor with diameter d is calculated using Equation 9.8c.

$$P_R = \pi \cdot d \cdot \varepsilon \cdot \frac{C_S}{10^8} \cdot [0.75 \cdot (T_1^4 - T_{21}^4) + 0.25 \cdot (T_1^4 - T_{22}^4)] \tag{9.8c}$$

As an example, the radiation of a new aluminum conductor ($A = 300\,\text{mm}^2$; $d = 22.5\,\text{mm}$; $\varepsilon = 0.08$) with conductor temperature 80 °C, ambient temperature 30 °C amounts to $3\,\text{W}\,\text{m}^{-1}$. With an ambient temperature of 0 °C, the heat radiation is ~$3.7\,\text{W}\,\text{m}^{-1}$.

Site tests of overhead line conductors confirmed that the heat dissipation occurs more via convection than through radiation. It is difficult to obtain a simple, manageable calculation method for the heat convection because of the nonlinear temperature dependences of the parameters of the surrounding air contributing to convection and because of the mutual influence of temperature and air-flow in the surrounding of the conductor. An approximation is given below.

The dependence on the physical parameters of air is described by the Nusselt factor (Nu), the Grashof factor (Gr), the Prandtl factor (Pr) and the Reynolds factor (Re) [37]. For convection in calm air, without any additional air flow, the heat dissipation resulting from convection P_K is given by Equation 9.9.

$$P_K = 1.163 \cdot \pi \cdot \lambda_L \cdot (T_1 - T_{21}) \cdot Nu \tag{9.9}$$

with

λ_L = heat conductivity of air
Nu = Nusselt factor as function of Grashof factor Gr according to Equation 9.10.

$$Nu \approx 0.436 + 0.55 \cdot Gr^{0.217} \qquad 1 \le Gr \le 10^4 \tag{9.10a}$$

$$Nu \approx 0.55 \cdot Gr^{0.228} \qquad 10^4 \le Gr \le 10^7 \tag{9.10b}$$

The Grashof factor Gr is determined by the diameter d, the air density γ, the volumetric expansion coefficient β, the temperature difference $T_1 - T_{21}$, the dynamic viscosity η and the acceleration due to gravity g according to Equation 9.11.

$$Gr = \frac{d^e \cdot \gamma^2 \cdot \beta \cdot (T_1 - T_{21})}{\eta^2 \cdot g} \tag{9.11}$$

Forced convection by wind arises likewise as a result of convection P_K according to Equation 9.9 above.

The Nusselt factor Nu is given as a function of the Reynolds factor Re by Equation 9.12.

$$Nu \approx 0.32 + 0.43 \cdot \mathrm{Re}^{0.52} \qquad 0.1 \le Re \le 10^3 \tag{9.12a}$$

$$Nu \approx 0.24 \cdot \mathrm{Re}^{0.6} \qquad 10^3 \le Re \le 5 \times 10^4 \tag{9.12b}$$

The Reynolds factor Re is calculated from the diameter d, the wind speed w perpendicular (crosswise) to the conductor rope, the air density γ, the dynamic viscosity η and the acceleration do to gravity g as in Equation 9.13.

$$\mathrm{Re} = \frac{d \cdot w \cdot \gamma}{\eta \cdot g} \tag{9.13}$$

For a new aluminum conductor, heat dissipation by convection, without forced convection by wind, amounts to $16\,\mathrm{W\,m^{-1}}$ ($A = 300\,\mathrm{mm}^2$; $d = 22.5\,\mathrm{mm}$; conductor temperature 80 °C; ambient temperature 30 °C). If the wind speed is assumed to be $0.6\,\mathrm{m\,s^{-1}}$ the heat dissipation by convection is approximately $41\,\mathrm{W\,m^{-1}}$. Figures 9.6–9.9 indicate the maximal permissible load current of an Aldrey 240 overhead line conductor ($d = 20.3\,\mathrm{mm}$; $R' = 0.138\,\Omega\,\mathrm{km}^{-1}$) for a number of parameters and conditions.

9.2.5 Examples for Permissible Thermal Loading

As can be seen from Figures 9.6 to 9.99, the wind velocity has a dominating influence on the maximal permissible load current of overhead line conductors in addition to the air temperature, whereas the emission coefficient and the reduction of the solar radiation due to atmospheric pollution can be almost neglected. It should be noted in this respect that a wind speed of zero in practice does not occur specifically in combination with high ambient temperatures. Solar radiation always results in a vertically aligned airflow (thermic), which is usually on the order of up to $0.8\,\mathrm{m\,s^{-1}}$.

9.3 Electric Field Strength

Design of conductors for overhead lines with nominal voltages above 110 kV has to take account of the electric field strength at the conductor surface [41], which is calculated according to Equation 9.14.

$$E_{\max} = \frac{U}{\sqrt{3}} \cdot \frac{C_1'}{2 \cdot \pi \cdot \varepsilon_0 \cdot n} \cdot \left(\frac{1}{r_\mathrm{T}} + \frac{2 \cdot (n-1) \sin \frac{\pi}{n}}{a} \right) \tag{9.14}$$

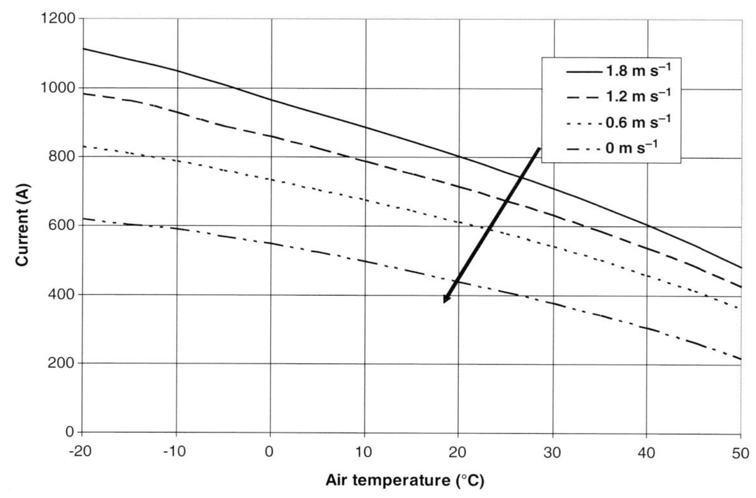

Figure 9.6 Maximal permissible load current of an Aldrey 240 overhead line conductor; $T_1 = 70\,°\text{C}$; $\varepsilon = 0.5$; clean atmosphere; parameter: wind velocity w.

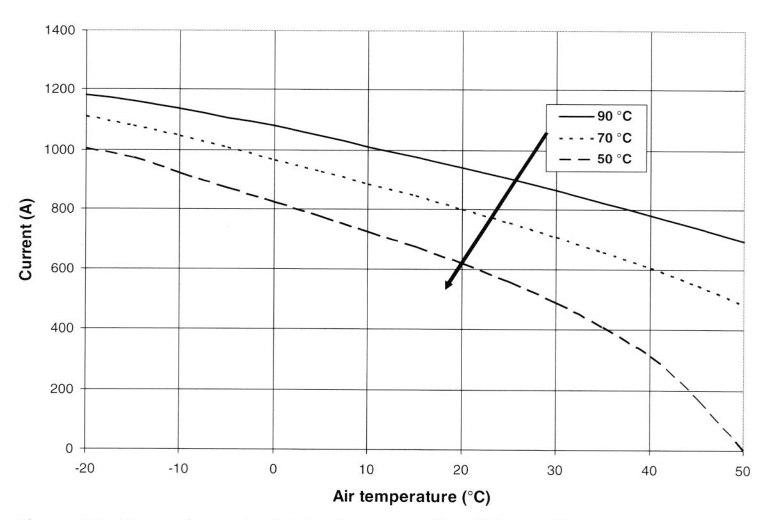

Figure 9.7 Maximal permissible load current of an Aldrey 240 overhead line conductor; $w = 1.8\,\text{m}\,\text{s}^{-1}$; $\varepsilon = 0.5$; clean atmosphere; parameter: conductor temperature T_1.

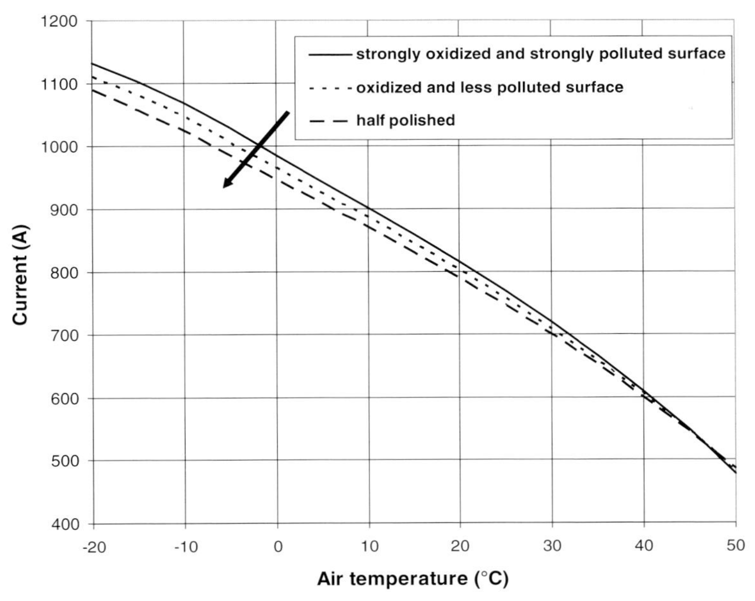

Figure 9.8 Maximal permissible load current of an Aldrey 240 overhead line conductor; $T_1 = 70\,°C$; $w = 1.8\,m\,s^{-1}$; clean atmosphere; parameter: emission coefficient ε.

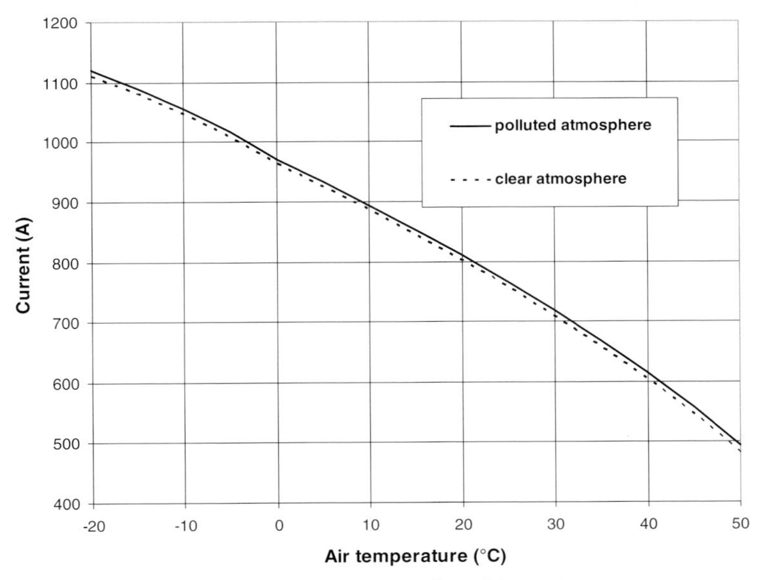

Figure 9.9 Maximal permissible load current of an Aldrey 240 overhead line conductor; $T_1 = 70\,°C$; $w = 1.8\,m\,s^{-1}$; $\varepsilon = 0.5$; parameter: atmospheric pollution.

with

U = maximal operating voltage
ε_0 = permittivity of air ($\varepsilon_0 = 8.854 \times 10^{-12}\,\mathrm{F\,m^{-1}}$)
r_T = radius of subconductor (in case of a multiple conductor type overhead line)
n = number of subconductors
a = distance of subconductors
C'_1 = capacitance in the positive-sequence component.

If the electric field strength (rotationally symmetric field) exceeds $16\,\mathrm{kV\,m^{-1}}$, corona discharges occur, which can cause disturbances of frequency-carrier signals and of medium-wave frequency transmitters apart audible noise.

9.4 Sag, Tensions and Minimum Distances

9.4.1 Minimal Length of Insulation

The mechanical design of overhead lines is determined by

- Minimum distances of the conductors to earth, to other conductors and to the tower
- Minimum clearances and minimum length of insulators, depending on the pollution class
- Sag, taking account of conductor and ambient conditions
- Tensile stress of the conductor
- Wind forces on the conductor, insulators and towers.

Minimum distances for nominal system voltage $U_n \geq 110\,\mathrm{kV}$ as indicated in Table 9.4 are defined in IEC 60071-1 (VDE 0111-1). The length of the insulators is determined by the minimum length of insulators and the specific creepage distance for different pollution classes according to IEC 60071-2 (VDE 0111-2).

The contamination classes are defined as follows below, whereas the specific creepage distance refers to the highest voltage for equipment U_m.

- Class 1 Lightly polluted areas without industry and with spread settlement (houses with exhaust from heating devices to be considered); areas with small industrial density or small populated areas, which are exposed to frequent wind and rain; areas far from sea shores or on large heights above sea level. The specific creepage distance can be kept at $1.6\,\mathrm{cm\,kV^{-1}}$.
- Class 2 Medium polluted area, industrial areas without any particular emissions, areas with medium population density with exhaust from heating devices to be considered; areas with high population density and/or industrial areas, which are exposed to frequent wind and rain; areas which are more than 1 km distant from the sea shore. The specific creepage distance shall be $2.0\,\mathrm{cm\,kV^{-1}}$.
- Class 3 Heavily polluted areas with high industrial density and suburbs of larger cities with considerable exhaust gases from heating; areas near sea shores

Table 9.4 Minimal distances for overhead transmission lines depending on the insulation strength.

Parameter	U_n (kV)			
	110	**380**	**750**	**±400**
Distance conductor–earth (line–earth) (m): $a_{LE} \geq 6+(U_n/150\,\text{kV}) \times 110/150$				
a_{LE} (m)	6.0	7.8	10.3	–8.0
Distance conductor–tower (line–tower) (m): $a_{LT} \geq U_n/150\,\text{kV}$				
a_{LT} (m)	0.74	2.56	5.0	–2.4
Distance conductor–conductor (line–line) a_{LL} (m): Determined by nominal lightning impulse withstand voltage (1.2/50 µs) and/or nominal switching impulse withstand voltage (250/2500 µs)				
1.2/50 (kV)	550	–	–	–
250/2500 (kV)	–	1550	2550	1300
a_{LL} (m)	1.1	2.65–3.35	6.8–7.95	2.05–2.65

which are exposed to relatively strong wind from the sea. The specific creepage distance shall be 2.5 cm kV^{-1}.

- Class 4 Very heavily polluted areas of limited extent, including areas which are exposed to thick, conductive deposits; areas of limited extent which are very close to the sea shore with occurrence of strong salty wind. The specific creepage distance shall be 3.1 cm kV^{-1}.

The classification of air pollution with respect to exhaust from heating devices should be done with care, since the increased use of gas for heating reduces the relevant emissions, with a corresponding effect on the creepage distance. Classification standards are lacking in some countries.

The minimal conductor–earth distance is determined by the maximal conductor sag and the maximal permissible continuous tensile stress of the conductor. According to DIN EN 50341-1 and DIN EN 50423-1 (VDE 0210), the following shall apply:

- The continuous tensile stress of the conductor must not be exceeded under any of the following condition: –5 °C with triple the normal load or double the increased additional load; –5 °C with normal additional load and wind load; or –5 °C with increased additional load and wind load.
- The horizontal component of the line tensile strength shall not exceed the values given in VDE 0210 Table 3 at the mean yearly average temperature (in Germany: +10 °C) without wind load.
- In order to calculate the maximal conductor sag, either +40 °C without additional load or –5 °C with normal or increased additional load are to be applied. The smallest sag always arises from –20 °C without additional load. In countries facing high current loading of the conductor during summer, higher temperatures have to be taken into account.

- Normal additional load in newtons per meter of conductor length is to be taken as $5+0.1d\ \mathrm{N\,m^{-1}}$, where d is the conductor diameter. For insulators, 50 newtons per meter of insulator chain should be used.
- Increased additional load can amount to a multiple of the normal additional load and its consideration should be based on observations spanning many years as well as on topographical and meteorological characteristics. Such additional load mainly has to be considered in cold climates because of ice and snow deposits.
- Additional load due to warning bowls or radar reflectors is to be considered as normal additional load with an additional 1 cm layer of ice on the entire surface (ice lining contributes $0.0075\ \mathrm{N\,cm^{-3}}$) for cold climates only.

9.4.2
Conductor Sag and Span Length

The site conditions during construction of the overhead line, especially ambient temperature during the pulling of the conductors, differ from those conditions defined for the determination of either the maximal conductor sag or the maximal permissible tensile stress of the conductor. It is therefore necessary to calculate the sag and the maximal tensile stress on the basis of the conditions during conductor pulling. If the suspension points of one span are at equal heights, the conductor sag is calculated using Equations 9.15a and 9.15b; if the suspension points are at different heights, the conductor sag is calculated using Equations 9.15c and 9.15d.

$$f=\frac{l^2\cdot\gamma}{8\sigma}+\frac{l^4\cdot\gamma^3}{384\cdot\sigma^3} \qquad \text{span length 500 m up to 1000 m} \tag{9.15a}$$

$$f=\frac{l^2\cdot\gamma}{8\sigma} \qquad \text{span length below 500 m} \tag{9.15b}$$

$$f=\left(\frac{l^2\cdot\gamma}{8\sigma}+\frac{l^4\cdot\gamma^3}{384\cdot\sigma^3}\right)\cdot\frac{1}{\cos\alpha} \qquad \text{span length 500 m up to 1000 m} \tag{9.15c}$$

$$f=\frac{l^2\cdot\gamma}{8\sigma}\cdot\frac{1}{\cos\alpha} \qquad \text{span length below 500 m} \tag{9.15d}$$

The parameters are explained below. The conductor sag for arbitrary temperatures and additional loads and/or the tensile stress for arbitrary sags and ambient temperatures are calculated by Equation 9.16,

$$\frac{1}{\sigma^2}\cdot\left(\frac{\gamma^2\cdot E\cdot l^2}{24}\right)-\sigma=\frac{E\cdot\gamma_0^2\cdot l^2}{24\cdot\sigma_0^2}-\sigma_0+E\cdot\varepsilon_t\cdot(\delta-\delta_0) \tag{9.16}$$

with parameters:

α = angle between horizontal and the suggested line between the suspension points

γ = relative density of the conductor
γ_0 = relative density of the conductor with additional load
l = span length
σ = applied tensile stress at temperature δ during conductor pulling
σ_0 = maximal permissible tensile stress at temperature δ_0
E = modulus of elasticity of the conductor
ε_t = thermal expansion factor of the conductor
δ = temperature of the conductor during conductor pulling
δ_0 = temperature of the conductor related to maximal permissible tensile stress
f = sag.

9.5 Short-Circuit Thermal Withstand Strength

The thermal withstand strength against short-circuits of any equipment is determined by the short-circuit duration, the initial short-circuit current and the conductor temperature prior to the short-circuit. The maximal permissible conductor temperatures during short-circuit are determined by VDE 0210, see Table 9.5. Manufacturers' specific data on the permissible conductor temperature are to be considered. The permissible temperature during short-circuits applies to a maximum short-circuit duration of $t_k = 5\,\mathrm{s}$ and the heating of the conductor is assumed to be adiabatic. Skin effect and proximity effect are generally neglected, the specific heat capacity is assumed constant, and the resistance temperature dependence is assumed linear.

The thermal equivalent short-time current I_{th} as reference value for the heat-production Q in a conductor with resistance R during the short-circuit duration T_k is calculated with Equation 9.17.

$$I_{th} = \sqrt{\frac{Q}{R \cdot T_k}} = \sqrt{\frac{\int_0^{T_k} i_k^2(t)\,dt}{R \cdot T_k}} \tag{9.17}$$

Table 9.5 Maximal permissible temperatures during short-circuit.

Conductor type	Material	Permissible conductor temperature during short-circuit (°C)
Single material	Cu	170
	Al	130
	E-AlMgSi (Aldrey)	160
	St	200
Bonding material	Al/St	160
	E-AlMgSi/St	160
	ACSR (STALUM)	260

The thermal equivalent short-time current I_{th} can also be calculated from the initial short-circuit current I''_k from Equation 9.18,

$$I_{th} = I''_k \cdot \sqrt{m+n} \tag{9.18}$$

with the factors m and n as in Figures 7.10 and 7.11 to take account for the thermal effects of the DC and the AC components of the short-circuit current. The thermal effect of several short-circuits occurring subsequently with arbitrary durations T_{ki} and r.m.s. values I_{thi} are calculated with a thermally equivalent short-time current according to Equation 9.19,

$$I_{th} = \sqrt{\frac{1}{T_k}\sum_{i=1}^{n} I_{thi}^2 \cdot T_{ki}} \tag{9.19}$$

with $T_k = \sum_{i=1}^{n} T_{ki}$

Calculation of the thermal short-time current density J_{th} of bare conductors (overhead conductors) is calculated based on the short-time current I_{th} and the conductor cross-section q_n according to Equation 9.20. The steel portion of Al/St-conductors is not considered.

$$J_{th} = \frac{I_{th}}{q_n} \tag{9.20}$$

The required thermal withstand strength to short-circuits is given, if Equation 9.21 is fulfilled. Data concerning the short-time current density of different materials are included in Table 7.3 and Figure 7.12.

$$J_{th} \le J_{thr} \cdot \sqrt{\frac{T_{kr}}{T_k}} \tag{9.21}$$

The short-time current density can also be calculated on the basis of an operating temperature of 20 °C in accordance with IEC 60865-1 (VDE 0103) according to Equation 9.22. If other temperatures are considered, Equation 9.22 must be modified accordingly.

$$J_{thr} = \frac{\sqrt{\dfrac{\kappa_{20} \cdot c \cdot \gamma}{\alpha_{20}} \cdot \ln \dfrac{1+\alpha_{20} \cdot (\delta_e - 20\,°\mathrm{C})}{1+\alpha_{20} \cdot (\delta_b - 20\,°\mathrm{C})}}}{\sqrt{T_{kr}}} \tag{9.22}$$

c = specific thermal capacity
γ = specific density
κ_{20} = conductivity (20 °C)
α_{20} = temperature coefficient (20 °C)
δ_e = permissible short-circuit temperature
δ_b = permissible temperature before short-circuit.

Parameter values are given in Table 9.6.

Table 9.6 Parameter for calculation of thermal short-time current density.

Parameter	Material			
	Cu	Al, AlMgSi, Al/St	St	STALUM
c (J kg^{-1} K^{-1})	390	910	480	607
γ (kg m^{-3})	8900	2700	7850	6590
κ_{20} (Ω^{-1} m^{-1})	56×10^6	34.8×10^6	7.25×10^6	39.8×10^6
α_{20} (K^{-1})	0.0039	0.004	0.0045	0.0036

9.6 Right-of-Way (ROW) and Tower Arrangement

When constructing overhead lines the route corridor for building the towers, stringing the conductors, taking account of maximal sag and sidewise swings of the conductors has to be granted as a legally free space, which is called the right-of-way (ROW). If the ROW cannot be obtained alternate routes must be devised.

The conductor sag of an overhead transmission line is different for different conductor types at constant tensile stress. This implies, that the tensile stress of different conductors is different if the conductor sag is kept constant. Therefore, the choice of an appropriate conductor has an influence on type, height and number of towers. The required route corridor for constructing the line has to be determined accordingly in a different way for different conductor types, as conductors with different sag have different sidewise oscillations. Since the conductor sag depends on the tower distance (span width), the overhead transmission line has to be designed by optimization procedures. The basic relations between the different tasks to be considered are represented in Figure 9.10.

The distance between the individual conductors on the same tower is determined by Equation 9.23.

$$a = k \cdot \sqrt{f + l_B} + s_U \tag{9.23}$$

with

f = conductor sag at 40 °C
l_B = overall length of the insulator
s_U = voltage-dependent value (according to VDE 0210 Table 16: reference value: $s_U \approx U_n/150$; U_n in kV)
k = factor for vertical and horizontal arrangement (according to VDE 0210 Table 17: reference range $k = 0.6$–0.95).

The conductor sag has an influence on the clearance between the conductors (Equation 9.23) and the distances between conductor and tower, thus having a direct impact on the design of the tower head and the tower height. It is clear

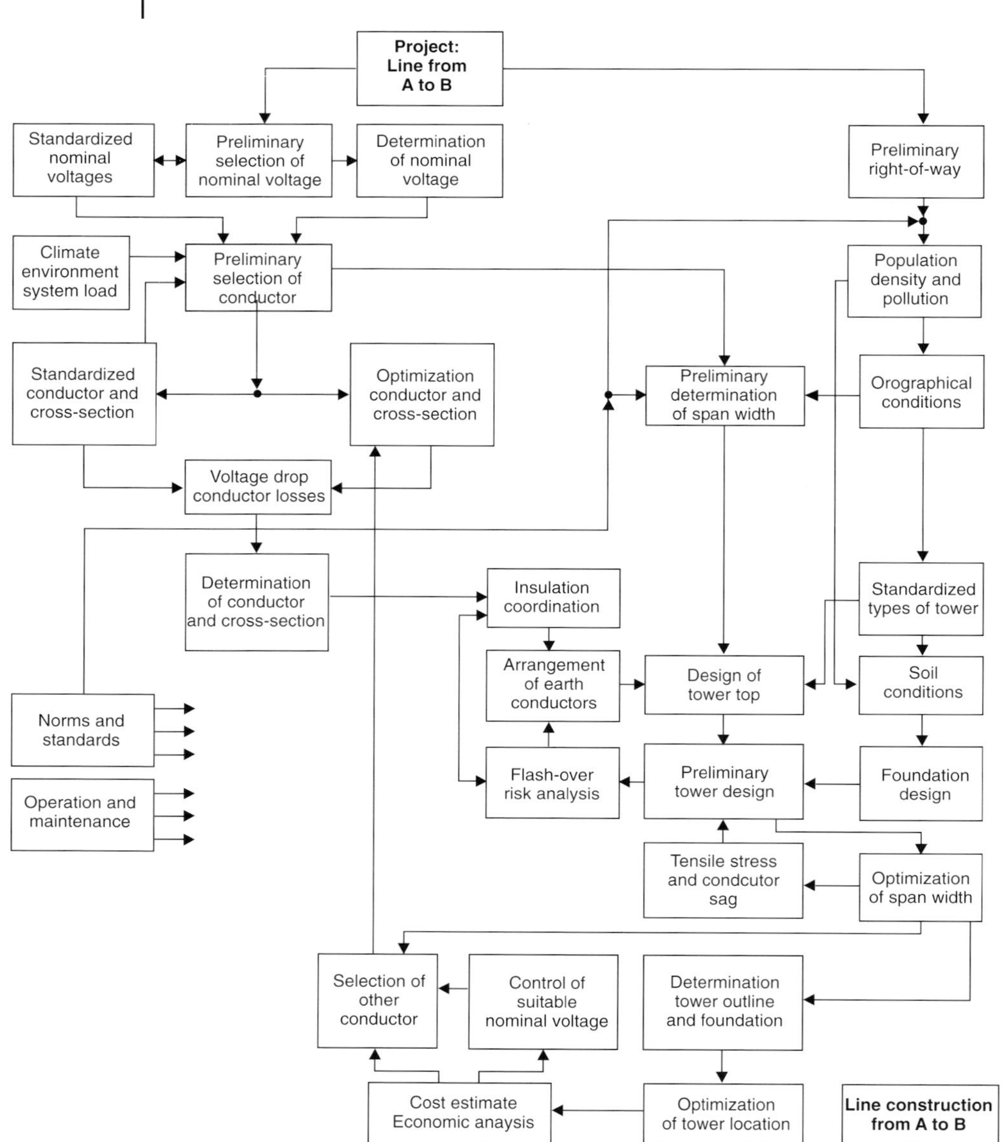

Figure 9.10 Basic relationships of different tasks for design of overhead transmission lines [43, 44].

therefore that it is not possible in some cases to replace a conductor with another type without changing the tower design.

Regarding the determination of the width of the route corridor b_{Tr}, the width of the tower head b_M, the sidewise oscillation amplitude by wind and an additional safety margin b_S have to be considered. The required width b_{Tr} can be established according to Equation 9.24. The first part states in each case the necessary width of the line corridor in the middle of the span width (due to wind aggravated oscillations). For span length up to 200 m Equation 9.24a applies, for span lengths from 200 m upwards Equation 9.24b applies.

$$b_{\mathrm{Tr}} = (f + l_{\mathrm{B}})\cdot\sin\left(\arctan\left(\frac{c\cdot p\cdot d}{g}\right)\right) + b_{\mathrm{M}} + b_{\mathrm{S}} \tag{9.24a}$$

$$b_{\mathrm{Tr}} = (f + l_{\mathrm{B}})\cdot\sin\left(\arctan\left(\frac{c\cdot p\cdot d\cdot(80+0.6\cdot l)}{g\cdot l}\right)\right) + b_{\mathrm{M}} + b_{\mathrm{S}} \tag{9.24b}$$

with

c = stagnation pressure factor of the conductor (according to VDE 0201 Table 6: reference range: $c = 1.0$–1.3)
p = stagnation pressure by wind (according to VDE 0201 Table 5: reference value: $p \approx w^2/1600$)
w = wind-speed in $\mathrm{ms^{-1}}$
d = diameter of conductor
l_{B} = overall length of insulator
l = span length
g = conductor weight per unit length
b_{M} = tower width
b_{S} = safety margin according to VDE 0201 Section 13.

The influence of different conductor types on the tower design is outlined in Table 9.7 for a 110 kV overhead line for a transmitted power S = 125 MVA. The

Table 9.7 Characteristic values of different conductor types for identical safety margin against conductor tearing ($\sigma_1 = 0.25\sigma_{\mathrm{B}}$).

Parameter	Unit	Conductor type		
		Al/St 240/40	**Alumoweld 254/28**	**Aldrey 300**
Specific weight	$\mathrm{kg\,km^{-1}}$	987	882	827
Thermal expansion factor	$10^{-5}\,\mathrm{K^{-1}}$	1.89	2.07	2.3
Modulus of elasticity	$\mathrm{N\,mm^{-2}}$	770	705	550
Maximal permissible current (calculation by program)	A	649	682	674
Breaking stress	N	864	832	854
Continuous tensile strength	$\mathrm{N\,mm^{-2}}$	2.08	2.31	1.96
Sag (60 °C; l = 420 m)	m	19.28	20.24	17.83
	%	108.1	113.5	100
Right-of-way	m	22.8	25.18	23.17
	%	98.32	108.7	100
Horizontal distance conductor-conductor at tower	m	3.36	3.42	3.40
	%	98.8	100.6	100
Vertical distance conductor-conductor at tower	m	3.91	3.99	4.18
	%	93.5	95.5	100

safety margin against conductor tearing is set to 300%, that is, the maximal tensile strength is 25% of the permissible breaking stress.

As can be seen from the parameters in Table 9.7, the conductor sag varies for different conductors by up to 13.5% relative to the Aldrey 300 conductor. This may result in the requirement to select another tower height or even another tower design. The width of the line corridor also varies clearly with different conductors. On the other hand, if one wishes the conductor sag to be constant, then different towers are necessary while using different conductors.

Last but not least, it is necessary to have as exact as possible knowledge of the topology and orography of the area for the design of transmission line routing. The detailed and final determination of tower locations is not a task of the planning procedure but has to be carried out prior to erection in the construction phase of the project.

9.7 Cost Estimates

Costs of overhead transmission lines consist of the annual capital costs and the annual cost of losses. The annual capital costs involve voltage-dependent and cross-section-dependent components and a fixed part according to Equation 9.25.

$$K_{\mathrm{i}} = \left[a + b \cdot U + c \cdot \sqrt[4]{n} \cdot n_{\mathrm{S}} \cdot q\right] \cdot e \cdot (r_n + \Delta k) \tag{9.25}$$

with

a, b, c = cost factors
e = factor for the consideration of cost increases
n = number of conductors per phase (subconductors)
n_{S} = number of AC systems per tower
q = cross-section of a conductor and/or sub-conductor
U_{n} = nominal voltage
r_{n} = annuity factor
Δk = additional cost per year.

The cost of losses are given by Equation 9.26.

$$K_{\mathrm{V}} = 3 \cdot I_{\max}^2 \cdot \frac{\rho}{q} \cdot \frac{n_{\mathrm{S}}}{n} \cdot [k_{\mathrm{F}} + k_{\mathrm{W}} \cdot \vartheta \cdot T_{\mathrm{Z}}] \cdot \mu \tag{9.26}$$

with μ and r as in Equations 9.27a and 9.27b:

$$\mu = \frac{q^n - r^{2n}}{q^n \cdot (q - r^2)} \tag{9.27a}$$

$$r = 1 + g \tag{9.27b}$$

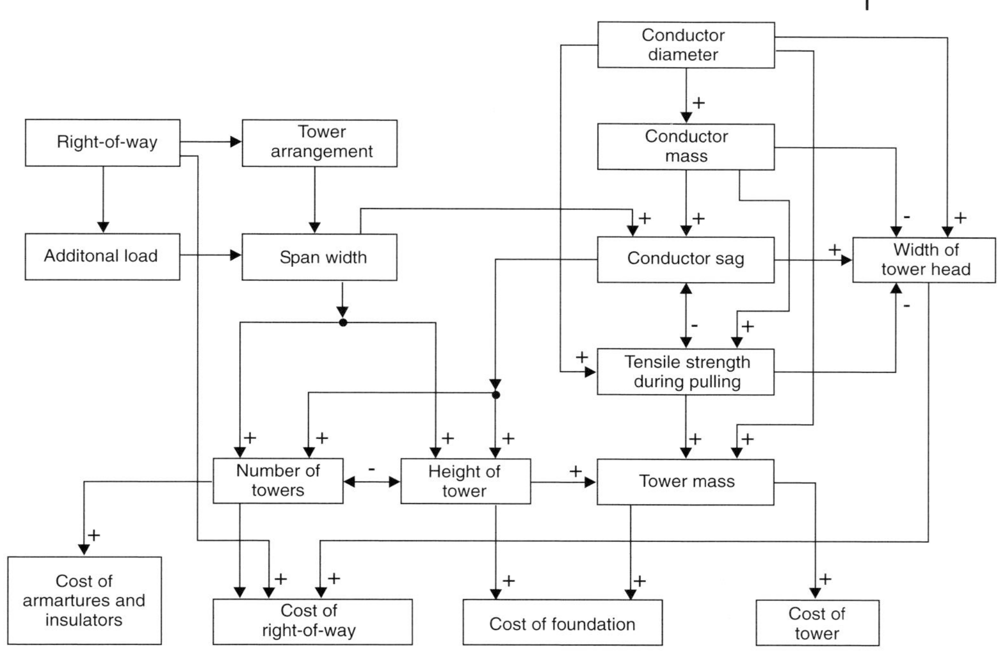

Figure 9.11 Simplified representation of the influence of different parameters on the costs of overhead transmission lines [44]. +, increase of cost; –, decrease of cost; some relations depend on the material.

The parameters are

k_P = annual specific cost related to transmittable power
k_W = energy cost
T_Z = period of analysis, typically 1 year
ϑ = degree of losses (see Section 4.2)
μ = factor for load increase over n years according to Equation 9.27a
q = interest factor
r = load factor r according to Equation 9.27b
g = factor of yearly load increase.

Various publications [45–47] describe detailed methods and equations for calculating the cost of losses. These methods are not suitable for planning purposes as the data and information needed are not available in the planning process but can only be obtained during operation.

Figure 9.11 gives a simplified representation of the parameters and relations which determine the costs of an overhead line. Other influences such as additional loads by wind and ice are not represented but may affect these costs significantly.

9.8
High-Temperature Conductors

9.8.1
General

Conductors for overhead transmission lines are operated under normal operating conditions up to 80 °C sometimes up to 90 °C. Conductors allowing operating temperatures above 80 °C (90 °C) are called high-temperature conductors; see CIGRE WG B2.12 (2004). They can roughly be divided into two categories [49]:

- Conventional high-temperature conductors (TAL), operating temperature up to 150 °C
- High-temperature conductors with operating temperatures up to 200 °C.

Conventional high-temperature conductors are state of the art and are used in special cases in the European transmission system and in other countries as well. High-temperature conductors operating at temperatures above 200 °C are not state of the art today. The high-temperature conductor costs are significantly higher than the costs of standard conductors; however, the permissible current carrying capacity of some conductor types is increased by more than 100%, whereas the sag is increased marginally only. The high costs limit the applications to special cases only. The different types of conductors and their technical properties are compared with standard Al/St-conductors in the following chapters. Detailed technical parameters, such as specific resistance, tensile strength, and permissible operating temperature are omitted in this book but must be obtained from manufacturers' data sheets.

9.8.2
Thermal Alloy Conductor Steel Reinforced (TACSR)

These conductors, called TAL-conductors, have construction details corresponding to Al/St-conductors according to DIN EN 50182. The aluminum layers are made from zirconium alloy (aluminum part with a degree of purity >99.7%). The recrystallization temperature of these conductors is much higher than that of Al/St-conductors, resulting in a sufficient mechanical strength of the conductor even at operating temperatures of 150 °C to 180 °C. The conductor core is made from aluminum-clad steel wires (STALUM). An increase of the current carrying capacity up to 150% at 150 °C can be achieved with TACSR-conductors as compared with standard Al/St-conductors operated at 80 °C. The permissible operating current can furthermore be increased by about 5% if the outer layer of the conductor is oxidized (black) in order to increase the heat radiation.

As the TACSR-conductor has nearly identical diameter and weight as compared with standard Al/St-conductors, towers, portals, and foundations are almost identical as compared with Al/St-conductors. The disadvantage, besides the high material costs, is the increase of ohmic resistance with increased temperature, thus increasing the ohmic losses. The sag at higher temperature is increased (approximately by 20% at 150 °C), requiring higher clearances and most probably an increase of the mast height.

9.8.3
Zirconium Alloy Conductor Invar Steel Reinforced (ZACIR)

The structure of this conductors, sometimes also named TACIR-conductor or ZTACIR-conductor, is similar to the TACSR-conductor; the steel core is made from a combination of steel wires and Invar-alloy (Ni-Fe-alloy), also called Permalloy or Nilo. Invar-alloy has a lower thermal expansion factor ($\approx 1.7 \cdot 10^{-5}$ to $2.0 \cdot 10^{-5}$ K^{-1}) as compared with Al/St-conductors. The conductor sag is slightly lower as against TACSR-conductors.

The operating temperature is given by some manufacturers as 200 °C, and the current carrying capacity is up to two times of that of Al/St-conductors. Besides higher price, the handling of the conductor at site is more complicated, as the Al-layers and the steel-alloy-layers have different temperature expansion factors, resulting in the need of pre-stretching of the conductors, which may require even stronger connection points.

9.8.4
Gap Thermal Resistant Aluminum Alloy Steel Reinforced (GTACSR)

These conductors, also called GAP-conductors, consist of an inner steel core and outer layers of zirconium alloy, similar to TACSR-conductors. The innermost aluminum layer, forming a circular tube, encloses the inner steel core nearly without mechanical contact. The gap between these layers is filled with a thermally stabilized lubricant. The outer aluminum layers are made from standardized zirconium alloy. The conductor behaves like a pure steel-cord with low thermal expansion with the aluminum layers sliding free on the steel core.

Conductor weight and diameter are almost identical to standard Al/St-conductors resulting in the same mechanical load at the tower. Due to the special thermal behavior, the expansion of the conductor is lower, resulting in similar sag at higher temperature as compared with standard Al/St-conductors at 80 °C. The disadvantage, besides the high material costs, is the increase of ohmic resistance with increased temperature, thus increasing the ohmic losses. Furthermore, the construction of the transmission line especially pulling of the conductor is complicated, as the steel core can be pulled only, thus requiring an increased number of suspension towers.

At operating temperature of 150 °C, an increase of the operating current up to 160% can be achieved as compared with standard Al/St-conductors at 80 °C. GTACSR-conductors are in use worldwide.

9.8.5
Annealed Aluminum Conductor Steel Supported (ACSS)

The conductor is made from an inner steel core with outer layers of thermally annealed aluminum. Due to the low tensile strength of the annealed aluminum, the conductor has a thermal behavior in principle similar to the GTACSR-conductor, that is, like a pure steel-conductor. The construction and installation comply with the standard Al/St- and GTACSR-conductors. The low breaking stress of the conductor as compared with TACSR requires a reduced tensile strength, resulting in higher sag. This may require higher towers or changes in the tower construction.

9.8.6
Aluminum Conductor Composite Core (ACCC)

In ACCC-conductors, the steel core is replaced by a composite plastic material. This supporting layer has a much higher breaking stress than steel cores and a significantly lower specific weight. The aluminum layers are of the same make as the TACSR-conductors. Due to the reduced weight of the conductor, the sag is significantly lower than that of standard Al/St-conductors. The long-term thermal and mechanical stabilities of the supporting plastic core under real conditions have not yet been tested. Conductor oscillations and line galloping due to the reduced weight are more probable as against Al/St-conductors. Sidewise wind loads will increase the side-swing of the conductors, thus requiring an increased safety distance.

9.8.7
Aluminum Conductor Composite Reinforced (ACCR)

The conductor is made from an inner core of aluminum-ceramic composite and outer layers of aluminum. The supporting core has a lower modulus of elasticity, resulting in less sag as compared with standard Al/St-conductors having the same operating temperature. The current capacity is significantly increased as the permissible operating temperature is increased. The long-term stability of the conductor is proved in several test installations; some hundred kilometers of installations are existing worldwide.

10
Flexible AC Transmission Systems (FACTS)

10.1
Basics of Transmission of Power through Lines

The transmission of electrical power by an overhead transmission line between two power systems is considered using the simplified equivalent circuit diagram according to Figure 10.1a with distributed longitudinal impedances $Z = R_L + jX_L$ and admittances $Y = G_E + jX_C$. With respect to voltage drop, reactive power balance and phase-angle, the model of the lossless line according to Figure 10.1b is adequate.

The active power P transferable over the line is calculated according to Equation 10.1,

$$P = \frac{U_A \cdot U_E}{Z_S \cdot \sin\theta} \cdot \sin\delta \tag{10.1}$$

with the characteristic impedance or surge impedance Z_S given by Equation 10.2a

$$Z_S = \sqrt{\frac{L}{C}} \tag{10.2a}$$

and the electrical length of the line θ according to Equations 10.2b and 10.2c:

$$\theta = \frac{2\pi}{\lambda} \cdot a \tag{10.2b}$$

$$\theta = 2\pi \cdot f \cdot \sqrt{L \cdot C} \cdot a \tag{10.2c}$$

with frequency f and line length a and δ the phase-angle of the line.

If natural power P_W according to Equation 10.3 is transmitted through the line,

$$P_W = \frac{U_A \cdot U_E}{Z_S} \tag{10.3}$$

Power System Engineering: Planning, Design and Operation of Power Systems and Equipment, Second Edition.
Jürgen Schlabbach and Karl-Heinz Rofalski.

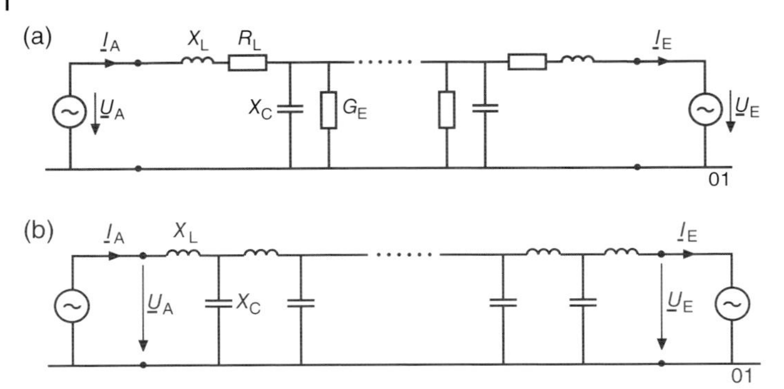

Figure 10.1 Equivalent circuit diagram of a line for quasi-stationary conditions. (a) Diagram with impedances and admittances; (b) simplified circuit diagram of the lossless line.

the reactive power of the line is balanced, as the reactive power produced by the capacitances (proportional to the square of the voltage) is equal to the reactive power needed by inductances (proportional to the square of the current). During higher loading the inductive voltage drop along the line becomes larger and thus the reactive (inductive) power becomes larger as compared with the reactive (capacitive) power produced by the capacitances, and the resulting voltage decreases.

The required reactive power must be made available or absorbed by the connected power systems at both ends of the line. In the circuit diagram of Figure 10.1 the voltage reaches its minimum in the middle of the line with same voltages at the sending and receiving ends of the line. When the line is loaded below the natural power, the capacitive part exceeds the inductive part and the voltage rises along the line length and reaches its maximum in the middle of the line. The unbalanced capacitive part of the reactive power in this case must be absorbed by the connected power systems.

If the line is assumed electrically short, then Equation 10.4 applies

$$\sin\theta \approx \theta = \omega \cdot \sqrt{L \cdot C} \cdot a \tag{10.4}$$

With this approximation Equation 10.5 is applicable.

$$Z_S \cdot \theta = \sqrt{\frac{L}{C}} \omega \cdot \sqrt{L \cdot C} \cdot a = \omega \cdot L \tag{10.5}$$

As a consequence Equation 10.1 is converted into Equation 10.6:

$$P = \frac{U_A \cdot U_E}{X_L} \cdot \sin\delta \tag{10.6}$$

The approximation of Equation 10.6 can be interpreted as neglecting the capacitance. If the capacitances are taken into consideration, the power transferable through the line is increased for quasi-stationary conditions.

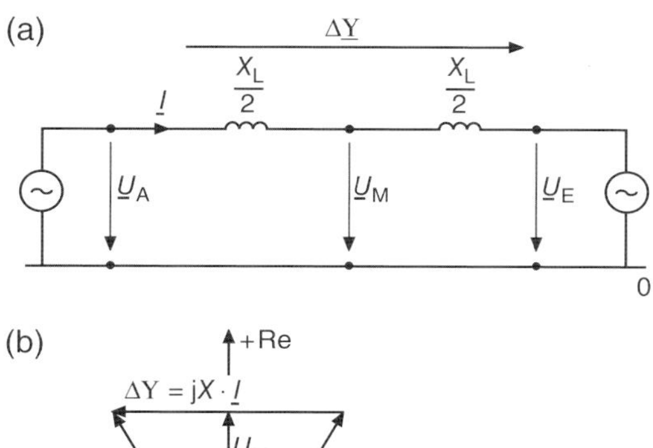

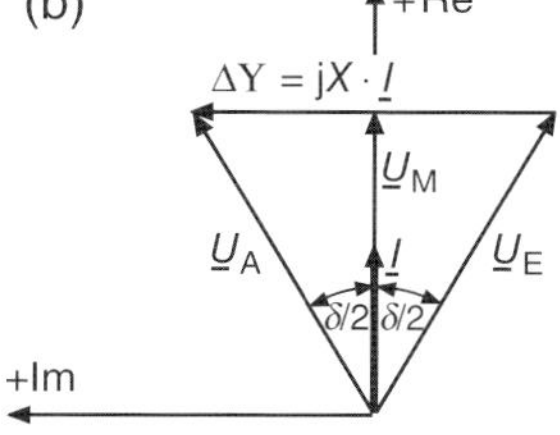

Figure 10.2 Equivalent circuit diagram for the determination of the transferable power. (a) Simplified equivalent circuit diagram of the lossless line; (b) vector diagram of the voltages and the current.

If the vector diagram of the voltages at the sending and receiving ends of the line is considered as in Figure 10.2 and if both voltages are assumed to have the same magnitude |U| = U then Equation 10.7 applies:

$$\underline{U}_A = U \cdot \left(\cos\frac{\delta}{2} + \mathrm{j}\sin\frac{\delta}{2}\right) \tag{10.7a}$$

$$\underline{U}_E = U \cdot \left(\cos\frac{\delta}{2} - \mathrm{j}\sin\frac{\delta}{2}\right) \tag{10.7b}$$

with the voltage in the middle of the line U_M given according to Equation 10.7c.

$$|\underline{U}_M| = \frac{|\underline{U}_A - \underline{U}_E|}{2} = U \cdot \cos\frac{\delta}{2} \tag{10.7c}$$

The active power P transferable through the line and the reactive power at the sending end Q_A and receiving end Q_E of the line are calculated form Equation 10.8,

$$P = U_M \cdot I = \frac{U^2}{X_L} \cdot \sin\delta \tag{10.8a}$$

$$Q_A = -Q_E = U \cdot I \cdot \sin\frac{\delta}{2} = \frac{U^2}{X_L}(1 - \cos\delta) \tag{10.8b}$$

with the current through the line according to Equation 10.9,

$$|\underline{I}| = \left| \frac{\underline{U}_{\rm A} - \underline{U}_{\rm E}}{{\rm j}X_{\rm L}} \right| = \frac{2 \cdot U}{X_{\rm L}} \sin \delta \tag{10.9}$$

The dependence of active and reactive power on the phase-angle of the line is outlined in Figure 10.3.

As can be seen from Figure 10.3, the current through the line is proportional the reactance of the line and increases with increasing phase-angle. The voltage in the middle of the line decreases with increasing phase-angle, as can be seen from Equation 10.7c. Increase of the transferable power can be achieved in accordance with Equation 10.8a by one the following means:

- Increase of the voltage in the middle of the line by capacitative compensation (parallel compensation with parallel capacitor)
- Reduction of the line reactance by serial compensation (serial compensation with serial capacitor)
- Decrease of the phase-angle of the line by phase-shifting equipment (phase-shifting transformer).

These three possibilities for increasing the transferable power can be realized both by conventional equipment such as capacitors or phase-shifting transformers and by electronic equipment to produce a flexible AC transmission system (FACTS). With electronic equipment, additional possibilities are available for improvement of the dynamic stability as well as further functions such as oscillation damping and voltage control; see also Section 10.5. The fundamentals of system operation, such as transmission of active power with low losses, operation of generators in the over-excited mode (leading power factor), the loading of equip-

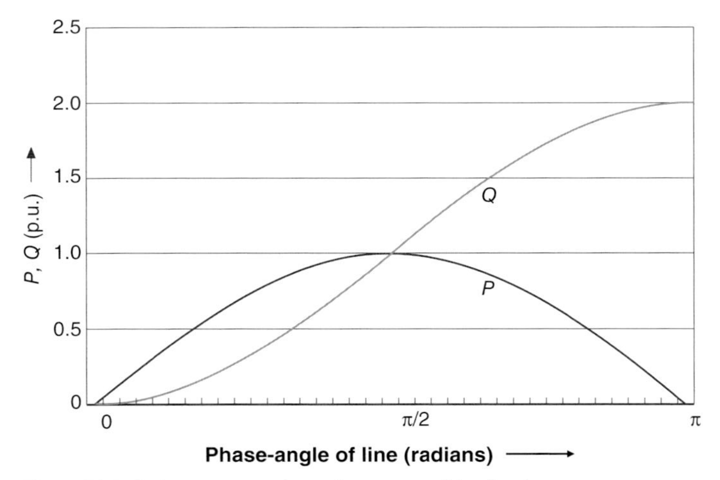

Figure 10.3 Active power and reactive power of the line in Figure 10.2 as a function of the phase-angle of the line.

ment below its permissible limits, voltage band according to IEC 60038, minimization of the transferable reactive power, limitation of the reactive power transferred through transformers and thus balanced reactive power of each voltage level, high lagging power factor of the load and other operational constraints have to be taken into account.

10.2 Parallel Compensation of Lines

Parallel compensation is the standard means of reactive power compensation in the MV and LV systems. The aim is to achieve a defined power factor of the load, limitation of the supplied reactive energy and a decrease of the voltage drop. Parallel compensation within the high-voltage range uses the same principle to increase the transferable power through lines. The system diagram of Figure 10.4 is employed. The lossless compensation (capacitor) is placed in the middle of the line. The voltage at the capacitor U_M is equal the voltages at the sending and receiving end of the line, U_A and U_E, if the maximal power is transmitted.

Similarly to Equation 10.7c, the voltages between the middle and the sending and receiving ends of the line are given by Equation 10.10:

$$U_{M1} = U_{M2} = U \cdot \cos\frac{\delta}{4} \tag{10.10}$$

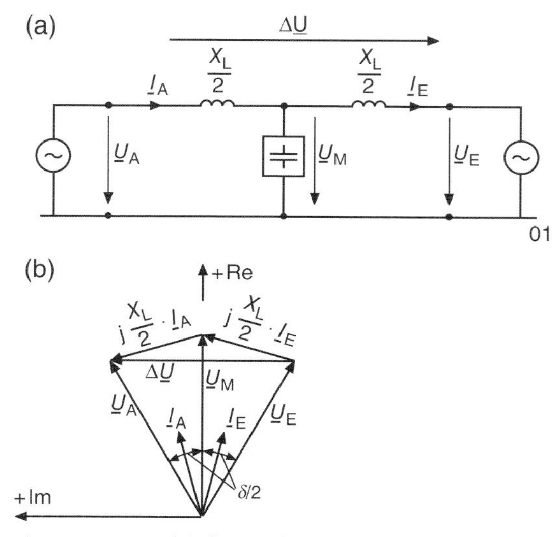

Figure 10.4 Simplified equivalent circuit diagram of the lossless line with parallel compensation in the middle of the line. (a) Equivalent circuit diagram; (b) vector diagram of the voltages and the current.

with the current I_A according to Equation 10.11:

$$I_A = I_E = \frac{U_M - U_A}{X_L/2} = \frac{4 \cdot U}{X_L} \cdot \sin\frac{\delta}{4} \tag{10.11}$$

Thus the active power P_K transferable over the line and the reactive power of the line Q_K are calculated according to Equation 10.12.

$$P_K = 2 \cdot \frac{U^2}{X_L} \cdot \sin\frac{\delta}{2} \tag{10.12a}$$

$$Q_K = 2 \cdot \frac{U^2}{X_L}\left(1 - \cos\frac{\delta}{2}\right) \tag{10.12b}$$

Figure 10.5 outlines the dependence of the active and reactive power on the phase-angle.

A parallel compensation in the middle of the line increases the transferable power to up to twice the value of the uncompensated line. The reactive power requirement is up to four times in this case and must be made available by the compensation plant and the connected power systems.

With compensation at the sending or receiving end of the line the conditions are worse because the compensation equipment does not here lead to an increase of the transferable power, assuming the same voltages at the line ends of the compensated and the uncompensated case. Compensation at one or both line ends leads to an increase of the transferable power only if higher voltages are achieved.

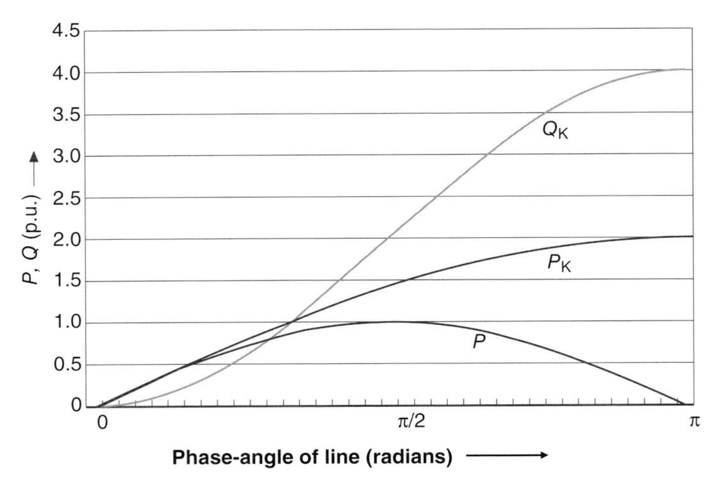

Figure 10.5 Transferable active power and reactive power as a function of the phase-angle of the line with parallel compensation in the middle of the line of Figure 10.4.

10.3 Serial Compensation of Lines

Serial compensation of lines is realized by reducing the line reactance by serial compensation, for example, a serial capacitor. The diagram of Figure 10.6 is employed. The line is assumed to be divided into two equal segments. Thus the voltage drop at the longitudinal reactance of the line is compensated with rising current partly by a voltage increase by the serial compensation, so that the total voltage drop of the line becomes smaller than without serial compensation.

The current of the line is calculated according to Equation 10.13a,

$$I = \frac{2 \cdot U}{(1-k) \cdot X_L} \cdot \sin\frac{\delta}{2} \qquad (10.13a)$$

with the compensation degree k of the compensated line given by Equation 10.13b,

$$k = \frac{X_C}{X_L} \qquad (10.13b)$$

Thus the transferable active power P_R and the reactive power Q_C produced by the serial compensation are given by Equations 10.14 a and b:

$$P_R = \frac{U^2}{(1-k) \cdot X_L} \cdot \sin\delta \qquad (10.14a)$$

$$Q_C = \frac{2 \cdot U^2}{X_L} \cdot \frac{k}{(1-k)^2} \cdot (1 - \cos\delta) \qquad (10.14b)$$

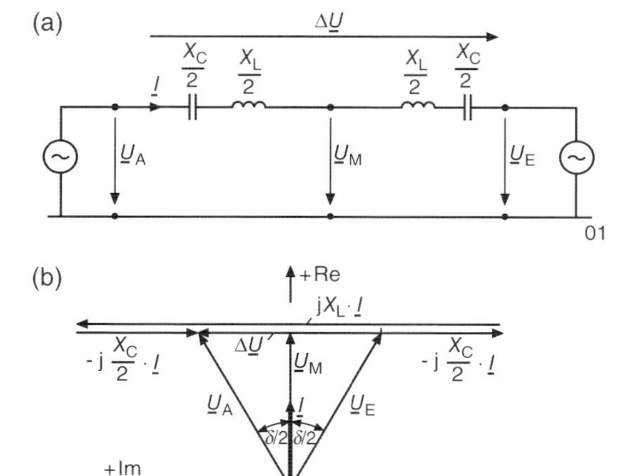

Figure 10.6 Equivalent circuit diagram of the lossless line with serial compensation. (a) Equivalent circuit diagram; (b) vector diagram of the voltages and the current.

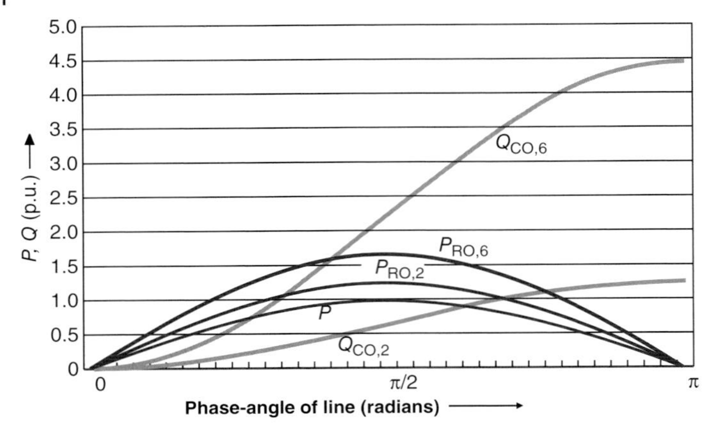

Figure 10.7 Transferable active power and capacitive reactive power of the serial compensated line of Figure 10.6 as a function of the line angle and compensation degree $k = 0.2$ and $k = 0.6$.

The active and reactive power depend on the phase-angle of the line as shown in Figure 10.7.

A serial compensated line leads to an increase of the transferable power, but the increase is inversely proportional to the degree of compensation. The reactive power of the serial compensation plant must be increased to increase the transferable active power.

10.4 Phase-Shifting Equipment

If the transferable power of a line is to be increased by changing of the phase-angle of the line, phase-shifting equipment–for example, a quadrature or phase-shifting transformer–has to be installed [53]. The aim of phase-shifting is to decrease the angle between the voltages at the sending and the receiving ends by addition of an auxiliary voltage U_Z with a defined phase-angle $\delta_Z > 0°$ related to the voltage U_A at the sending end of the line according to Equation 10.15 and represented in Figure 10.8. With the installation at the receiving end of the line, the phase-angle has to be related to the voltage U_E at the receiving end of the line. The voltage at the sending end of the line, relevant for the phase-angle of the line, can be tuned and thus the phase-angle of the line can be adjusted by changing either the angle or the magnitude of the auxiliary voltage.

$$\underline{U}_Z = U_Z \cdot e^{j\alpha} \tag{10.15}$$

where α is the phase-angle between the voltage at the sending end and the auxiliary voltage. The most favorable effect is achieved if an auxiliary voltage with a phase-

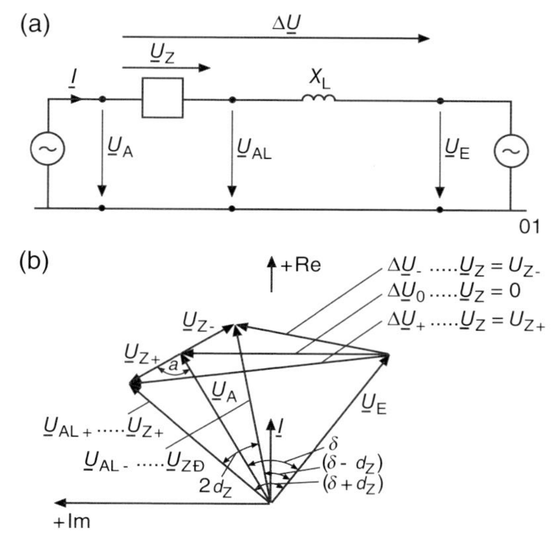

Figure 10.8 Equivalent circuit diagram of the lossless line with phase-shifting (quadrature booster) transformer. (a) Equivalent circuit diagram; (b) vector diagram of the voltages and the current.

angle of 90 is added, as the phase-angle is changed substantially in this case without affecting the magnitude of the voltage.

The active power P_S transferable through the line and the reactive power Q_S at the sending and receiving end of the line are calculated according to Equation 10.16,

$$P_S = \frac{U^2}{X_L} \cdot \sin(\delta - \delta_Z) \tag{10.16a}$$

$$Q_S = \frac{U^2}{X_L} [1 - \cos(\delta - \delta_Z)] \tag{10.16b}$$

with δ_Z the phase-angle of the additional phase-shift. The dependency of the active and reactive power from the phase-angle is outlined in Figure 10.9.

The concept of phase-shifting can be applied for example, for balancing the loading of two parallel lines in the power system with different impedances [50] and [51].

The transferable power can be kept nearly constant for a wide range of phase-angles by adaptation of the magnitude of the auxiliary voltage. Thus the mode of operation of the phase-shifting equipment can be adapted to changed system conditions.

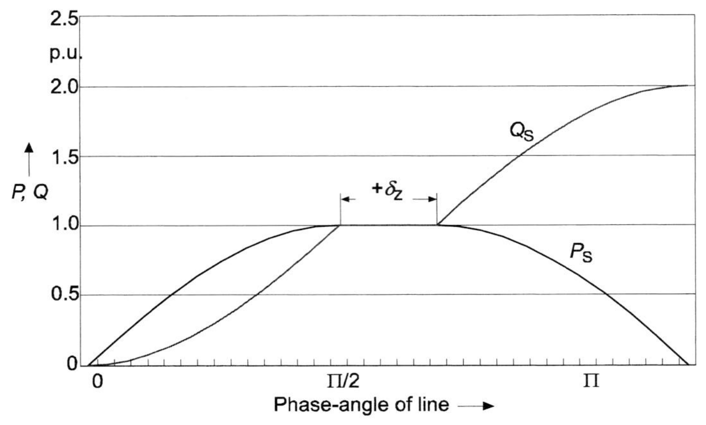

Figure 10.9 Transferable active power and reactive power of the line of Figure 10.8 as a function of the phase-angle of the line.

10.5
Improvement of Stability

The mode of operation of phase-shifting equipment for increasing the transferable power as described in the previous sections can be applied to improve the stability of the generators and the power system. The stability limit of a synchronous generator under changing loading conditions, for example, by short-circuits, can be determined graphically in accordance with Figure 10.10. The generator is operated in the quasi-stationary mode with active power P_1 followed by a stepwise change in load (active power P_2). If the system voltage is assumed to be constant, the phase-angle δ_1 corresponds to the rotor phase-angle. The active power P_1 and P_2 are therefore related to the rotor phase-angle δ_1 and δ_2. In case of a load change from P_1 to P_2 the rotor phase-angle will attune with a damped oscillation to the new operating point δ_2 with an overshoot up to the rotor phase-angle δ_3 corresponding to an active power P_3. The maximal attainable rotor angle is determined from the equal-area criterion, that is, the balance of the accelerating and the decelerating torques as represented in Figure 10.10.

As an example, consider a short-circuit on the line between the generator and the power system according to Figure 10.11a. The generator is operated with the active power P_1, the assigned rotor phase-angle is δ_1 (curve 1 in Figure 10.11c). During the short-circuit the transferable active power is reduced substantially (curve 2 in Figure 10.11c) as the voltage at the generator decreases and/or the impedance between generator and system increases. The generator is accelerated, as the acceleration torque remains constant during the short-circuit. The rotor phase-angle increases at the same time and would probably exceed the limit at which the generator slips and runs asynchronously. It is assumed in the example that the short-circuit is switched off before the generator slips. At this time the

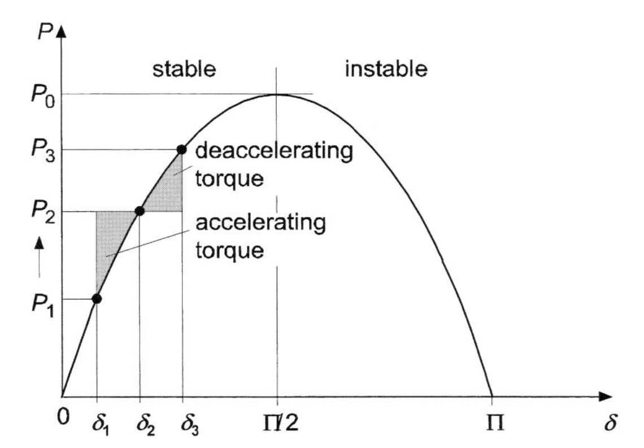

Figure 10.10 Graphic determination of the stability of a synchronous generator during stepwise load change in the power system using the equal-area criterion [52].

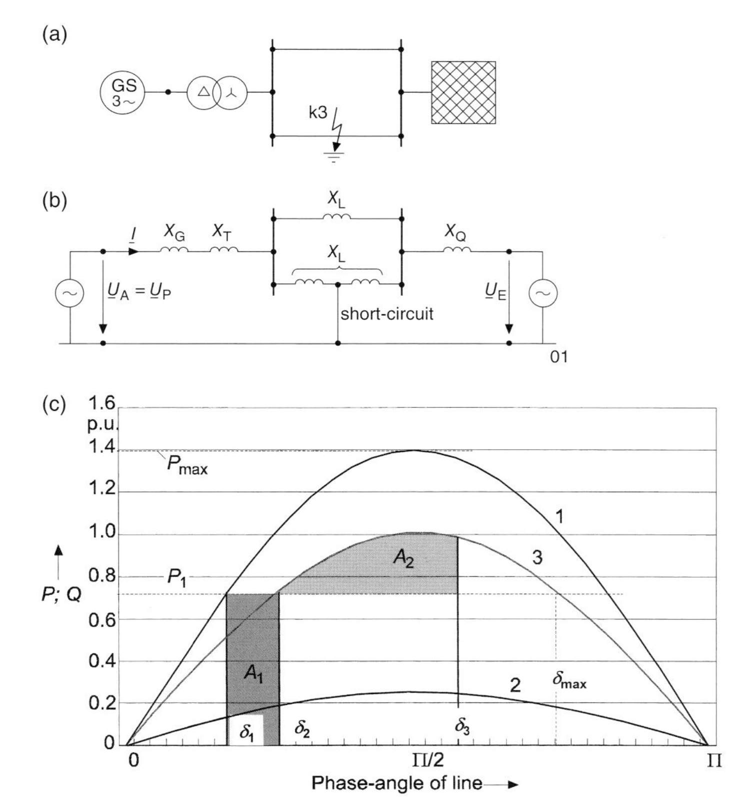

Figure 10.11 Diagram for the determination of the stability of a generator in the case of short-circuits. (a) Connection diagram; (b) equivalent circuit diagram in the positive-sequence component; (c) transferable active power as function of the phase-angle of the line for different operating conditions. **1**, prior to short-circuit; **2**, during short-circuit; **3**, after short-circuit.

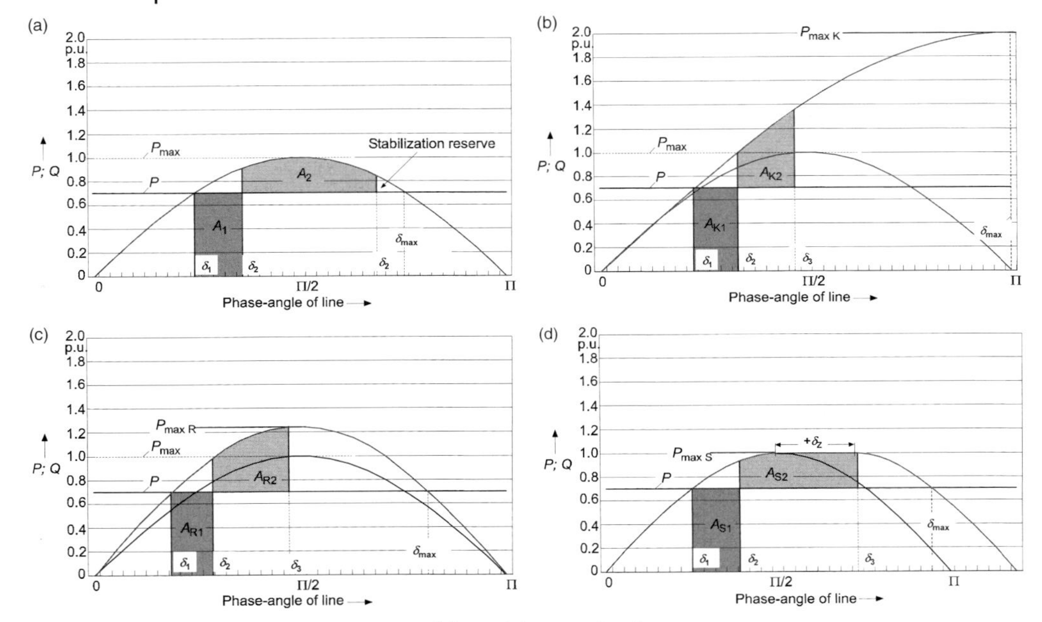

Figure 10.12 Improvement of the stability margin of a generator [52]. (a) Without measures; (b) parallel compensation; (c) serial compensation; (d) phase-shifting.

rotor phase-angle will be between δ_2 and δ_3. After disconnection of the short-circuit the transferable active power increases again, though not to the previous value P_1, as the impedance between the generator and the power system is increased due to the shut-down of the faulted line (curve 3 in Figure 10.11c). It is further assumed that under quasi-stationary operating conditions the rotor phase-angle of the generated active power P_1 will be approximately δ_3. The stability limit with the rotor phase-angle δ_{max} is determined by the equal-area criterion (areas A_1 and A_2). In case of larger phase-angles the generator will slip.

The influence of the different measures for the increase of the transferable active power on the stability limits is represented in Figure 10.12 on the basis of the equal-area criterion. It is assumed that the active power P_1 is fed by a generator through the line into the power system. The rotor phase-angle δ_1 according to the individual Figures 10.12a–d depends on the different measures applied. Additionally, the dependence of the transferable active power without additional measures is also given in the diagrams to indicate the improvement. During short-circuit the transferable active power decreases (not represented in the diagrams); at the time of the disconnection of the short-circuit the rotor phase-angle will be δ_2, since the generator is accelerated during the short-circuit duration.

Due to transient oscillations, as explained in Figure 10.11, for the instant of short-circuit clearing (disconnection) the rotor phase-angle reaches the value δ_3, determined by the equal-area criterion. If the rotor phase-angle exceeds the value

δ_{max}, the generator will slip and stability will no longer be achieved. The difference between the angles δ_3 and δ_{max} represents the reserve margin for stability.

As can be seen from Figure 10.12, the transferable power through the line and/or the stability margin is significantly increased in all cases compared with the case without any measures.

10.6 Basics of Flexible AC Transmission Systems (FACTS)

The different methods to increase the current carrying capacity of transmission lines, as explained in Sections 10.2–10.5 can be realised with power-electronic equipment. Modern installations offer more operational options such as damping of oscillations, voltage control, and reduction of harmonics and interharmonics. These equipment are summarized under the expression FACTS, which is a generic term for technologies related to the control of active and reactive power flow and voltage stability by power-electronic equipment. Due to their high flexibility and dynamic, they are perceived as a key technology for the challenges of decentralized, renewable energy systems. FACTS either use thyristors (first generation) or inverse gate bipolar transistors (IGBT) (second generation) as semiconductor devices. An overview of the different types of FACTS is indicated in Figure 10.13.

In the case of the serial compensation thyristor switched serial capacitor (TSSC), the reactive power can be switched off via a thyristor-bypass, thus allowing a more flexible power flow control. With an additional reactor controlled by a thyristor, a thyristor controlled series capacitor (TCSC) is reached. FACTS with thyristors connected in parallel are summarized under the term static-var compensator (SVC), switched capacitors (thyristor switched capacitor [TSC]), and switched reactors (thyristor switched reactor [TSR]) are possible. A faster response time is achieved with controlled reactors (thyristor controlled reactor [TCR]), with the reactor to be controlled continuously by the thyristor. The required capacitive

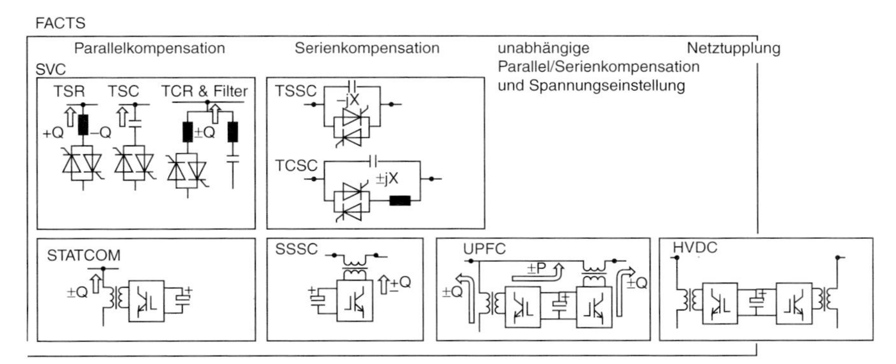

Figure 10.13 Overview of Flexible AC Transmission Systems (FACTS) [54].

reactive power is provided by the switched capacitor banks. As harmonic currents are generated by the TCR, the capacitor bank is usually designed as a filter circuit.

FACTS of the second generation have a pulse inverter (IGBT) with voltage control. A device of high importance is the static synchronous compensator (STATCOM). The reactive power control is independent from the system voltage; capacitive and inductive reactive power can be controlled with high dynamic.

SVC and STATCOM play an important role for the dynamic voltage support. They are therefore operated at critical locations in the power system to correct dynamic voltage fluctuations due to changes in load or to damp load flow oscillations. Even in distribution systems, SVC and STATCOM can protect consumer installations against voltage dips and surges as well as reduce harmonics, thus improving the power quality.

Basically, modern FACTS can be installed in serial connection as static synchronous series compensator (SSSC). Of high practical importance is the combination of series and parallel compensation, the unified power flow controller (UPFC), which allows not only parallel and serial compensation but also an active power exchange between the two inverters. The world's first UPFC, each with +160 Mvar for the parallel and series compensator, was commissioned in 1998 in Eastern Kentucky on the 138-kV level.

10.7 HVDC-Transmission (High-Voltage-Direct-Current)

10.7.1 General

HVDC-transmission is in several cases superior to the HVAC-transmission of electrical energy for the following reasons:

- Coupling of asynchronous networks (HVDC links between regions in the US)
- Coupling of networks of different frequency (such as supply of railway power supply systems from public three-phase systems)
- Connection of large power plants without increasing the short-circuit power (Kingsnorth power station in London)
- High power transmission over large distances (Three Gorges Dam in China)
- Offshore connection of generation (offshore wind farms in the North Sea)
- Connection of separate power systems by submarine cables (France–UK, Norway–Netherlands, Sweden–Finland).

In HVDC-transmission, the transmittable power is determined by the DC-current and the DC-voltage only as per Equation 10.17. No reactive power is required.

$$P = U_{\mathrm{DC,A}} \cdot I_{\mathrm{DC}} = U_{\mathrm{DC,A}} \cdot \frac{U_{\mathrm{DC,A}} - U_{\mathrm{DC,E}}}{R} \tag{10.17}$$

with

$U_{DC,A}$	DC-voltage at sending end of the line
$U_{DC,E}$	DC-voltage at receiving end of the line
R	Resistance of the DC line

10.7.2 Converter Stations and Related Equipment

The basic scheme of an HVDC link consists of two converter stations, either equipped with thyristors or inverse gate bipolar transistor (IGBT), which are connected via transformers to the AC-systems. The DC-connection between the two converter stations is realized by an overhead transmission line of by a cable.

Converter station with thyristors are so-called line-commutated converters, which require an AC-voltage and reactive power for commutation, power control, and harmonic's reactive power and distortion power. Changing the direction of power flow is only possible by reversing the voltage, resulting in remaining space charges in the DC-system, which may increase the risk for partial discharges in XLPE-cables (cross-linked polyethylen).

Converter station with IGBT are self-commutated converters, controlled as constant voltage source. The IGBT can be controlled by pulse width modulation; reactive power is not required. Active and reactive power can be controlled in any combination from the viewpoint of the AC-system. Self-commutated converters can be operated without any AC-voltage signals; they are capable for black-starts, which are a suitable measure to restart a power system after blackout. Changing the direction of power flow is possible by reversing the direction of current flow; this enables the use of XLPE-cables for the DC-connection.

10.7.3 Breakers, Reactors, Electrodes and other Equipment

For smoothing of the DC-current, smoothing reactors are required in converter stations with thyristors at the beginning and the end of a DC line. They also serve to limit the short-circuit currents for faults on the DC line. Regarding converters with IGBT (mainly voltage source converters), the DC-voltage has to be smoothed by capacitors. Filters can be installed alternatively on the DC-side.

Circuit breakers at the AC-side shall be capable to switch large capacitive currents. Circuit breakers on the DC-side are not required for two-terminal connections as DC-faults can be cleared by lowering the DC-voltage. In multiterminal connections, DC-breakers are necessary. The latest technology consists of a hybrid DC-switch made of a mechanical breaker (disconnector) in parallel to an electronic breaker.

Figure 10.14 outlines the layout of an HVDC-station with thyristors for transmission of 1000 MW, DC-voltage of ±400 kV.

HVDC overhead lines require only one single conductor (unipolar or monopolar transmission). The return line is by the earth return. No redundancy for power

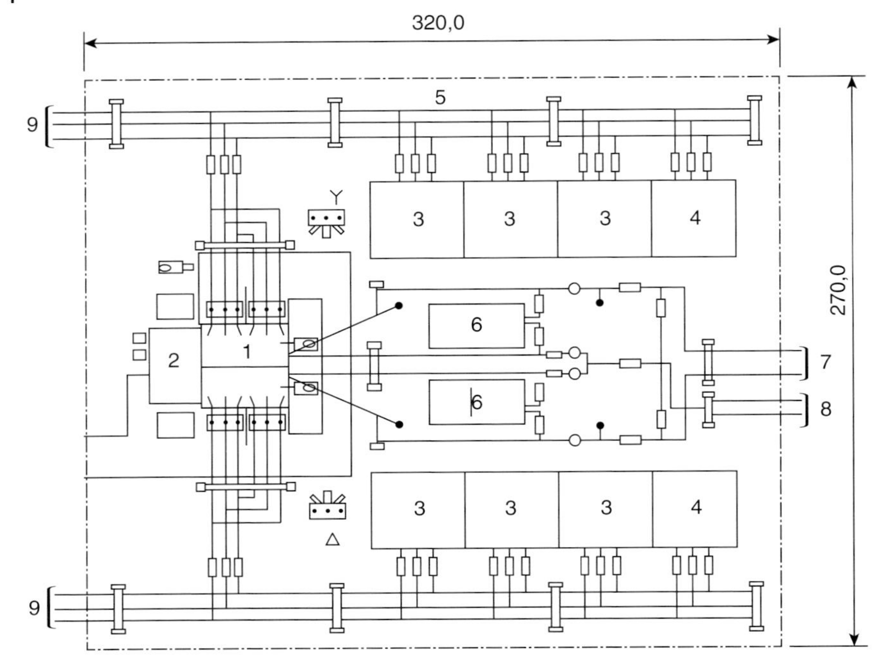

Figure 10.14 Arrangement of an HVDC station. Source: ABB Switchgear manual. 1, valve hall; 2, control building; 3, phase filter; 4, capacitors; 5, AC switchgear; 6, DC filter; 7, DC line; 8, conductor to earthing electrode; 9, AC line.

transmission in case of failure of the line is given. In case of bipolar transmission, the return line is by the second conductor. The transmission line itself consists of either two monopolar lines or one bipolar line, depending on the required redundancy in case of damage of the tower.

HVDC cables are used as single-phase cables, in exceptional cases as two-phase cables. Today, mainly XLPE-cables are in use; however oil-immersed cables and gas-pressure cables are installed as well. It should be noted that XLPE-cables can only be used in case of converters with IGBT technology.

Earthing electrodes are required at both ends of the DC-transmission in case of monopolar transmission. In case of bipolar transmission, earthing electrodes are installed to increase the availability of the transmission in case of outage of one pole. The environmental effects of earthing electrodes and currents through earth, such as electrolysis of water, magnetic fields, corrosion, and step voltages shall be considered.

11
Load-Flow and Short-Circuit Current Calculation

11.1
Load-Flow Calculation

An important tool for the planning of electrical power systems is the load-flow analysis or load-flow calculation. The objective is the determination of significant parameters of the power system for normal operation and under emergency conditions, such as

- Voltages at grid-stations, substations and busbars in terms of kilovolts or percent of nominal voltage and in terms of phase angle
- Currents and/or active and reactive power-flow on overhead lines, in cables and through transformers and other equipment, as well as the resultant relative loading
- Active and reactive power losses of equipment, power systems and subsystems
- Exchange of active and reactive power between power systems or between groups of power systems
- Balance of generation and load in subsystem areas
- Power transfer through power systems or subsystems
- Required voltage control range of transformers and generators
- Reactive power needs and/or compensation needs at busbars, made available by generators, compensation equipment or flexible AC transmission systems (FACTS)
- Further characteristics, which can be determined from the knowledge of current and voltages in the power system.

Generally load-flow calculation assuming symmetrical, three-phase operation is adequate. On the basis of the equations for admittance Y and power S according to Equation 11.1,

$$\underline{Y} = \frac{\underline{I}}{\underline{U}} \tag{11.1a}$$

$$\underline{S} = \sqrt{3} \cdot \underline{U} \cdot \underline{I}^{*} \tag{11.1b}$$

Power System Engineering: Planning, Design and Operation of Power Systems and Equipment, Second Edition.
Jürgen Schlabbach and Karl-Heinz Rofalski.

the voltages are calculated at given infeed power and/or load at the busbars. If the current is substituted in Equation 11.1b by Equation 11.1a nonlinear, quadratic equations are obtained, for the solution of which mathematical procedures are available like such as the Newton–Raphson method, the current iteration method or the decoupled load-flow calculation. The differences between the different methods are seen in the calculation time, in the number of necessary iterations and in the convergence behavior, that is, whether a meaningful solution can be found at all.

The choice of suitable representation of the consumer load and the definition of the busbar nodes also influence the convergence behavior and the calculation time. If the nodes are defined as PQ busbars, then the voltages at the busbar are calculated for given active power P and reactive power Q. If the node is defined as a PV busbar, then the voltage U (symbolized V) at the node is kept constant with given active power P and the reactive power Q, needed to maintain the voltage, is calculated. A change of the type of node within a calculation run is supported by all load-flow programs. The voltage at a PQ node is kept constant if the voltage exceeds a defined bandwidth and the node is then regarded as a PV node. For the balance of the active and reactive power in the power system including the losses, a balance node, called slack, is needed. The differences of the summarized losses, loads and infeeds are balanced at the slack, that is, the balanced power is placed by the program at the slack as an additional infeed or load. The indication of subsystem balance (zone slack) is supported by most load-flow programs. It has to be ensured that the slack can also actually feed or absorb the calculated balance as generation or load.

In principle, loads in the power system can be defined in different ways. In case of modeling of the load as *constant impedance,* the active and reactive parts of the load must be defined. The apparent power changes with the square of the voltage. The representation of the load as *constant current* requires as input data the current and the power factor, leading to a linear interdependence of power and voltage. For load-flow calculation in high-voltage transmission systems, the load is assumed as *constant power* with active and reactive parts. Thus, there is no voltage-dependent consumer load in this case.

As the system load consists of a high number of different consumers, an exact indication of the kind of load representation is hardly possible, as explained below for different applications. As an example, the active power of a synchronous generator remains constant in the voltage range 90–110% of the rated voltage; the active power of lighting equipment in this voltage range changes with the exponent p = 1.5, whereas the active power of electric heating devices changes almost quadratically. Measurement in MV systems [55] indicated that the system load changes within the voltage range $U = 0.85\,U_n$ to $1.15\,U_n$ according to Equation 11.2.

$$P(U) = P_n \cdot \left(\frac{U}{U_n}\right)^p \tag{11.2a}$$

$$Q(U) = Q_n \cdot \left(\frac{U}{U_n}\right)^q \tag{11.2b}$$

The exponents p and q depend on the load to a great extent. An increase in motor load results in a decrease of the exponent p, which in industrial systems can be within the range $p = 0.5–1.1$ and in purely residential areas up to $p = 2.1$. The exponent q of the reactive power (Equation 11.2b) is usually in the range $q = 3.0–5.8$, which is due mainly to the reactive power of the transformers. The recommended exponents for the active power, $p = 1$ (constant load), and the reactive power, $q = 2$ (constant impedance), given in various publications and program handbooks, are therefore to be handled with caution. At least for the analysis of the low-voltage system, the active load has to be represented as constant impedance ($p = 2$). More details can be found in references [56] and [57].

The time dependent pattern of load and generation has also to be taken into account for load-flow analysis. Power systems normally have pronounced load variations and different maximum load portions of the individual loads. The representation of all loads as maximum load will yield wrong results, and, being on the safe side, it will lead to economically unfavorable investment decisions, see also Chapters 2 and 4. Load-flow calculations with actual load variation curves, as also recorded, for example, for the medium-voltage range by load dispatch centers, are more reliable.

11.2 Calculation of Short-Circuit Currents

11.2.1 General

AC and DC power systems have to be planned and designed in such a way that installations and equipment are able to withstand the expected short-circuit currents. Both the thermal and the electromagnetic (electromechanical) effects of the short-circuit current have to be considered for the determination of the maximal short-circuit currents. The protection devices must detect the short-circuit definitely, and for this the knowledge of the minimal short-circuit current is necessary. Breakers, switches and fuses must cut off short-circuit currents safely and rapidly. Short-circuit currents through earth can lead to interference with communication circuits and nearby pipelines as well as resulting in impermissible step, touch and grounding voltages. The treatment of the neutral point of a power system is an important factor in the determination of the voltage of the neutral point in the case of asymmetrical short-circuits. Short-circuits may cause mechanical oscillations of the generator shaft and may also cause oscillations of the active and reactive power with the consequence of possible loss of stability of the power system. The calculation of short-circuit currents and their effects is dealt with in the various parts of IEC 60909, IEC 60865 and IEC 61660 (VDE 0102 and VDE 0103). IEC 61660 applies to DC Installations. In three-phase AC systems, different types of short-circuits may occur as outlined in Figure 11.1.

In the case of single-phase faults, the currents through earth are indicated depending upon the type of neutral earthing as:

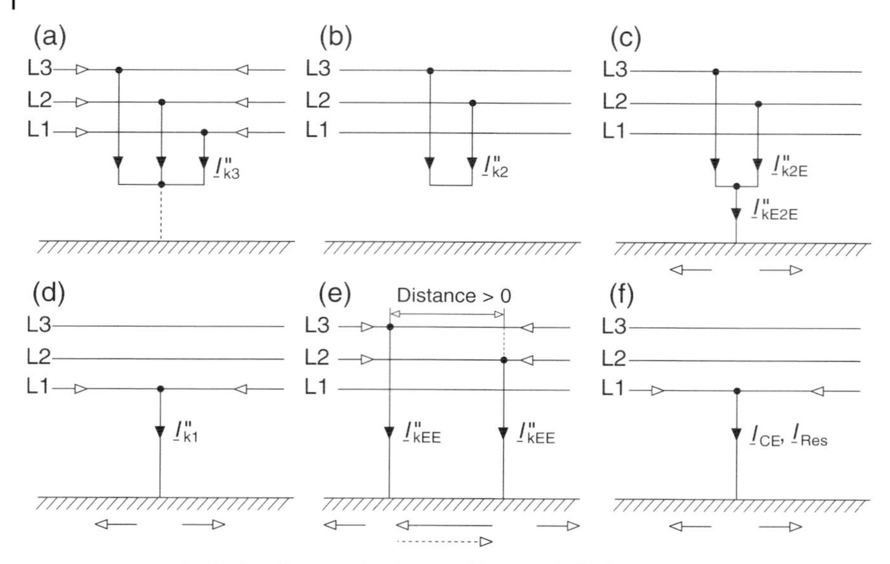

Figure 11.1 Types of short-circuits in three-phase AC systems. (a) Three-phase short-circuit; (b) two-phase (double-phase) short-circuit without earth connection; (c) two-phase short-circuit with earth connection; (d) single-phase (line-to-ground) short-circuit; (e) double-earth fault (only in systems with isolated neutral or with resonance earthing; (f) line-to-ground fault (only in systems with isolated neutral or with resonance earthing).

- Single-phase short-circuit current I''_{k1} (low-impedance or direct earthing)
- Capacitive earth-fault current I_{CE} (isolated neutral) or
- Ground fault residual current I_{Res} (resonance earthing).

Short-circuit currents can be of the near-to-generator or far-from-generator types. In near-to-generator short-circuits the AC part of the short-circuit current decays during the short-circuit duration, whereas in far-from-generator short-circuits the AC part of the short-circuit current remains constant. A short-circuit is considered as near-to-generator if the short-circuit current contribution in a three-phase short-circuit is twice the rated current of at least one generator in the system or if the contribution to the short-circuit current of all synchronous and asynchronous motors exceeds 5% of the short-circuit current without motors. The time characteristics of near-to-generator and far-from-generator short-circuit currents are outlined in Figure 11.2. The calculation of the short-circuit currents according to IEC 60909 determines significant parameters, that is, the initial short-circuit current I''_k, the peak-short-circuit current i_p and so on, and not the time curve of the current.

Two different procedures for the calculation of short-circuit currents are described in IEC 60909; the *method of the equivalent voltage source at the short-circuit location* is described in more detail in this chapter. The voltages of all system feeders, gen-

(a)

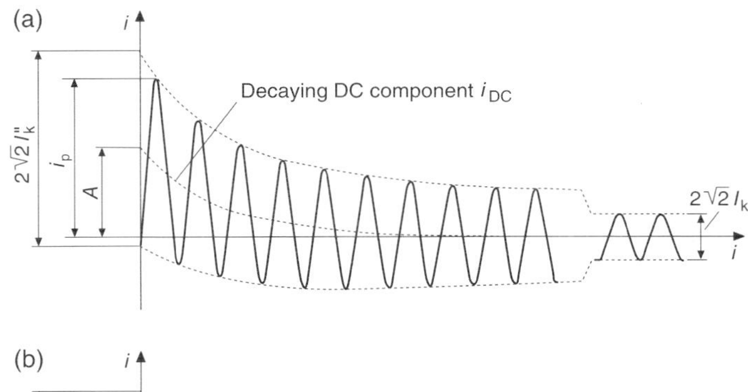

(b)

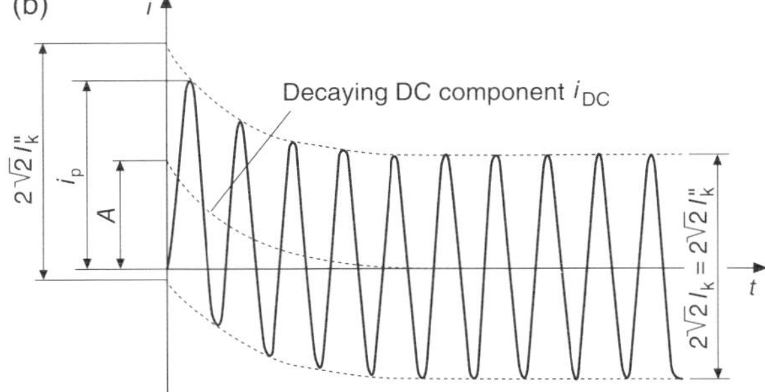

Figure 11.2 Time curve of short-circuit currents and parameters. (a) Near-to-generator short-circuit; (b) far-from-generator short-circuit. I''_k, initial (symmetrical) short-circuit current (AC); i_p, peak short-circuit current; I_k, steady short-circuit current; i_{DC}, DC component; A, initial value of the DC component i_{DC}.

erators and motors are represented by one equivalent voltage at the short-circuit location causing the same short-circuit current as the superposition of all individual short-circuit currents of the individual sources. The equivalent voltage source has the magnitude $cU_n/\sqrt{3}$, using a voltage factor c as in Table 11.1.

The calculation of maximal short-circuit currents has to take the following into account:

- The voltage factor c_{max} according to Table 11.1 has to be applied. National standards can specify other values.
- For the representation of system feeders the smallest equivalent impedance Z_{Qmin} is to be considered, so that the contribution to the short-circuit current becomes maximal.
- The contribution of motors must be accounted for (see Section 11.2.6).
- The resistances of lines are to be calculated for a temperature of 20 °C.
- The operation schedule of power stations and system feeders is to be selected in such a way as to achieve the largest contribution to the short-circuit current.
- The operating conditions of the power system are to be selected assuming the largest contribution to the short-circuit current.

Table 11.1 Voltage factor c according to IEC 60909.

Nominal system voltage (U_n)	Voltage factor c for calculation of	
	Maximal short-circuit current c_{max}	Minimal short-circuit current c_{min}
Low-voltage: 100 V up to 1000 V inclusive (see IEC 60038/05.87, Table I)		
(a) 400 V/230 V	1.00	0.95
(b) Other voltages	1.05	1.00
(c) Tolerance +6%	1.05	0.95
(d) Tolerance +10%	1.10	0.95
Medium-voltage >1 kV to 35 kV (see IEC 60038/05.87, Table III)	1.10	1.00
High-voltage >35 kV to 230 kV (see IEC 60038/05.87, Table IV)	1.10	1.00
380 kVa	1.10	1.00

a cU_n shall not exceed the highest voltage for equipment U_m according to IEC 60071.

The calculation of the minimal short-circuit current is based on the following:

- The voltage factor c_{min} according to Table 11.1 has to be applied.
- For the representation of system feeders the largest equivalent impedance Z_{Qmax} is to be considered, so that the contribution to the short-circuit current becomes minimal.
- Motors are to be neglected; exceptions will be considered in special cases.
- The resistance of lines is to be calculated taking the temperature of lines at the end of the short-circuit duration; see Table 8.8 for cables.
- The operation schedule of power stations and system feeders is to be selected in such a way as to achieve the smallest contribution to the short-circuit current.
- The operating conditions of the power system are to be selected assuming the smallest contribution to the short-circuit current.

The calculation of symmetrical short-circuit current is based on the impedances of the positive-sequence component of the equipment. In the case of asymmetrical short-circuits the impedances of the negative-sequence component and the zero-sequence component are needed additionally. Impedances of equipment are to be calculated at the voltage level of the short-circuit location taking account of the transformers' rated transformation ratios.

11.2.2
Initial Short-Circuit Current (AC)

The equations for calculation of the initial short-circuit currents of the different types of short-circuits are outlined in Table 11.2. Impedances of generators, unit transformers in power stations and high-voltage transformers in the power system are subject to correction factors according to Table 11.3 [58].

11.2.3
Peak Short-Circuit Current

The peak short-circuit current i_p is the maximal instantaneous value of the short-circuit current, to be calculated according to Equation 11.3:

$$i_P = \kappa \cdot \sqrt{2} \cdot I_k'' \tag{11.3}$$

with the peak factor κ according to Equation 11.4:

$$\kappa = 1.02 + 0.98 \cdot e^{-3R/X} \tag{11.4}$$

Table 11.2 Calculation equations for initial short-circuit current.

Type of short-circuit	Equation	Remark
Three-phase	$I_{k3}'' = \dfrac{cU_n}{\sqrt{3}\lvert\underline{Z}_1\rvert}$	
Two-phase without earth connection	$I_{k2}'' = \dfrac{cU_n}{\lvert 2\underline{Z}_1\rvert}$	
Two-phase with earth connection		
General	$I_{k2EE}'' = \left\lvert\dfrac{-\sqrt{3}cU_n\underline{Z}_2}{\underline{Z}_1\underline{Z}_2 + \underline{Z}_1\underline{Z}_0 + \underline{Z}_2\underline{Z}_0}\right\rvert$	Current through earth
	$I_{k2EL2}'' = \left\lvert\dfrac{-jcU_n(\underline{Z}_0 + \underline{a}\underline{Z}_2)}{\underline{Z}_1\underline{Z}_2 + \underline{Z}_1\underline{Z}_0 + \underline{Z}_2\underline{Z}_0}\right\rvert$	Current in phase L2
	$I_{k2EL3}'' = \left\lvert\dfrac{jcU_n(\underline{Z}_0 - \underline{a}^2\underline{Z}_2)}{\underline{Z}_1\underline{Z}_2 + \underline{Z}_1\underline{Z}_0 + \underline{Z}_2\underline{Z}_0}\right\rvert$	Current in phase L3
Far-from-generator ($Z_1 = Z_2$)	$I_{k2EE}'' = \dfrac{\sqrt{3}cU_n}{\lvert\underline{Z}_1 + 2\underline{Z}_0\rvert}$	Current through earth
	$I_{k2EL2}'' = \dfrac{cU_n\left\lvert\dfrac{\underline{Z}_0}{\underline{Z}_1} - \underline{a}\right\rvert}{\lvert\underline{Z}_1 + 2\underline{Z}_0\rvert}$	Current in phase L2
	$I_{k2EL3}'' = \dfrac{cU_n\left\lvert\dfrac{\underline{Z}_0}{\underline{Z}_1} - \underline{a}^2\right\rvert}{\lvert\underline{Z}_1 + 2\underline{Z}_0\rvert}$	Current in phase L3
Single-phase		
General	$I_{k1}'' = \dfrac{\sqrt{3}cU_n}{\lvert\underline{Z}_1 + \underline{Z}_2 + \underline{Z}_0\rvert}$	
Far-from-generator ($Z_1 = Z_2$)	$I_{k1}'' = \dfrac{\sqrt{3}cU_n}{\lvert 2\underline{Z}_1 + \underline{Z}_0\rvert}$	

Table 11.3 Impedance correction factors for short-circuit current calculation according to IEC 60909.

Equipment	Correction factor	Remark
Synchronous generator	$K_G = \frac{U_{nQ}}{U_{rG}(1+p_G)} \cdot \frac{c_{max}}{1+x_d'' \cdot \sin\varphi_{rG}}$	If the voltage is kept constant at U_{rG}, $p_G = 0$ To be applied for positive-, negative- and zero-sequence components
Generators and unit transformers with tap-changer in power stations	$K_{KWo} = \frac{U_{nQ}}{U_{rG}(1+p_G)} \cdot \frac{U_{rTLV}}{U_{rTHV}} \cdot (1 \pm p_T) \cdot \frac{c_{max}}{1+x_d'' \cdot \sin\varphi_{rG}}$	To be applied for positive-, negative- and zero-sequence components
Generators and unit transformers without tap-changer in power stations	$K_{KWs} = \frac{U_{nQ}^2}{[U_{rG}(1+p_G)]^2} \cdot \frac{U_{rTLV}^2}{U_{rTHV}^2} \cdot \frac{c_{max}}{1+\lvert x_d'' - x_T \rvert \cdot \sin\varphi_{rG}}$	If the voltage is kept constant at U_{rG}, $p_G = 0$ To be applied for positive-, negative- and zero-sequence components only in case of over-excited operation (leading power factor)
Transformers installed in the power system	$K_T = \frac{U_{nQ}}{U_{bmax}} \cdot \frac{c_{max}}{1 + x_T \frac{I_{bmaxT}}{I_{rT}} \sin\varphi_{bT}}$ Approximation $K_T = 0.95 \cdot \frac{c_{max}}{1+0.6 x_T}$	In case of lagging power factor at transformer and $U_{bLV} > 1.05\,U_n$ To be applied for positive-, negative- and zero-sequence components

Symbols: HV and LV are to be understood as the higher and the lower voltage side of the transformer.
U_{rG} = rated voltage of generator
U_{nQ} = nominal system voltage
U_{rTHV} = rated voltage of transformer at HV side
U_{rTLV} = rated voltage of transformer at LV side
U_{bmax} = maximal voltage prior to short-circuit
I_{bmaxT} = maximal operating current prior to short-circuit
I_{rT} = rated current of transformer
x_T = relative impedance voltage of transformer tap-changer in middle position, $x_T = u_{xT}$.
x_d'' = relative subtransient impedance of generator
p_T = control range of tap-changer, for example: $p_T = 15\%$
p_G = voltage control range of generator, for example: $p_T = 5\%$
φ_{bT} = Phase-angle prior to short-circuit
φ_{rG} = phase-angle between U_{rG} and I_{rG}

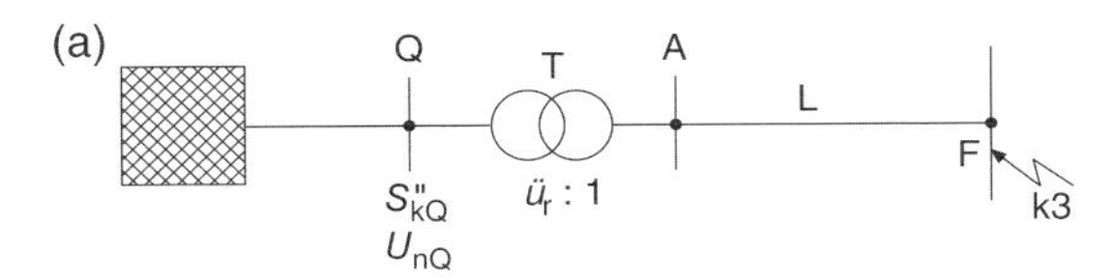

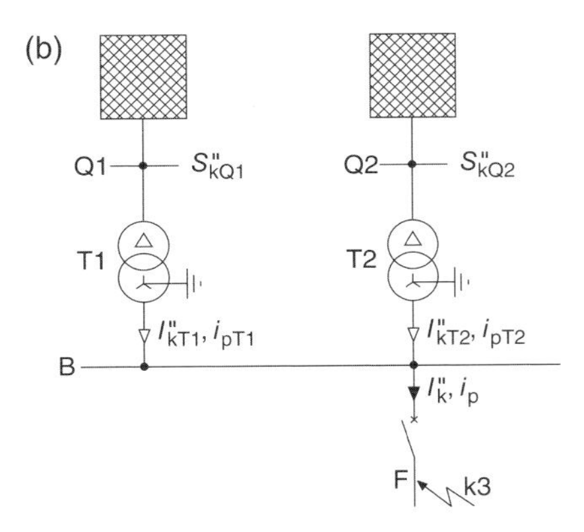

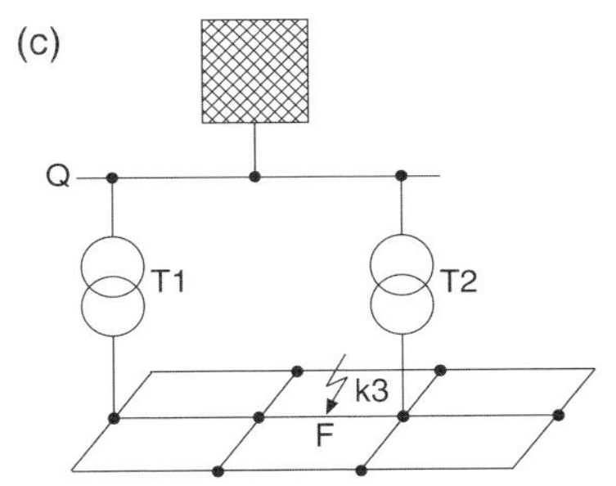

Figure 11.3 **system** diagram for calculation of peak short-circuit current. (a) Simply fed system; (b) multilateral simply fed system; (c) meshed system.

Different restrictions are to be considered depending on the kind of feed into the short-circuit. In the case of the single-fed short-circuit (Figure 11.3a), the peak-factor κ can be calculated from the short-circuit impedance. In the case of multilateral single-fed short-circuits (Figure 11.3b), the impedance ratio R/X of the individual transformers or lines (branches) is calculated. The partial peak short-circuit currents of the individual branches are also calculated, and finally the total peak short-circuit current is obtained by superposition of the partial peak short-circuit currents according to Equation 11.5:

$$i_p = \sum_i \kappa_i \cdot \sqrt{2} \cdot I''_{ki} \tag{11.5}$$

If the short-circuit occurs in a meshed power system (Figure 11.3c), the peak short-circuit current cannot be calculated by superposition, but other methods, described below, can be applied depending on the system conditions.

11.2.3.1 Uniform or Smallest *R*/*X* Ratio (Method A)

The peak-factor κ is calculated with the smallest ratio R/X of all branches (network components) of the power system. Only those branches are to be considered which contribute more than 80% of the short-circuit current and which have the same voltage level as the short-circuit location. The peak factor is to be limited to κ = 1.8 in LV systems. The results are always on the safe side if all system branches feeding the short-circuit are considered. The error can, however, amount to almost 100%. If one considers only those branches contributing 80% of the short-circuit current and if the ratios R/X of the branches are significantly different, the results cannot be considered on the safe side. The method should therefore be applied only with relatively uniform R/X ratio and if $R/X < 0.3$.

11.2.3.2 *R*/*X* Ratio of the Short-Circuit Impedance at the Short-Circuit Location (Method B)

The peak-factor κ is calculated from the ratio R/X of the short-circuit impedance at the short-circuit location and multiplied by a safety factor of 1.15. In LV systems the factor $1.15 \times \kappa$ is limited to the value 1.8, in HV systems to 2.0. The error of the results can be positive or negative. If the application of the method is restricted to those cases where $0.005 \leq R/X \leq 1.0$, the error is limited to +10% to −5%. The safety factor 1.15 is not to be applied for $R/X < 0.3$.

11.2.3.3 Equivalent Frequency f_c (Method C)

With this procedure the peak-factor κ is calculated from the R/X ratio at an equivalent frequency according to Equation 11.6. The equivalent frequency in 50 Hz systems is $f_c = 20\,\text{Hz}$ and in system with nominal frequency 60 Hz the equivalent frequency is $f_c = 24\,\text{Hz}$.

$$\frac{R}{X} = \frac{R_c}{X_c} \cdot \frac{f_c}{f} \tag{11.6}$$

R_c and X_c are the equivalent resistance and reactance at the short-circuit location at equivalent frequency. The peak-factor is limited to $\kappa = 1.8$. With the method of equivalent frequency, the error is less than ±5% if the R/X ratio is in the range $0.005 \leq R/X \leq 5.0$. The method represents a good compromise between security (result not on the unsafe side) and economy (error as small as possible).

11.2.4 Symmetrical Short-Circuit Breaking Current

The calculation of the symmetrical short-circuit breaking current I_b, often called the breaking current or the partial breaking current of a generator, is carried out using Equation 11.7.

$$I = \mu \cdot I_K'' \tag{11.7}$$

The factor μ depends on the minimum time delay of the breaker (including time delay of the protection relays) and can be calculated according to Equations 11.8. The equations are valid for HV synchronous generators with rotating exciters or static inverters. Further restrictions are described in IEC 60909. If parameters are unknown, the factor μ is set equal to 1.0.

$$\mu = 0.84 + 0.26 \cdot e^{-0.26 I''_{kG}/I_{rG}} \quad \text{with} \quad t_{min} = 0.02\,s \tag{11.8a}$$

$$\mu = 0.71 + 0.51 \cdot e^{-0.30 I''_{kG}/I_{rG}} \quad \text{with} \quad t_{min} = 0.05\,s \tag{11.8b}$$

$$\mu = 0.62 + 0.72 \cdot e^{-0.32 I''_{kG}/I_{rG}} \quad \text{with} \quad t_{min} = 0.1\,s \tag{11.8c}$$

$$\mu = 0.56 + 0.94 \cdot e^{-0.38 I''_{kG}/I_{rG}} \quad \text{with} \quad t_{min} > 0.25\,s \tag{11.8d}$$

For far-from-generator short-circuits, the breaking current I_b is equal to the initial short-circuit current I''_k for all types of short-circuits.

11.2.5 Steady-State Short-Circuit Current

In case of near-to-generator short-circuits, the steady-state short-circuit current I_k depends on (among other things) the saturation of the generator and the operating status of the power system (tripping of circuits due to protection during the time elapsed of the short-circuit) and can only be calculated with some inaccuracy. The method proposed in IEC 60909 therefore indicates upper and lower limit values for those cases where the short-circuit is fed by one synchronous generator only. Maximal excitation of the generator leads to the maximal steady-state short-circuit current I_{kmax} according to Equation 11.9a. For the calculation of the minimal steady-state short-circuit current, a constant no-load excitation of the generator is assumed. The minimal steady-state short-circuit current I_{kmin} is calculated according to Equation 11.9b.

$$I_{kmax} = \lambda_{max} \cdot I_{rG} \tag{11.9a}$$

$$I_{k\,min} = \lambda_{min} \cdot I_{rG} \tag{11.9b}$$

The factors λ_{max} and λ_{min} for turbo and salient-pole type generators are to be taken from Figures 11.4 and 11.5 with x_{dsat} the reciprocal of the short-circuit ratio. The remarks to be found in IEC 60909 should to be noted.

For far-from-generator short-circuits, the steady-state short-circuit current I_k is equal to the initial short-circuit current I''_k for all types of short-circuits.

11.2.6 Influence of Synchronous and Asynchronous Motors

Synchronous motors are treated for the calculation of short-circuit current as synchronous generators. Asynchronous (induction) generators are treated like asynchronous (induction) motors. Asynchronous motors are considered only in the following cases:

- The sum of the rated currents of the asynchronous motors is more than 1% of the initial short-circuit current without motors.

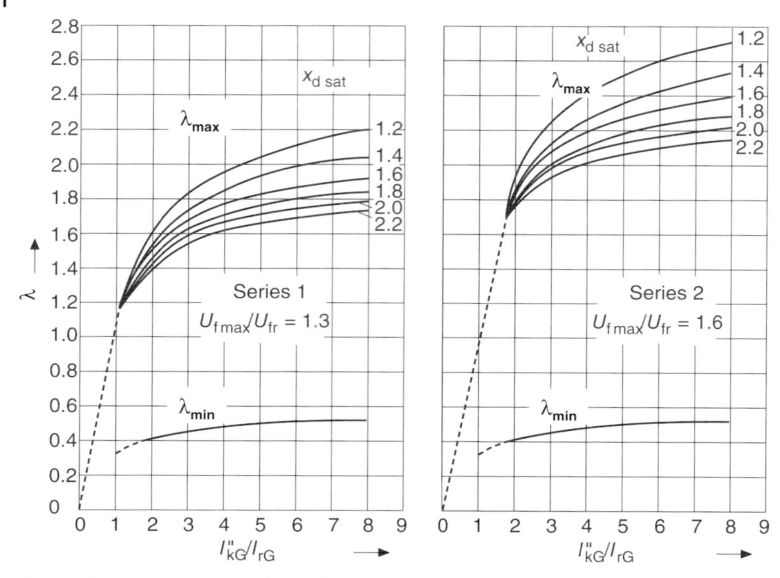

Figure 11.4 Factors λ_{min} and λ_{max} for turbo generators.

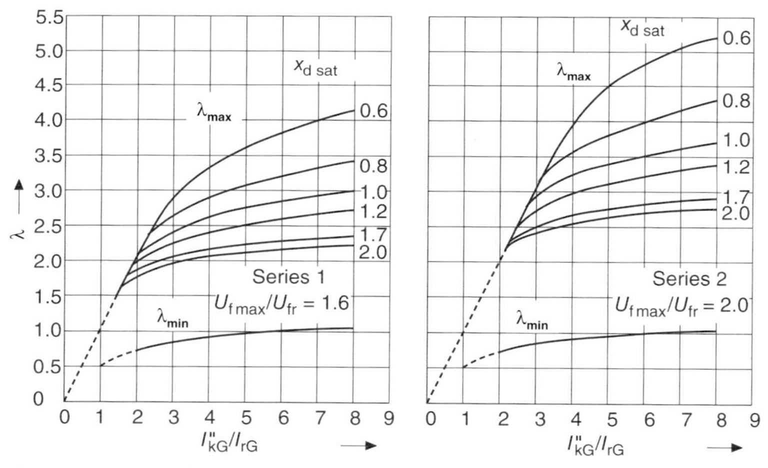

Figure 11.5 Factors λ_{min} and λ_{max} for salient pole generators.

- The contribution of the asynchronous motors to the initial short-circuit current without motors is more than 5%.
- Equation 11.10 is satisfied when the asynchronous motors are fed through two-winding transformers.

$$\frac{\sum P_{rM}}{\sum S_{rT}} > \frac{0.8}{\left| \frac{100c\sum S_{rT}}{\sqrt{3}\cdot U_{nQ}\cdot I_k''} - 0.3 \right|} \tag{1.10}$$

where S_{rt} is the rated apparent power of the transformer, U_{nQ} is the nominal voltage, P_{rM} is the rated active power of the motor and I''_{k} the initial short-circuit current without motors.

Asynchronous motors contribute to the initial short-circuit current, to the peak short-circuit current, to the breaking current and in the case of asymmetrical short-circuits to the steady-state short-circuit current as well. Asynchronous motors in auxiliary supply systems of power stations and in industry (e.g. steel processing, chemical industry, pumping plants) are considered in any case. Motors which cannot be operated in parallel due to interlocking are to be neglected for the calculation. Motors in public LV supply systems are also neglected. Static inverter-fed drives are considered for the calculation of the initial short-circuit current and the peak short-circuit current only in case of three-phase short-circuits and only in those cases where there is the possibility of an inverter operation.

The equations for calculation of the different parameters of the short-circuit current are outlined in Table 11.4. The factor q for the calculation depends on the minimum time delay t_{min} of the breaker (including time delay of the protection) and can be calculated according to Equations 11.11.

$$q = 1.03 + 0.12\ln(m) \qquad t_{min} = 0.02\,\mathrm{s} \tag{11.11a}$$

$$q = 0.79 + 0.12\ln(m) \qquad t_{min} = 0.05\,\mathrm{s} \tag{11.11b}$$

$$q = 0.57 + 0.12\ln(m) \qquad t_{min} = 0.10\,\mathrm{s} \tag{11.11c}$$

$$q = 0.26 + 0.10\ln(m) \qquad t_{min} \geq 0.25\,\mathrm{s} \tag{11.11d}$$

The factor m is the rated active power of the motor per pair of poles in units of MW. No values greater than 1.0 are to be used for the factor q.

11.3 Short-Circuit Withstand Capability

The calculation of the thermal and electromagnetic effects of AC short-circuit currents is defined in IEC 60865-1 (VDE 0103). Short-circuit currents cause heating of conductors by ohmic losses (thermal effects). Conductors carrying currents, in this case short-circuit currents, cause forces against other conductors (electromagnetic effects). The basics of calculation of the thermal effects of short-circuit current are outlined in the individual chapters of this book, dealing with transformers, cables and overhead lines.

The calculation of the electromagnetic effects of conductors carrying currents requires a more detailed analysis based on detailed knowledge of the constructional details and the design of the equipment and installations. The exact determination of the electromagnetic effects therefore cannot be a task of power system planning but is closely related to the design and construction of equipment and installations.

Table 11.4 Calculation of short-circuit currents of asynchronous motors.

Parameter	Type of short-circuit		
	Three-phase	**Two-phase**	**Single-phase**
Initial short-circuit current	$I''_{k3M} = \frac{cU_n}{\sqrt{3}Z_M}$	$I''_{k2M} = \frac{\sqrt{3}}{2} I''_{k3M}$	$I''_{k1M} = \frac{c\sqrt{3}U_n}{Z_{1M} + Z_{2M} + Z_{0M}}$ in systems with low-impedance earthing only
	$i_{p3M} = \kappa_M \sqrt{2} I''_{k3M}$	$i_{p2M} = \frac{\sqrt{3}}{2} i_{p3M}$	$i_{p1M} = \kappa_M \sqrt{2} I''_{k1M}$
	MV motors: $\kappa_M = 1.65$ ($R_M/X_M = 0.15$) active power per pair of poles <1 MW $\kappa_M = 1.75$ ($R_M/X_M = 0.10$) active power per pair of poles ≥1 MW LV motors including connection cables: $\kappa_M = 1.30$ ($R_M/X_M = 0.42$)		
Symmetrical short-circuit breaking current	$I_{b3M} = \mu q I''_{k3M}$	$I_{b2M} \approx \frac{\sqrt{3}}{2} I''_{k3M}$	$I_{b1M} \approx I''_{k1M}$
	μ according to Equation 11.8 q according to Equation 11.11		
Steady-state short-circuit current	$I_{k2M} = 0$	$I_{k2M} \approx \frac{\sqrt{3}}{2} I''_{k3M}$	$I_{k1M} \approx I''_{k1M}$

Three-phase short-circuits and two-phase short-circuits without earth connection generally cause maximal short-circuit currents and therefore maximal electromagnetic forces on other conductors. As the short-circuit currents are AC currents, the resulting forces also vary with time. The parallel arrangement of conductors at distance a is taken as an example. In case of a two-phase short-circuit, the force on the two conductors (here conductors S and T in flat formation) with length l is calculated using Equation 11.12. The variable amplitude of the force always acts in the same direction.

$$F_{k2} = \frac{\mu_0}{2\pi} \cdot \frac{l}{a} i_{L2} i_{L3} \tag{11.12}$$

For a three-phase short-circuit and assuming cables in flat formation, the middle conductor is stressed most by the electromagnetic forces. The magnetic fields caused by the currents in the outer conductors (here conductors L1 and L3) are subtracted. The force on the middle conductor is calculated according to Equation 11.13 and acts in different directions.

$$F_{k3} = \frac{\mu_0}{2\pi} \cdot \frac{l}{a} i_{L2}(i_{L1} - i_{L3}) \tag{11.13}$$

The maximal electromagnetic forces occur shortly after the short-circuit initiation due to the decreasing DC component. The peak short-circuit current is the maximal instantaneous value of the current, resulting in the maximal electromagnetic force according to Equations 11.4 and 11.5.

$$F_{k2\max} = \frac{\mu_0}{2\pi} \cdot \frac{l}{a} i_{p2} \tag{11.14}$$

$$F_{k3\max} = \frac{\mu_0}{2\pi} \cdot \frac{l}{a} \frac{\sqrt{3}}{2} i_{p3} \tag{11.15}$$

The equations given above are valid only if the distance a between the conductors is small compared with the conductor length l but large compared with the conductor radius r ($l/a > 10$; $a/r > 10$). It can be assumed that these requirements are fulfilled in HV equipment and installations. Detailed information and examples of the calculation of effects of short-circuit currents can be found in reference [59].

11.4 Limitation of Short-Circuit Currents

11.4.1 General

The expansion of electrical power systems by addition of new power stations and new lines (overhead transmission lines and cable circuits) results in an increase of short-circuit currents due to an increase in the number of sources feeding the short-circuit and to a reduction of system impedance. Existing installations have to be improved and equipment has to be replaced or reinforced to prevent the permissible short-circuit current being exceeded. Measures to limit the short-circuit currents can also be implemented which might be more economic than the replacement of equipment and installations. Various measures have to be taken into account, such as those affecting the whole system (e.g. higher voltage level), those concerning installations and substations (e.g. separate operation of busbars) and those related to equipment (e.g. I_p limiter).

All measures have an influence on the system reliability as well, which must be guaranteed under outage conditions of the equipment when the measures for limitation of short-circuit currents have been put in place. Measures for short-circuit current limitation decrease the voltage stability, increase the reactive power requirement, reduce the dynamic stability and increase the complexity of operation. Furthermore, some measures to limit short-circuit currents will contradict requirements for a high short-circuit level; for example, in the case of connection of an arc-furnace the short-circuit level has to be kept high to reduce voltage flicker.

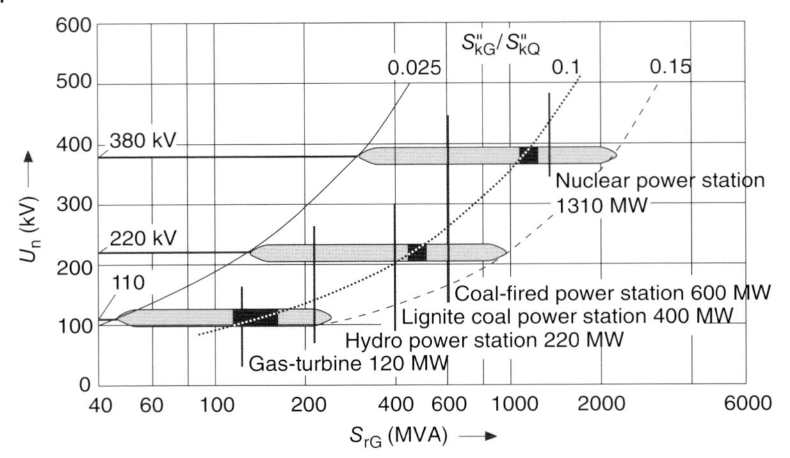

Figure 11.6 Selection of suitable voltage levels for the connection of power stations.

Decisions on the location of power stations are determined by, in addition to other criteria, the availability of primary energy (lignite-coal fired power stations are built near the coal mine), requirements for cooling water (thermal power stations are placed near the sea or by large rivers), geological conditions (hydro power stations can only be built if water reservoirs are available), the requirements of the power system (each power station requires a system connection at suitable voltage level) and the closeness to consumers (combined heat and energy stations need heat consumers nearby).

The connection of large power stations is determined by the contribution to the short-circuit current. Figure 11.6 outlines considerations for selecting the suitable voltage level for the connection of power stations to the power system. It is assumed that more than one power station is connected to the system.

As generation of electrical energy for which there are no consumers makes no sense, a suitable power system has to be planned and constructed accordingly. Because the number of combined heat and power stations (distributed generation) with connection to medium-voltage and even to low-voltage system is increasing, additional considerations of protection, operation and short-circuit level in the different voltage levels [60] are necessary; see also Chapter 12.

11.4.2 Measures in Power Systems

11.4.2.1 Selection of Nominal System Voltage

A higher nominal system voltage at constant rated power of feeding transformers will reduce the short-circuit level proportionally. The selection of nominal system voltage must take into account the recommended voltages according to IEC 60038 and also the common practice of the utility itself and perhaps of the

Table 11.5 Selection of recommended voltage according to IEC 60038.

	Nominal voltage	Application	Remarks
Low voltage	400 V/230 V	Private consumers Small industrial consumers	According to IEC Table I
	500 V	Motor connection in industry	Not listed in IEC
Medium voltage	6 kV	HV motors in industry, auxiliary supply in power stations	According to IEC Table III
	10 kV	Urban distribution systems, industrial systems	According to IEC Table III
	20 kV	Industrial systems, rural distribution systems	According to IEC Table III
	30 kV	Electrolysis, arc furnace, rectifiers	Not listed in IEC
High voltage	110 kV	Urban transport systems	According to IEC Table IV
	220 kV	Transport system with regional task	According to IEC Table IV
	380 kV	Transmission system country-wide	According to IEC Table V The highest voltage of equipment $U_{bmax} = 420\,kV$ is specified

whole country. Table 11.5 lists a selection of recommended voltages. The table also includes information on typical applications in Europe.

The short-circuit current is directly proportional to the voltage level or to the voltage ratio of feeding transformers if all other parameters are constant. The selection of a new nominal system voltage normally is only possible when new electrification projects are considered. As the impedance voltage of transformers increases with increasing voltage, an additional positive effect on the reduction of short-circuit currents is seen. It should be noted that as a by-effect the transmittable power of overhead lines and cables increases with increasing voltage without increasing in the cross-section of the conductor. On the other hand, the voltage drop of the transformer increases with increase of impedance voltage.

11.4.2.2 Operation as Separate Subsystems

The power system is operated as several subsystems, which are connected at a higher voltage level. Figure 11.7 outlines the general structure of a 132 kV cable system (total load approximately 1500 MW). The system is supplied from the 400 kV system and by power stations connected to the 132 kV level. Assuming a meshed system operation, that is, the 132 kV system is operated as one system

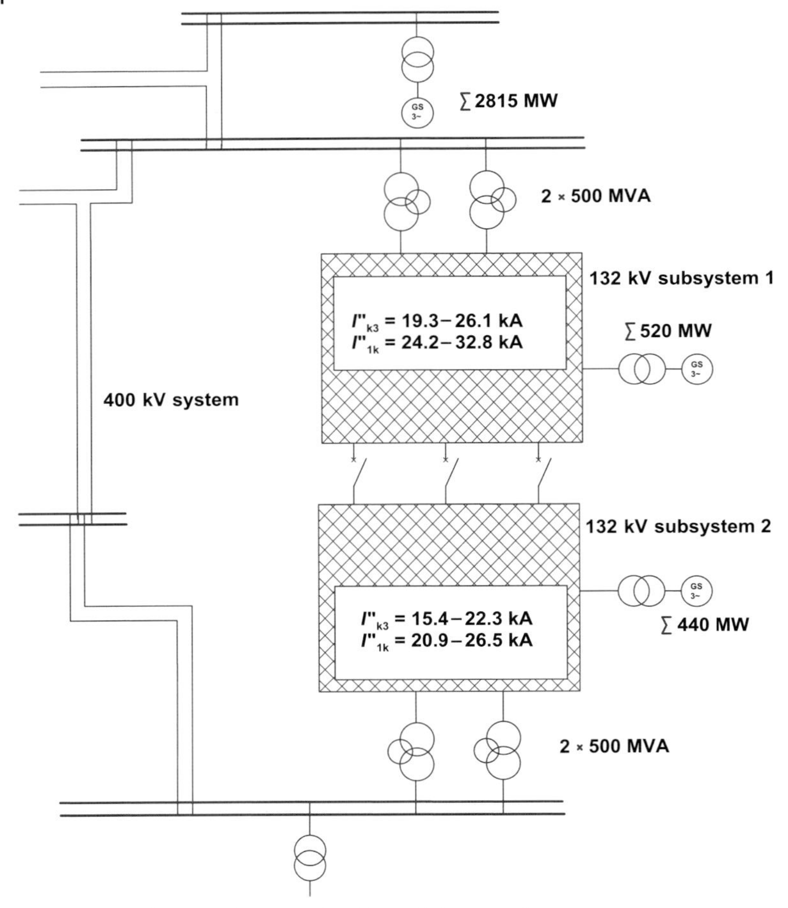

Figure 11.7 Schematic diagram of a 400/132 kV system for urban load; values of short-circuit currents.

with all breakers closed, the short-circuit currents in case of three-phase and single-phase short-circuits are $I''_{k3} = 26.0 - 37.4\,\text{kA}$ and $I''_{k1} = 37.3 - 45.7\,\text{kA}$. Operating the 132 kV system as two separate subsystems coupled only on the 400 kV level, the short-circuit currents will be reduced to the values as indicated in Figure 11.7.

Operating the 132 kV system as two separate subsystems will require additional cable circuits and an extension of the switchgear to fulfill the $(n - 1)$-criteria for a reliable power supply, see Chapter 3.

11.4.2.3 Distribution of Feeding Locations

Power stations and system feeders from higher voltage levels are to be connected to several busbars in the system. This measure was realized in the power system of Figure 11.7 as a by-effect of the system separation. A further example is outlined

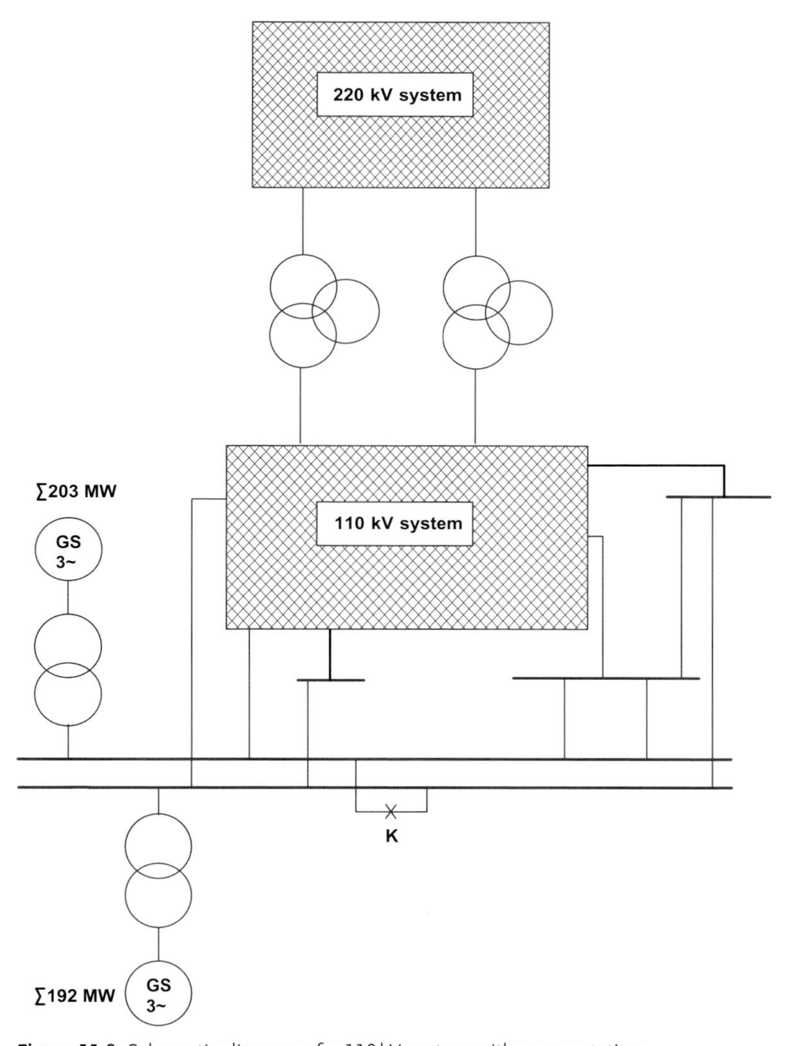

Figure 11.8 Schematic diagram of a 110 kV system with power station.

in Figure 11.8. A power station of 395 MW is connected to the 110 kV system, which has a second supply from the 220 kV system. The 110 kV system is a pure cable network and the shortest cable length between any two substations is 11.2 km. If the busbar-coupler K in the power station is closed, the three-phase short-circuit current at the busbar is $I''_{k3} = 37.6\,\text{kA}$; the short-circuit currents at the busbars in the 110 kV system remain below $I''_{k3} = 33.5\,\text{kA}$. If the busbar-coupler K is operated open, the short-circuit currents at the busbars in the power station are $I''_{k3} = 28.0\,\text{kA}$ and $I''_{k3} = 29.3\,\text{kA}$. For short-circuits at the busbars in the system itself, the short-circuit currents are reduced significantly.

The generators and the 110 kV cables in the power stations need to be switched on to the busbars in such a way that the generated power can be transferred to the power system without overloading any of the cable even under outage conditions.

11.4.2.4 Coupling of Power System at Busbars with Low Short-Circuit Level

Different parts of the power system should be connected only at busbars with low short-circuit level. Figure 11.9 outlines a 30 kV system with overhead lines, which

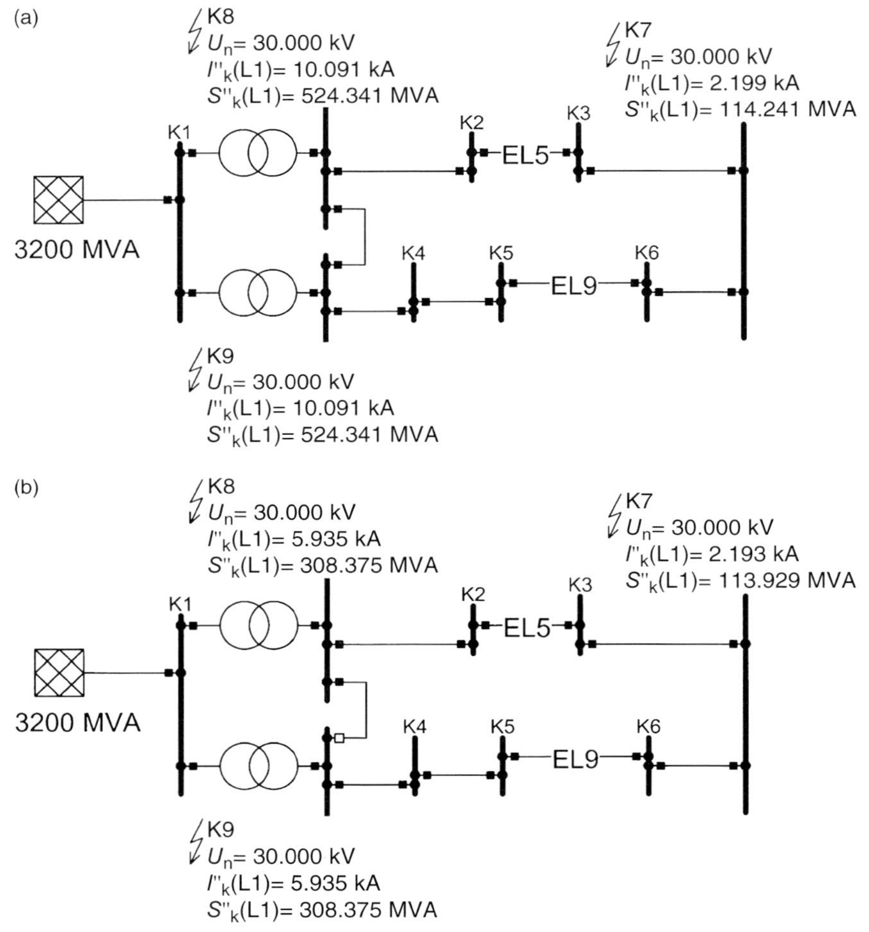

Figure 11.9 Equivalent circuit diagram of a 30 kV system with feeding 110 kV system. Result of three-phase short-circuit current. $S''_{kQ} = 3.2\,GVA$; $S_{rT} = 40\,MVA$; $u_{krT} = 12\%$; $t_{rT} = 110/32$; OHTL 95Al; $l_{tot} = 56\,km$. (a) Operation with transformers in parallel; (b) limitation of short-circuit current, transformers are not operated in parallel.

is fed from the 110 kV system by two transformers operated in parallel. The three-phase short-circuit current is $I''_{k3} = 10.09\,\text{kA}$. If the transformers are not operated in parallel and the system is coupled at busbar K7 (right-hand side of Figure 11.9) the short-circuit current at the feeding busbar is $I''_{k3} = 5.94\,\text{kA}$.

It should be noted that the short-circuit level at busbar K7 is affected only to a minor extent. If the transformers are loaded only up to 50% of their rated power and if the lines have sufficient thermal rating, both system configurations have the same supply reliability.

11.4.2.5 Restructuring of the Power System

Restructuring of power systems is comparatively costly and complicated. In medium-voltage systems, restructuring is in most cases only possible together with the commissioning of new primaries, loop-in and loop-out of cable (overhead line) circuits and the operation of the system as a radial system. In high-voltage systems, restructuring requires a totally different system topology. Figure 11.10 compares two system topologies – a meshed system and a radial system.

As can be seen from Figure 11.10 the short-circuit currents are reduced from $I''_{k3} = 23.5\,\text{kA}$ to $I''_{k3} = 22.7\,\text{kA}$ (3.8%) with the new topology. The reduction of the short-circuit currents is comparatively small, but will be more significant if an increased number of feeders (or generators) are to be connected [61].

11.4.3 Measures in Installations and Switchgear Arrangement

11.4.3.1 Multiple Busbar Operation

The connection of lines and feeders to more than one busbar per substation is advantageous compared with the operation of the substation with single busbar or with bus-coupler closed. Figure 11.11 gives a schematic diagram of a 110 kV system. The 110 kV substation is equipped with a double busbar and one additional spare busbar (transfer busbar). The substation is fed from the 220 kV system; outgoing 110 kV cables are connected to each of the two busbars in operation.

Operation with two separate busbars reduces the three-phase short-circuit current from $I''_{k3} = 16.3\,\text{kA}$ to $I''_{k3} = 14.9\,\text{kA}$ (8.6%) at SS1 and $I''_{k3} = 15.3\,\text{kA}$ (6.1%) at SS2. Each of the two busbars SS1 and SS2 can be switched-on to the transfer busbar without coupling the busbars SS1 and SS2.

11.4.3.2 Busbar Sectionalizer in Single-Busbar Switchgear

Single busbars can be equipped with a busbar sectionalizer, so that an operation mode similar to double-busbar operation is possible. The outgoing feeders and the feeding transformers need to be connected to the busbar section in such a way that the loading of feeders is approximately equal. Figure 11.12 illustrates an industrial system with nominal voltage 6 kV which is fed from the 30 kV system.

The short-circuit current at the feeding busbar is reduced by 16.8% from $I''_{k3} = 11.4\,\text{kA}$ to $I''_{k3} = 9.48\,\text{kA}$ if the busbar sectionalizer is kept open. The outgoing

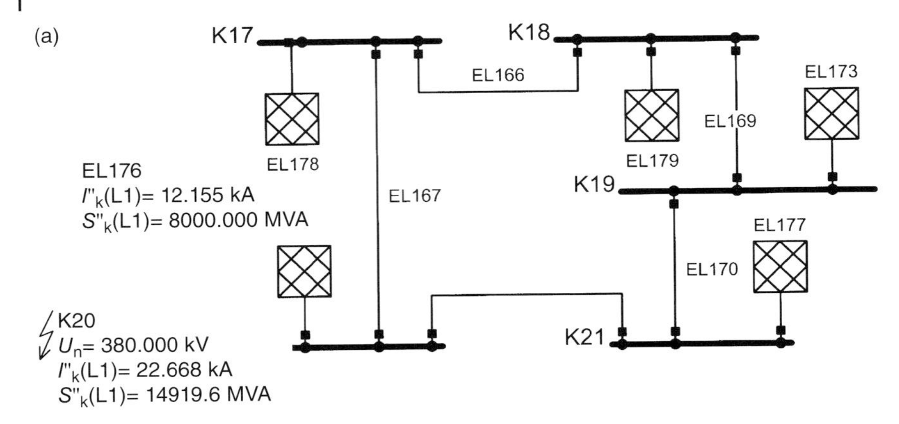

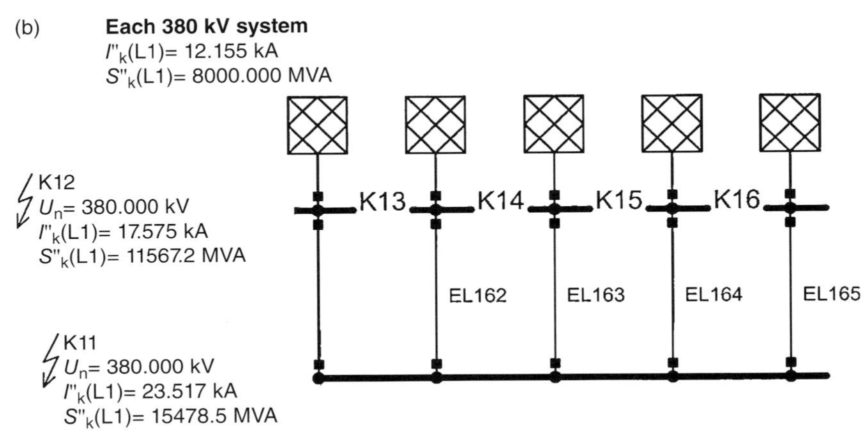

Figure 11.10 Equivalent circuit diagram of a 380 kV system and results of three-phase short-circuit current calculation. $S''_{kQ} = 8\,\text{GVA}$; OHTL ACSR/AW 4 × 282/46; $l_i = 120\,\text{km}$. (a) Ring fed system; (b) radial fed system.

feeders have to be arranged in so that the loading will be approximately equal for both busbar sections K3 and K4.

11.4.3.3 Short-Circuit Current Limiting Equipment

Short-circuit current limiting equipment and fuses (medium-voltage and low-voltage systems) can be installed to reduce the short-circuit level in parts of the installations. In medium-voltage installations, an I_p limiter can be installed. Figure 11.13 is a schematic diagram of an industrial system. The existing switchgear A with low short-circuit rating is extended with the busbar section B, which is fed by an additional system feeder Q2. The maximal permissible short-circuit current $I''_{kA\,max}$ of busbar section A is exceeded by this extension.

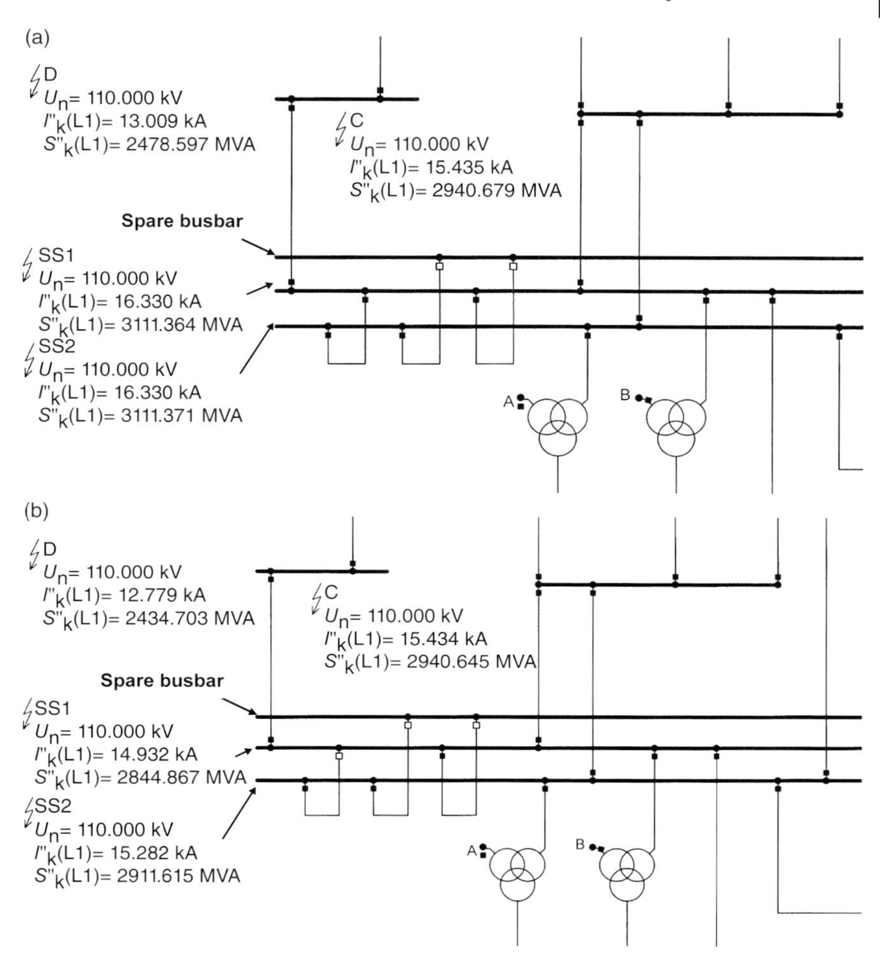

Figure 11.11 Schematic diagram of a 110 kV substation fed from the 220 kV system. Result of three-phase short-circuit current calculation. (a) Operation with two bus-couplers closed, transfer busbar not in operation; (b) operation with all bus-couplers open.

The total short-circuit current from both system feeders should be limited to the permissible short-circuit current I''_{kAmax} of busbar section A in case of a short-circuit at busbar A. If the relation $I''_{\text{kQ1}}/I''_{\text{kQ2}}$ depends on the ratio $Z_{\text{Q1}}/Z_{\text{Q2}}$ of the feeders Q1 and Q2, it is sufficient to measure the partial short-circuit current through the I_{p} limiter. The current ratio is defined according to Equation 11.16,

$$\frac{I_1}{I_2} = \frac{I''_{\text{kQ1}}}{I''_{\text{KQ2}}} \tag{11.16}$$

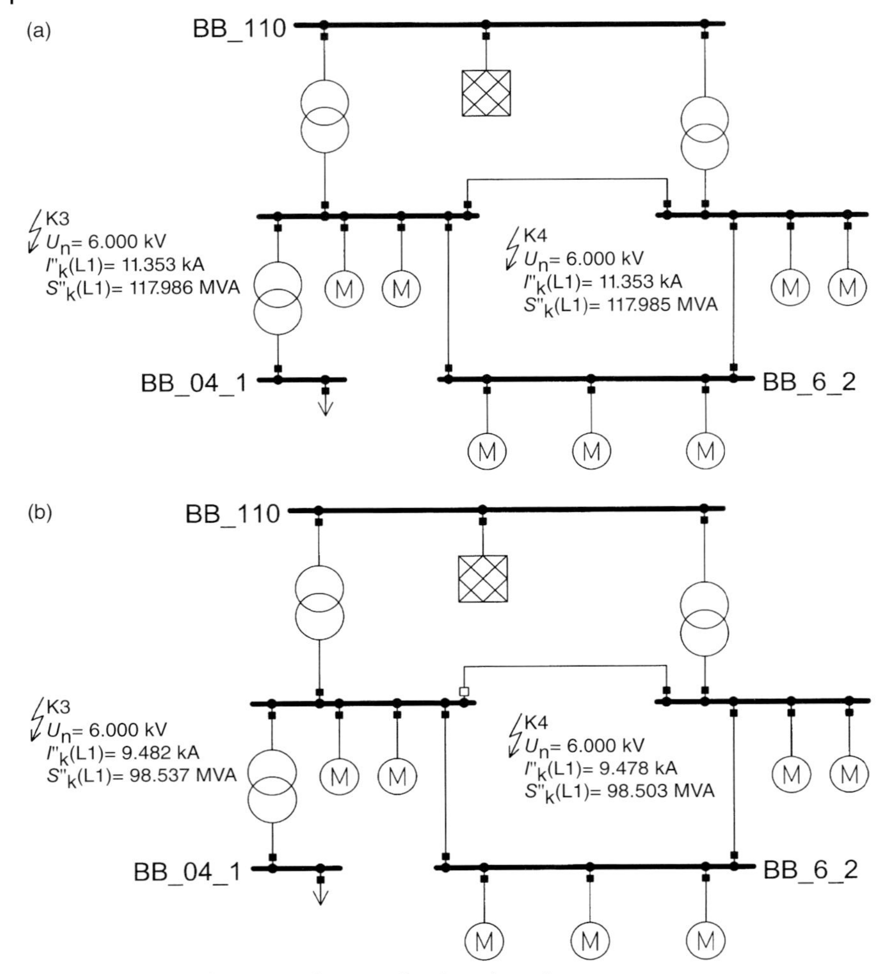

Figure 11.12 Equivalent circuit diagram of a 6 kV industrial system. Results of three-phase short-circuit current calculation. (a) Busbar sectionalizer closed; (b) busbar sectionalizer open.

and the total short-circuit current according to Equation 11.17,

$$I_3 = I_2 \cdot \left(1 + \frac{I''_{kQ1}}{I_{kQ2}}\right) \leq I''_{kA\,max} \tag{11.17}$$

The threshold value I_{2an} of the I_p limiter is given by Equation 11.18a.

$$I_{2an} = I''_{kA\,max} \cdot \frac{I''_{kQ2}}{I''_{kQ1} + I''_{kQ2}} \tag{11.18a}$$

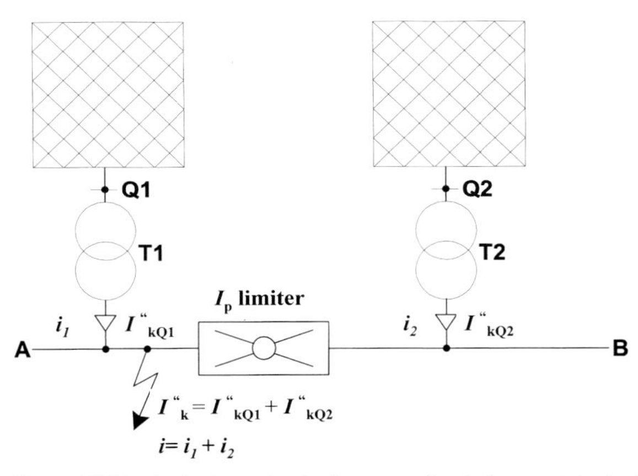

Figure 11.13 Equivalent circuit diagram of switchgear with single busbar.

In case the permissible short-circuit currents I''_{kAmax} and I''_{kBmax} of both busbar sections A and B are exceeded, the threshold value $I_{1\text{an}}$ of the I_{p} limiter for short-circuits at busbar section B is also needed according to Equation 11.18b.

$$I_{1\text{an}} = I''_{\text{kBmax}} \cdot \frac{I''_{\text{kQ1}}}{I''_{\text{kQ2}} + I''_{\text{kQ1}}} \tag{11.18b}$$

The threshold value I_{an} of the I_{p} limiter is set to the minimum of both values according to Equation 11.19.

$$I_{\text{an}} = \text{MIN}\{\text{I}_{1\text{an}}; I_{2\text{an}}\} \tag{11.19}$$

The detailed design and determination of the settings are determined by different topologies of the power system, different phase-angles of the branch short-circuit currents and different rating of the switchgear in the system, as well as other factors.

Figure 11.14 represents the time progression of short-circuit currents at section A in Figure 11.13. The branch short-circuit current i_2 from system feeder Q2 is switched off by the I_{p} limiter within approximately 7 ms, thus reducing the peak short-circuit current significantly.

The technical layout of one phase of an I_{p} limiter is shown in Figure 11.15. Inside an insulating tube (1) the main current conductor (3) with a breaking element actuated by a triggerable explosive loading (2) is located. If the threshold value is exceeded, the tripping circuit triggers the explosive loading; the arc inside the insulating tube cannot be quenched and is commutated to the fuse element (4), which will blow according to the fuse-characteristic. The main element, that is, the insulating tube with main conductor and fuse element, needs to be replaced after operation of the I_{p} limiter. A measuring unit (5) with

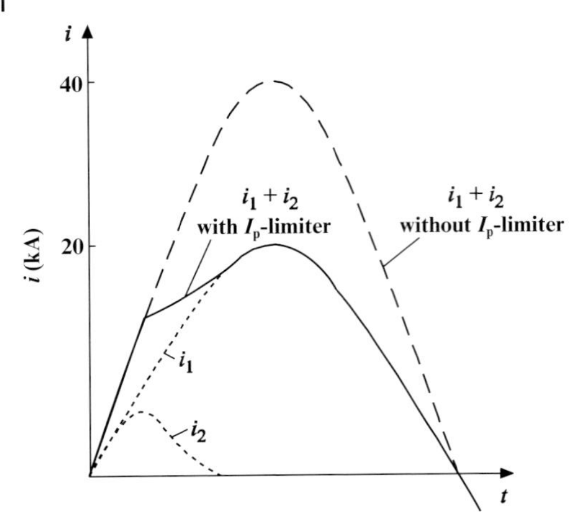

Figure 11.14 Time progression of short-circuit current in installations with and without I_p limiter.

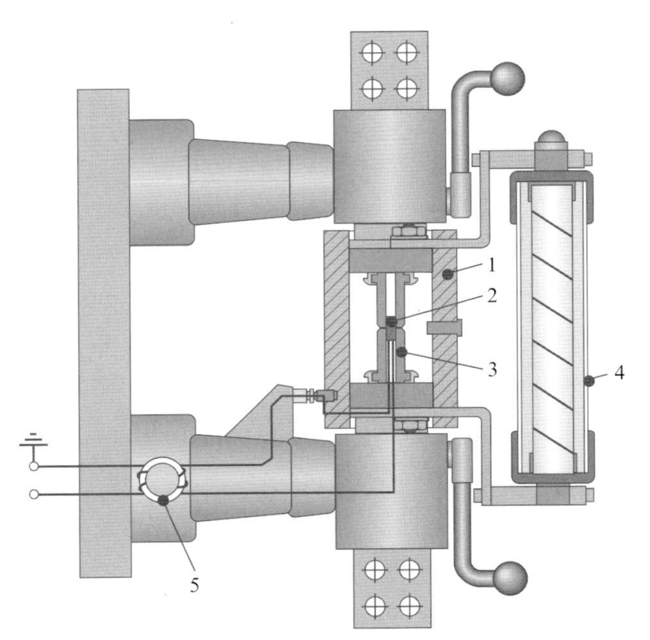

Figure 11.15 Cutaway view of an I_p limiter support. **1**, insulating tube; **2**, explosive loading; **3**, main conductor; **4**, fuse element; **5**, transducer. Source: ABB Calor Emag Schaltanlagen AG.

tripping circuit is needed to compare the actual current value with the threshold value.

I_p limiters are nowadays available with thyristor technology. The short-circuit current can be limited within 1 ms to 2 ms after initiation of the fault. The I_p limiter is back in operation after fault clearing and replacement of main conductor and fuse is not necessary. Additional operational functions, such as limitation of start-up current of large motors, can also be realized. Superconducting I_p limiters are undergoing laboratory tests [62]; first applications were tested in MV-systems in the course or the European Ecoflow-Project.

11.4.4
Measures Concerning Equipment

11.4.4.1 Impedance Voltage of Transformers

Transformers with high impedance voltage reduce the short-circuit level, but the reactive power losses are increased and the tap-changer need to be designed for higher voltage drops. Figure 11.16 shows the equivalent circuit diagram of a 10 kV system fed from a 110 kV system by three transformers, $S_{rT} = 40\,MVA$. The system load is $S_L = 72\,MVA$, cos $\varphi = 0.8$. The short-circuit power of the 110 kV system is $S''_{kQ} = 2.2\,GVA$; the voltage at the 10 kV busbar is to be controlled within a bandwidth of ±0.125 kV around $U = 10.6\,kV$.

The relevant results of load-flow and short-circuit analysis are outlined in Table 11.6. As can be seen, the increase of the impedance voltage from 13% to 17.5% reduces the short-circuit current but increases the reactive power losses and increases the number of steps at the tap-changer to control the voltage.

Table 11.6 Result of load-flow and short-circuit analysis according to Figure 11.16.

u_{krT} (%)	I''_{k3} (kA)	I''_{k1} (kA)	Tap-changer position $U \approx 10.6\,kV$	Reactive power losses of one transformer (Mvar)
13	35.2	22.5	+6	2.61
17.5	28.9	20.7	+8	3.58

11.4.4.2 Short-Circuit Limiting Reactor

The application of short-circuit limiting reactors can be considered a measure related to switchyards or one related to equipment. Figure 11.17 outlines the equivalent circuit diagram of a 10 kV system in the paper-processing industry with direct connection to an urban 10 kV system. Two reactors are installed to limit the short-circuit currents. The three-phase short-circuit current without local generation in the industrial system at the coupling busbar between industry and utility is $I''_{k3} = 20.43\,kA$.

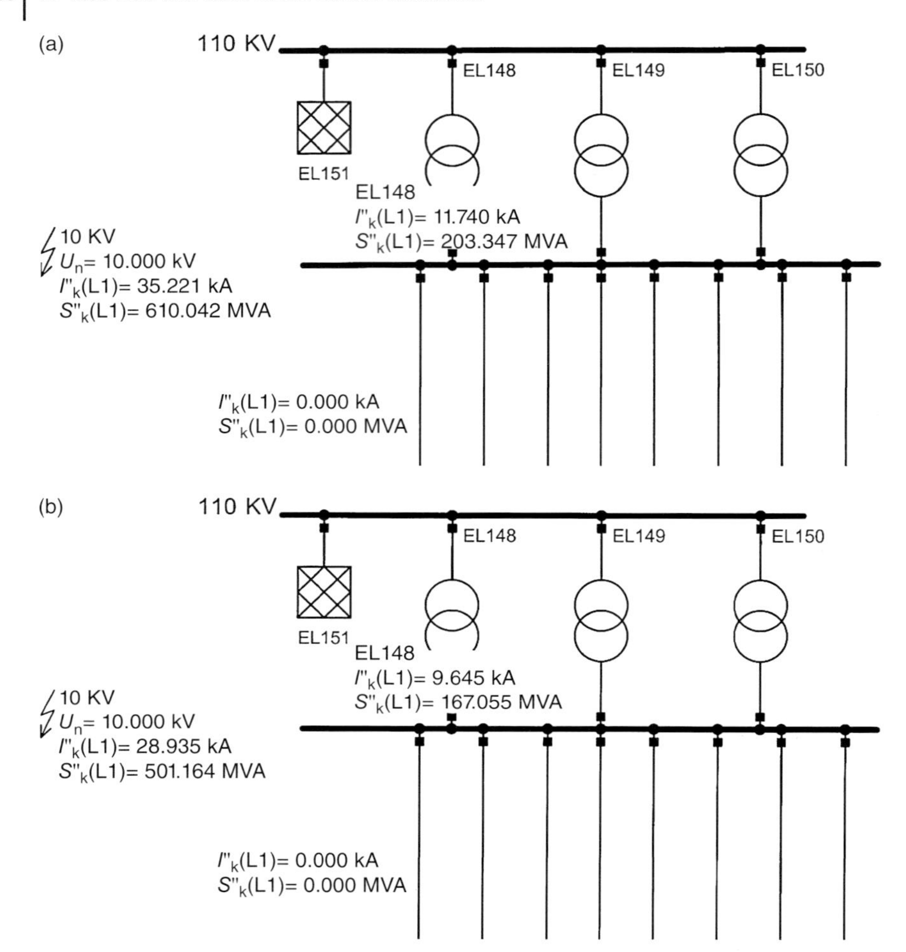

Figure 11.16 Equivalent circuit diagram of a 10kV system with incoming feeder. Results of three-phase short-circuit current calculation. (a) Impedance voltage 13%; (b) impedance voltage 17.5%.

The industrial system is connected to a combined-cycle power plant with four generators of 6.25 MVA each; three out of four generators are allowed to be in operation at the same time. The short-circuit current is increased by this to 25.6 kA. To limit the short-circuit current to $I''_{k3} \le 21.5\,\text{kA}$, reactors with $I_n = 1600\,\text{A}$; $u_k = 20\%$ were installed. The short-circuit current is reduced to $I''_{k3} = 21.1\,\text{kA}$.

11.4.4.3 Earthing Impedances

Single-phase short-circuit currents can be reduced significantly by the installation of earthing impedances in the neutral of transformers or at artificial neutrals

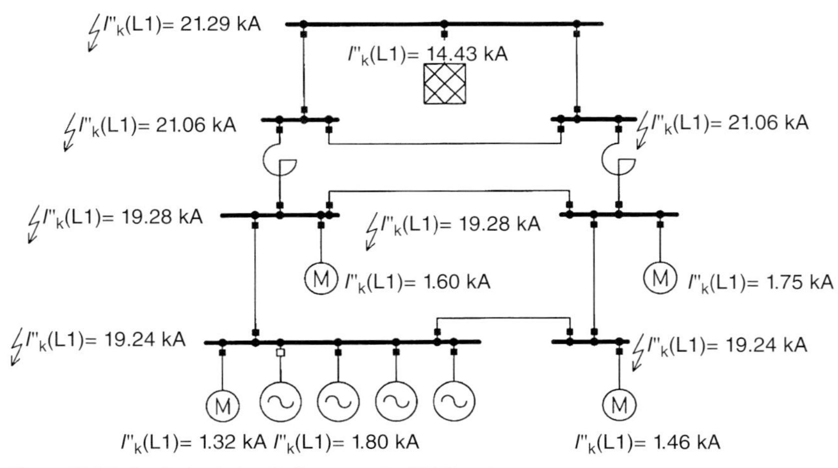

Figure 11.17 Equivalent circuit diagram of a 10 kV system with short-circuit limiting reactors. Results of three-phase short-circuit current calculation.

without affecting the three-phase short-circuit currents. Figure 11.18 represents a 132/11.5 kV substation with four transformers ($S_r = 40\,\text{MVA}$, $u_k = 14\%$). The 132 kV system has direct neutral earthing; the short-circuit currents are $I''_{k3} \approx 29.3\,\text{kA}$ and $I''_{k1} \approx 37.3\,\text{kA}$ without contribution from the 132/11.5 transformers.

The permissible short-circuit current in the 11.5 kV system is 25 kA. The single-phase short-circuit currents at the 11.5 kV busbar are $I''_{k1} = 15.04\,\text{kA}$ if one transformer is in operation and $I''_{k1} = 29.27\,\text{kA}$ if two transformers are operated in parallel.

In order to limit the single-phase short-circuit current on the 11.5 kV side to $I''_{k1} \leq 25\,\text{kA}$ (two transformers in parallel), an earthing resistance of $R_E = 0.31\,\Omega$ or an earthing reactor of $X_E = 0.1\,\Omega$ needs to be installed in the 11.5 kV neutral of each of the transformers [63].

11.4.4.4 Increased Subtransient Reactance of Generators

Generators are the direct sources for short-circuit currents; the contribution of one generator to the short-circuit current is inversely proportional to the subtransient reactance X''_d if rated power and rated voltage are not changed. An increased subtransient reactance reduces the branch short-circuit current and by this the total short-circuit current. Figure 11.19 indicates the results of short-circuit current calculation for a power station. Generators of different transient reactance ($x''_d = 12 - 17.8\%$) but identical rating $S_{rG} = 150\,\text{MVA}$ are installed. The three-phase branch short-circuit currents are in the range $I''_{k3} = 2.32 - 2.75\,\text{kA}$ depending on the subtransient reactance.

High subtransient reactance of generators has a negative impact on the dynamic stability of the generators. For short-circuits on the transmission line with

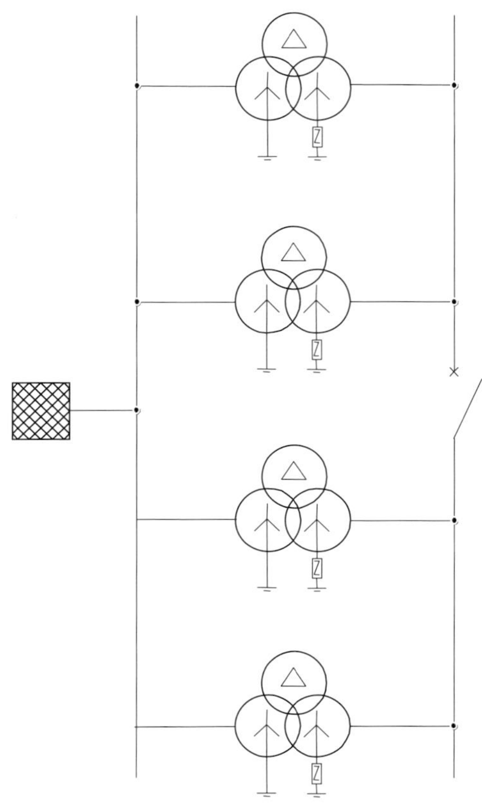

Figure 11.18 Equivalent circuit diagram of 11.5 kV system fed from the 132 kV system.

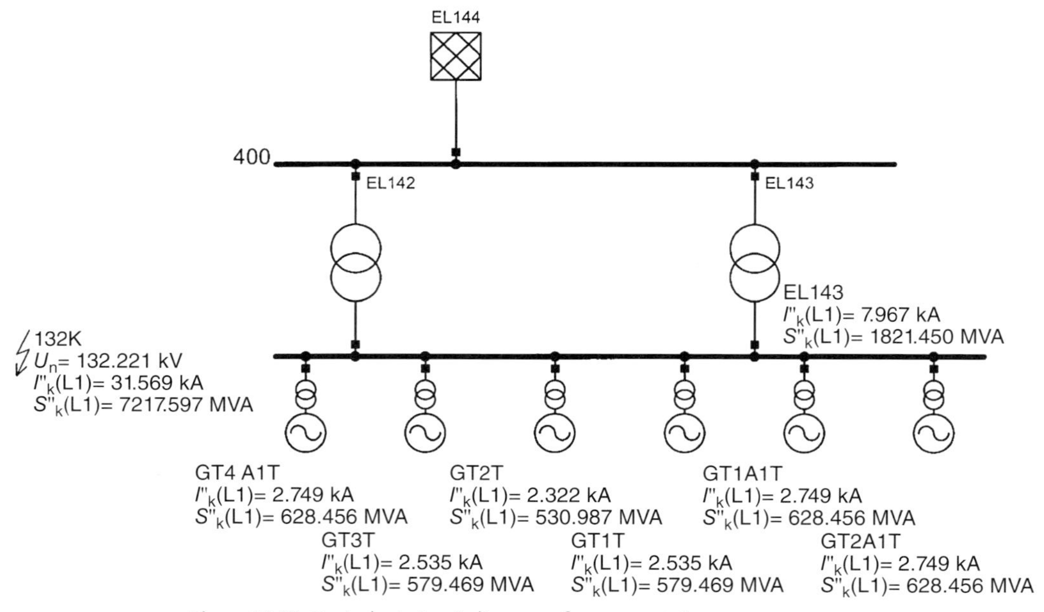

Figure 11.19 Equivalent circuit diagram of a power station with 132 kV busbar. S_{rG} = 150 MVA; $x_d'' = 12 - 17.8\%$. Results of three-phase short-circuit current calculation.

subsequent fault clearing, the transmittable power from a power station is reduced if the fault clearing time of the protection is kept constant or the fault clearing time must be reduced to keep the transmittable power constant. Details can be obtained from [59] and [61].

12 Connection of "Green-Energy" Generation to Power Systems

12.1 General

Power plants must be connected to the power system in such a way as to avoid negative effects on the operation of the power system and equipment. The following should be noted:

- The rated power of equipment in the power system must be sufficient to enable the transfer of the produced power from the generation plant into the power system and to the consumers.
- The short-circuit currents of the power system must not be inadmissibly increased by the power plants.
- The voltage rise at the connection point (point of common coupling PCC) and in the power system must remain below the permissible limits.
- Voltage changes by switching of the generation must remain within permissible limits.
- The power quality with respect to harmonics, interharmonics, asymmetry and flicker must be properly maintained and not degraded by the connection of the generating system.

Due to the intensified development of "green-energy" generation plants, special aspects need to be considered such as use of power electronics for connection to the AC systems (e.g. with photovoltaic and wind energy), sudden increase or decrease in generation (e.g. operation of wind energy plants), and distributed generation (decentralized) with small generation units (e.g. co-generation units producing heat and electricity, biogas powered plant). The existing standards, regulations and guidelines usually deal with the effects of the connection of one unit or a limited number of generation units to the power system. If a great number of generation units are connected to one power system level, as will be the case for the connection of photovoltaic or co-generation units to the low-voltage system or connection of wind parks to the high-voltage transmission system, the assessment of the connection and the operation of the generation units will need to be carried out differently. This is because the connection of one unit probably can fulfill the requirements of the standards, but the limits are exceeded

Power System Engineering: Planning, Design and Operation of Power Systems and Equipment, Second Edition.
Jürgen Schlabbach and Karl-Heinz Rofalski.

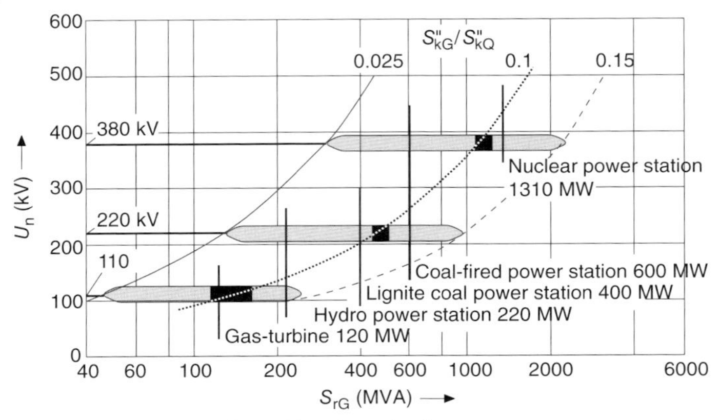

Figure 12.1 Selection of suitable voltage levels (HV system) when connecting large generation units. Data and type of typical power plants and generation units.

with connection of an increased number of units due to the uniform operation characteristic of all units. The increased usage of small co-generation units and the connection of photovoltaic installations in low-voltage systems require new considerations regarding system protection, system operation, power quality and limitation of short-circuit currents concerning these voltage levels. For high-voltage transmission systems the connection of wind parks is achieved either directly by three-phase AC connection or in case of "offshore-plants" by means of HVDC.

The connection of large generation plants is determined by, among other factors, the contribution of the generation to the short-circuit current. Figure 12.1 indicates the increase of short-circuit power with increasing system voltage and the suitable range of the generator rating which can be connected to the indicated voltage level.

In medium- and low-voltage systems the short-circuit power is lower than in high-voltage systems. The permissible share of the short-circuit level due to the connection of generation is therefore lower as well. Table 12.1 indicates typical short-circuit levels S''_{kQ} of MV and LV systems, the contribution of generation units S''_{kG} and the permissible rated power S_{rG} of generation units.

It should be noted that the contribution to the short-circuit current of generation units connected to the system through power electronics is lower; for example, in the range of the rated current of the generation unit.

In small hydro plants up to rated power of 500 kW, asynchronous generators are used; for generation with higher rated power, synchronous generators are applied. The scheme for connection of small hydro power plants therefore does not differ much from the schemes used for other conventional forms of generation. Small hydro units are connected to LV and MV systems, sometimes in remote areas, having low short-circuit power.

Table 12.1 Contribution of generation units to the short-circuit level in MV and LV systems.

U_n (kV)	S''_{kQ} (MVA)	S''_{kG} (contribution) (MVA)	$S_{rG}(S''_{kG}/S''_{kQ} = 0.1)$ (MW)
0.4	50	1–2.5	Up to 0.1
6	300	2–5	Up to 1
10	500	5–50	Up to 20
20	1000	15–100	Up to 30
30	1500	30–190	Up to 50

Wind energy plants are equipped either with asynchronous generators (also with static frequency converter in the rotor circle, so-called static inverter cascades) or with synchronous generators with or without intermediate DC circuit, provided to control the rotor speed independently of the frequency of the power system. Reactive power is needed both for the magnetization of asynchronous generators and for the commutation of static frequency converters. Power electronic devices generate harmonic and interharmonic currents, which need to be considered for the assessment of power quality. Generation units in general can cause voltage fluctuations due to the fluctuating generation. This fact must be given special attention in the case of fluctuating wind power in terms of voltage flicker. Wind energy plants and wind parks with total power above ~40 MW must be allocated to the high-voltage transmission system.

Photovoltaic installations are used either with comparatively low rating of a few kilowatts installed locally on building roofs and connected to the LV system, or as large-scale plants in the capacity range above some hundred kilowatts, to be connected through transformers to the MV system. The electrical energy produced as DC voltage is converted into the desired AC voltage by an inverter, adjusting frequency and voltage level. DC/AC inverters emit harmonic and interharmonic currents.

In case of the connection of fuel-cells, where the electricity is produced at DC voltage level, the same technical conditions apply as for photovoltaic installations.

The production of electricity using biomass can be implemented by means of conventional steam turbines, screw expansion machines or micro gas-turbines, using predominantly synchronous generators or inverter installations. Asynchronous generators are also employed, however.

Table 12.2 outlines planning tasks and engineering studies that are recommended to be carried out when connecting "green-energy" production plants to LV and MV systems. When connecting production plants to high-voltage transmission systems, extensive system studies (load-flow analysis, short-circuit current calculation, harmonics, stability, etc.) are recommended in order to ensure trouble-free and reliable operation. Norms, standards and recommendations of utilities can only offer guidelines and cannot substitute for a careful engineering study and assessment of the results.

Table 12.2 Recommended system studies for the connection of "green-energy" generation plants.

Type of generation unit		System studies to be carried out				
	Generator, inverter	**Short-circuit**	**Load-flow**	**Harmonics, interharmonics**	**Voltage asymmetry**	**Voltage fluctuations, flicker**
General	Synchronous	Y	Y	—	—	—
	Asynchronous	Y		—	—	—
Hydro power	Synchronous	Y	Only in case of high rating	—	—	—
	Asynchronous	Y		—	—	—
Wind energy	Synchronous	Y	Connection to MV and HV systems	—	—	Y
	Asynchronous	Y		—	—	Y
	Inverter			Y	—	(Y)
Co-generation plant	Synchronous	Y	—	—	Y	—
	Asynchronous	Y	—	—	Y	—
Photovoltaic	Inverter	(Y)	(Y)	Y	Y	Y
Fuel-cell	Inverter	(Y)	(Y)	Y	Y	(Y)

Y, strongly recommended; (Y), recommended in certain cases; —, not necessary.

12.2 Conditions for System Connection

12.2.1 General

When connecting renewable energy sources (green energy) and other generation facilities to the public power systems, the technical conditions and regulations of the utilities, the relevant international and national standards and norms (IEC, DIN, EN, VDE, ANSI, BS, etc.), and guidelines of technical associations (in Germany e.g., VDE/FNN; BDEW; VDN etc.) have to be taken into consideration, such as:

- ENTSO: Network code for requirements for grid connection applicable to all generators, Jun. 2012 [65]
- VDN: Transmission Code 2007, Network and System Rules for the operation of the German transmission system, Aug. 2007 [66]

- VDN-technical guideline: Technical conditions for connection of renewable generation plants to high and extra-high-voltage systems, Aug. 2004[67]
- E VDE-AR-N 4130: Technical conditions for connection of generation units to EHV-systems, expected 2014 [68]
- E VDE-AR-N 4120: Technical conditions for connection of customer equipment to high-voltage systems (110 kV), Nov. 2012 [69]
- BDEW-technical guideline: Connection of generation units to medium-voltage systems, June 2008, supplements Feb. 2011 [70]
- VDE-AR-N 4105: Connection of generating units to low-voltage systems, Jul. 2011 [71]
- VDN: DA-CH-CZ Technical rules for the assessment of power system perturbations, 2007 [72]

Definitions of major terms as used in the Technical Rules and Guidelines are presented as follows:

Point of common coupling (PCC) is the location within the power system, closest to the customer system; other customer facilities (load or generation) can be connected. The PCC is important for the assessment of power system perturbations and for all other effects and interactions between the generation plant and the power system.

Generation unit is a single unit for generating electrical power; Type 1 is given in case of direct coupling of a synchronous generator (with or without unit transformer) to the grid, all other systems are of Type 2.

Connecting circuit includes all equipment such as circuit breaker, cables, transformers, reactive power compensation, flexible AC transmission system (FACTS) that are required to connect one or more generation units to the power system.

Generation plant comprises one or more generation units and all equipment (connecting circuit) necessary for the connection to the power system.

Connected apparent power is the total sum of the maximum apparent power of the generation plant.

Agreed apparent power is the apparent power, which results from the agreed active power and the lowest agreed power factor cos φ of the generation plant or generation unit.

Installed active power is the sum of the rated active power of generation units within a power generation plant.

Operable installed active power (of generation plant) is the sum of the rated active power of generation units, except those in revision or out of order due to faults.

Available active power is the maximum possible value of the active power of generation plant at PCC.

Maximum active power is the highest active power of a generation plant, measured as maximum mean value within an interval of 10 minutes. The maximum active power can be obtained from the unit-certificate.

Instantaneous active power is the instantaneous value of the active power of the generation plant at PCC.

Available reactive power is the maximum possible value of the reactive power that a generation plant can provide, both overexcited (leading) and underexcited (lagging), depending on the instantaneous active power and the voltage at the PCC.

Instantaneous reactive power is the instantaneous value of the reactive power of the generation plant at the PCC.

12.2.2 Calculation of Power System Impedance at Point of Common Coupling

12.2.2.1 Structure of Power System

To assess the connection of generation plants, especially regarding the assessment of network disturbances, it is necessary to include the impedance of the feed-in network at the PCC. Harmonics and other emissions occur at all voltage levels of electrical power supply systems. The basic structure of an electric power supply system or scheme is shown in a simplified form in Figure 12.2.

When considering the different voltage levels (LV, MV, HV), it is assumed that generation plants are connected to the 380-kV level. Power generation at other levels solely increases the short-circuit power but has no effect on the fundamental consideration.

In LV-system harmonic currents are assumed to be fed into the grid, for example, by PV-inverter. These currents cause a voltage drop at the impedance of the supplying transformer. At the MV-level, other LV-networks are assumed to be connected as well as further current sources of harmonics, for example, in the form of inverters of wind energy units. The emitted currents are superimposed currents and cause a voltage drop at the impedance of the HV/MV-transformer. This is the fact also at the 110-kV level; thus, it can be stated that the emission of harmonic currents and all other disturbance emissions are added up from the LV- to the HV-level. An exception is those harmonics with orders dividable by three, since they form a zero-sequence system. They are blocked by transformers having a vector group Yd or Dy (LV-transformers) or transformers with isolated neutral.

The resulting harmonic voltages at the different voltage levels are transferred in accordance with the ratios of the transformers from the higher to the lower voltage systems; thus, the harmonic voltages are summed up from the HV-level to the LV-level.

Considering typical values for the short-circuit power of the power systems as indicated in Figure 12.2, the short-circuit power of the individual voltage levels differ by about one order of magnitude from HV- to LV-level. Permissible levels of disturbance voltages must be shared between the different voltage levels taking account of the typical ratio of impedances of the individual voltage levels. A first approximation of the impedance ratios provides the following:

$$Z_{\mathrm{HV}} : Z_{\mathrm{MV}} : Z_{\mathrm{LV}} = (16\% \ldots 32\%) : (32\% \ldots 57\%) : (27\% \ldots 43\%)$$

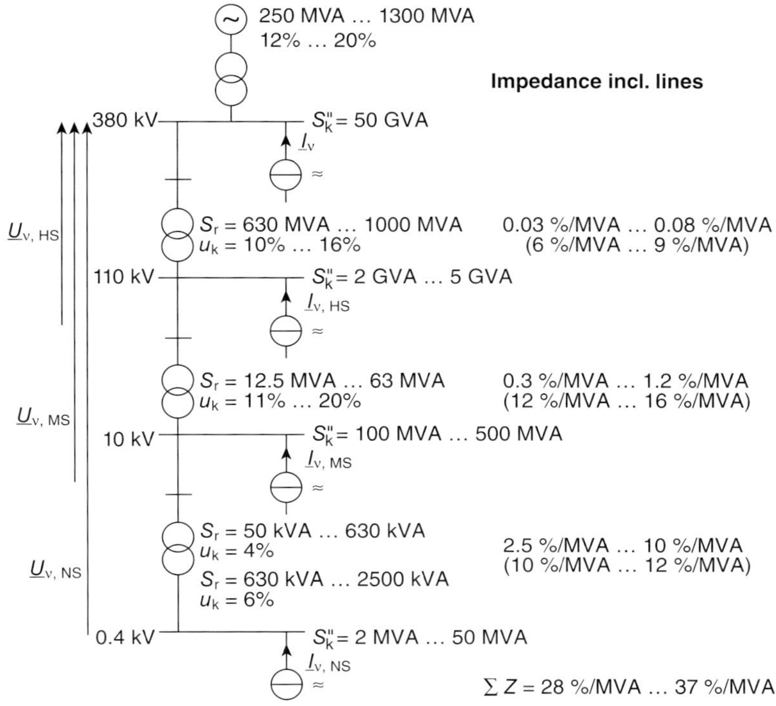

Figure 12.2 Basic structure of an electrical power system; example: emission of harmonic currents I_v. Impedance values in %/MVA: upper value: transformer impedance; lower value in brackets: impedance including lines.

The calculation of the power system impedance for the assessment of system perturbations have to be carried out according to VDN Technical Rules [72]. Normal operating conditions that determine the minimal short-circuit power are assumed. Temporarily operational special switching conditions are not considered. The results of this calculation differ from the results of short-circuit current calculations, carried out in accordance with DIN EN 60909-0 (VDE 0102) due to the different assumptions to be made, reference Table 12.3.

For the assessment of the connection of generation units to the different voltage levels, the short-circuit power $S_{k,PCC}$, calculated as per the last column of Table 12.3, is used.

12.2.2.2 Parallel Resonances in Electrical Power Systems

A typical structure of an MV-power system is outlined in Figure 12.3 consisting of a supply via a transformer from the 110-kV system. A wind energy plant, load, MV-cable circuits, and compensation equipment are connected at PCC. The cable capacitances and also the capacitors, if any, must be taken into account. In

Table 12.3 Calculation of minimal short-circuit currents by IEC 60909-0 (VDE 0102). Calculation of maximal impedance for the assessment of system perturbations (VDN Technical Rules [72]).

	Calculation of minimal short-circuit currents IEC 60909-0 (minimal impedance)	**Calculation of maximal system impedance VDN Technical Rules (maximal impedance)**
Voltage	Voltage factor c and nominal system voltage, see Section 11.2.1	MV: agreed supply voltage LV: nominal system voltage
Impedance of lines	MV: temperature at end of short-circuit duration LV: 80 °C	MV: normal operating temperature LV-cables: 70 °C
Impedance correction factors	Generators, power stations, and transformers (see Table 11.3)	No correction factors
System condition	Defined conditions as per IEC 60909-0	Normal operating conditions
Frequency dependency	None (impedance at 50 Hz)	Higher frequencies to be considered
Motor load	Not applicable for calculation of minimal short-circuit currents	Normal operating conditions to be considered
Consumer load	Not to be considered	To be considered
System impedance	$Z_{\mathrm{k,PCC}} = \dfrac{c \cdot U_{\mathrm{n}}^2}{S''_{\mathrm{k,PCC,min}}}$	$Z_{\mathrm{k,PCC}} = \dfrac{U_{\mathrm{PCC}}^2}{S_{\mathrm{k,PCC}}}$

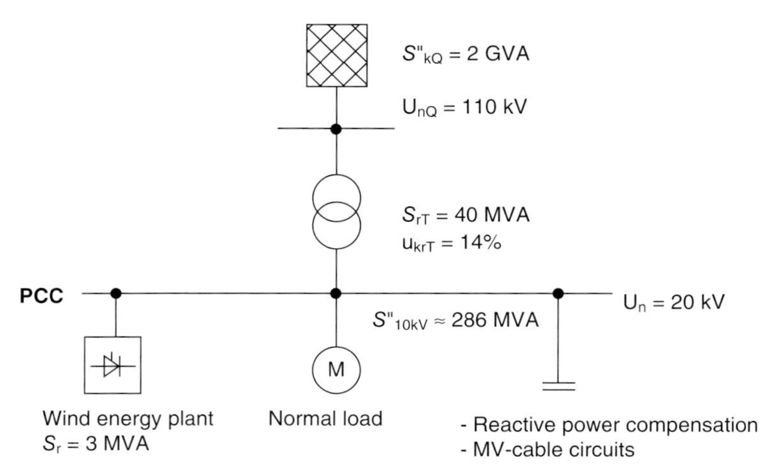

Figure 12.3 Simplified structure of a MV-power system with generation, load and compensation.

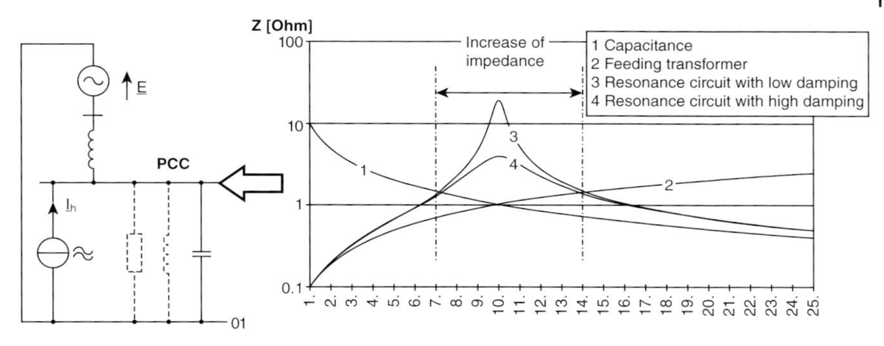

Figure 12.4 Electrical diagram of a parallel resonance circuit and impedance as a function of frequency.

LV-systems, a similar configuration can be assumed; mainly, photovoltaic generation is connected.

For further investigation of the supply scheme with regard to system resonances, the equivalent circuit of the network in the positive-sequence system, as per Figure 12.4, is used.

The reactance of the feeding transformer (and of the power system at the higher voltage level) and the cable capacitances (capacitance of the compensation as well) form a parallel resonant circuit, seen from the PCC, reference Figure 12.4. The resistances of the transformer and of the load are acting as attenuation (damping) of the resonance circuit. The resonance frequency f_{res} is calculated using Equation 12.1:

$$f_{res} = \frac{1}{2\pi * \sqrt{L * C}} \tag{12.1a}$$

$$f_{res} = f_1 * \sqrt{\frac{S_{k,PCC}}{Q_C}} \tag{12.1b}$$

$$f_{res} \approx f_1 * \sqrt{\frac{S_{r,T}}{u_{kr,T} * Q_C}} \tag{12.1c}$$

with

$S_{k,PCC}$	short-circuit power at PCC
Q_C	reactive power of the capacitor
$S_{r,T}$	apparent rated power of the transformer
$u_{kr,T}$	rated impedance voltage of the transformer

The damping d of the resonance circuit is calculated by Equation 12.2:

$$d = \frac{1}{R} \cdot \sqrt{\frac{L}{C}} \tag{12.2}$$

The reciprocal of the attenuation is referred to as the resonance quality Q. Figure 12.4 shows the impedance versus the frequency of a parallel resonance circuit. The impedance of the resonance circuit is increased in a typical bandwidth $(f_{\text{res}}/\sqrt{2} < f < f_{\text{res}} \cdot \sqrt{2})$ as compared with the impedance at the PCC without resonance. The damping is higher in case of high load conditions in the system. The motor loads, represented by their inductivity, lead to a shift of the resonant frequency to lower frequencies. This effect is relatively minor, taking account of the impedance values of the transformers and the motor loads.

12.2.2.3 Typical Resonances in Power Systems

In a 110/30-kV system shown in Figure 12.5, a 12-pulse converter is to be connected at node B2 (industrial power system 30 kV) to supply a medium-frequency converter in an industrial installation.

Two system operating scenarios were analyzed, that is,

- Meshed 30-kV network and
- 110/30-kV transformer T124 switched off at B3.

Reference a system configuration as per Figure 12.5 calculations of resonances reveal at node B2 a series resonance frequency $f_{\text{resR}} \approx 750$ Hz and two parallel resonances at frequencies $f_{\text{resP1}} = 650$ Hz and $f_{\text{resP2}} = 850$ Hz. These resonances occur due to the parallel circuits of the capacitance of the 30-kV cables B2-B3 and B2-A9 and

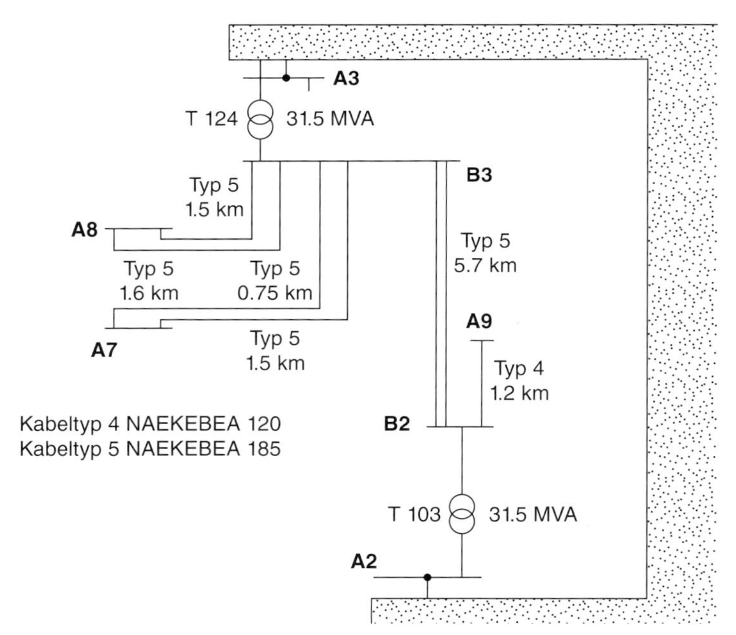

Figure 12.5 Power system diagram of a 30-kV system fed from a 110-kV system at nodes A2 and A3 [73].

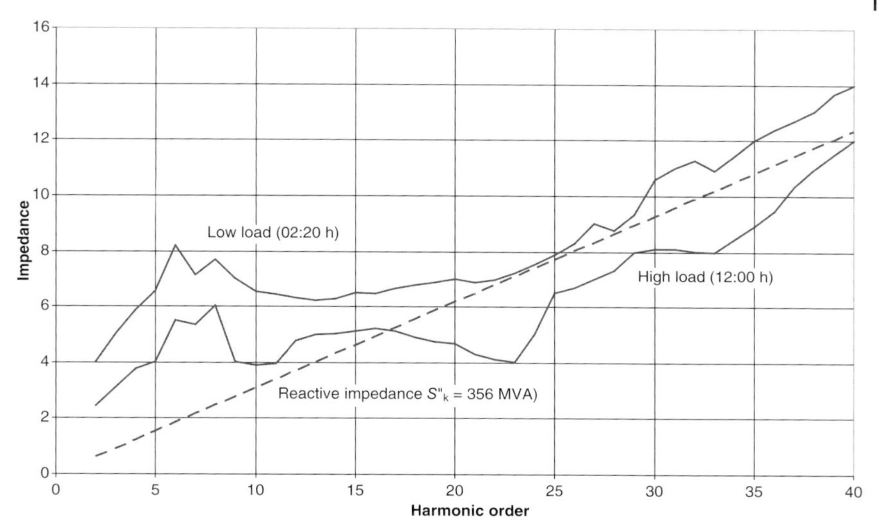

Figure 12.6 Impedance versus frequency of an urban 10-kV system at high-load and low-load conditions.

the inductances of the 110/30-kV transformers. These, in turn, are to be regarded as in series with a parallel circuit of the capacitances of the 110-kV network (mainly cables) and the inductance of the HV-supply-network feeding the 110-kV system. The series resonance frequency of the 30-kV network at node B2 is more or less maintained for the second operational scenario; that is, the cables B2-B3 are switched off at B3. The impedance of the parallel resonance would, of course, be substantially greater and would lead to a significant rise in the voltage harmonics for this operating condition.

Figure 12.6 outlines the measuring results of the frequency-dependent impedances of an urban 10-kV system with a short-circuit power $S''_k = 356\,\text{MVA}$ at low load and high load conditions. A parallel resonance frequency occurs in the range of 300–400 Hz. The resonance frequency is more or less independent from the system load conditions. The system impedance for all frequencies is generally lower at high load as compared with low load condition.

Parallel resonances occur at each voltage level. One has to observe that the impedance of a resonance circuit is increased in case of parallel resonance in a certain bandwidth $((f_{\text{res}}/\sqrt{2} < f < f_{\text{res}} \cdot \sqrt{2}))$ near to the resonance frequency. Typical values of resonances are

- LV-system $f_{\text{res}} > 2\,\text{kHz}$
- MV-system (mainly cables) $f_{\text{res}} \approx 300\,\text{Hz} \ldots 700\,\text{Hz}$
- MV-system (mainly overhead lines) $f_{\text{res}} > 1.5\,\text{kHz}$
- HV-system $f_{\text{res}} \approx 1\,\text{kHz} \ldots 2\,\text{kHz}$ (difficult to estimate)

A detailed analysis using a suitable program for the determination of resonance frequencies is strongly recommended in any case.

12.2.3 Short-Circuit Currents and Protective Devices

All installations must be designed to withstand the foreseeable short-circuit stress and the evidence of adherence to the standards has to be documented in a suitable way. If the short-circuit level in the system is increased, for example, by the connection of generation, which is generally the case, suitable measures are to be planned and installed by the contractor to limit the short-circuit current. If data concerning the expected short-circuit current are unknown, multiples of the rated current of the generator are to be used as contribution to the short-circuit current as outlined in Table 12.4.

Short-circuit current calculations are necessary to determine the thermal and mechanical strength as parameters for the design of the generation as well as the short-circuit current contribution to the generation.

Protection devices such as short-circuit protection, overload protection, and measures against touch voltage must be installed. The design of necessary protection devices depends on the kind of connection (single-phase, three-phase), the type of generator (inverter, synchronous, asynchronous), and the type of switching interface. In addition, voltage and frequency control devices are needed, which actuate switching devices when exceeding the limit values. The setting ranges of the relevant protection device shall be as stated below:

- Voltage increase (three-phase; LV only) 100–115% of U_n
- Voltage increase (three-phase; MV and HV) 100–130% of U_n

Table 12.4 Estimates for the contribution to short-circuit currents of generation units; exact values can be found from the certificate of the generation unit.

Type of generation unit	Initial symmetrical s.-c. current I''_k	Peak s.-c. current I_p	Symmetrical s.-c. breaking current I_b	Steady state s.-c. current I_k
Synchronous generator	$8 \cdot I_r$	$20 \cdot I_r$	$5 \cdot I_r$	$5 \cdot I_r$
Asynchronous generator	$6 \cdot I_r$	$12 \cdot I_r$	$5 \cdot I_r$	$5 \cdot I_r$
Double-fed asynchronous generator	$3 \cdot I_r$	$8 \cdot I_r$	$1 \cdot I_r$	$1 \cdot I_r$
Full-scale electronic converter	$1 \cdot I_r$	$2 \cdot I_r$	$1 \cdot I_r$	$1 \cdot I_r$

s.-c., short-circuit.

- Fast voltage increase (three-phase; MV and HV) 100–130% of U_n
- Under-voltage (three-phase; all voltage levels) 70–100% of U_n
- Fast under-voltage (three-phase; all voltage levels) 70–100% of U_n
- Reactive power and low voltage (single-phase; MV only) 70–100% of U_n
- Frequency increase (single-phase; all voltage levels) 50–52 Hz
- Frequency decrease (single-phase; LV and HV) 47–50 Hz

Generation units connected to power systems with auto-reclosing measures has to be provided with instantaneous tripping of the generation plant (in case of auto-reclosure), if successful auto-reclosing is to be achieved.

12.2.4 Voltage Control and Reactive Power Supply under Steady-State Conditions

12.2.4.1 Generation Connected to Low-Voltage Systems

Requirements for generation units connected to low-voltage networks are defined in VDE-AR-N 4105 [71]. Generation connected to LV-systems must contribute to the voltage control under normal operating conditions. Generators, which cannot regulate reactive power, such as asynchronous generators, for example, used for small cogeneration plants, are exempted from the regulations, whereas a fixed power factor can be defined. All other generation facilities have to supply reactive power to control the voltage at the PCC for normal operating conditions and for active power above 20% of the rated active power, for example,

$\Sigma S_r \leq 3.68\,\text{kVA}$	(cos φ=0.95 lag. . .0.95 lead) as per DIN EN 50438
$3.68\,\text{kVA} < \Sigma S_r \leq 13.8\,\text{kVA}$	(cos φ=0.95 lag. . .0.95 lead) with characteristic curve cos $\varphi = f(P)$
$\Sigma S_r > 13.8\,\text{kVA}$	(cos φ=0.9 lag. . .0.90 lead) with characteristic curve cos $\varphi = f(P)$

It shall be possible to adjust the reactive power during operation within 10 s within the specified range. For the startup-sequence of the generation, a transitional period of 10 minutes is allowed. Characteristic curves cos $\varphi = f(P)$ are defined by the distribution system operator depending on the real system requirements; see Figure 12.7. The characteristic curves shall have less than (3. . .4) patterns.

12.2.4.2 Generation Connected to Medium-Voltage Systems

The Technical Guideline of BDEW "Connection of generation units to medium-voltage systems" [70] describes in detail the requirements for generation facilities connected to MV-systems. At PCC, a power factor between cos φ=0.95 lag. . .0.95 lead is required. The defined range of the power factor must be achievable for normal operating conditions.

As additional requirement, derived from the Transmission Code 2007 [66], the required range of power factor can be defined to be cos φ=0.975 lag. . .0.9 lead or cos φ=0.95 lag. . .0.925 lead.

The reactive power has to be adjustable within several minutes. The reactive power set point resp. characteristic curve can be defined by

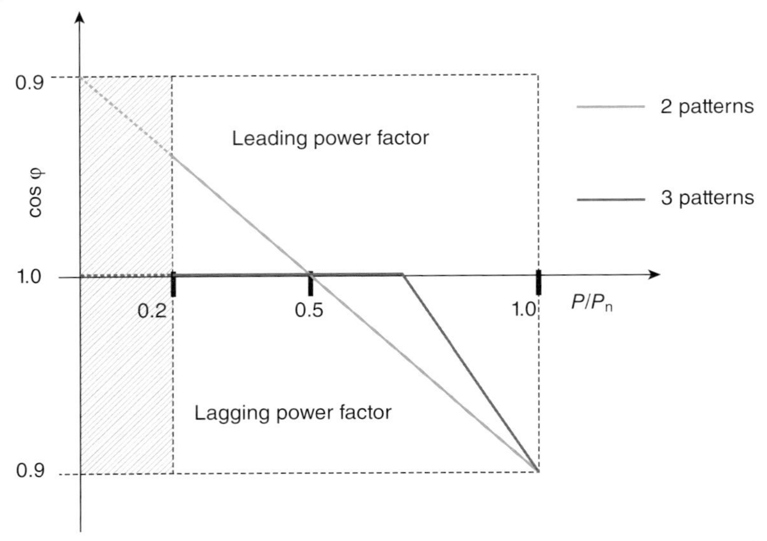

Figure 12.7 Sample characteristic curve of power factor and active power (cos $\varphi=f(P)$) curve.

- Fixed value of the power factor cos φ
- Fixed determined reactive power Q
- Characteristic curve (cos $\varphi=f(P)$); each set point has to be achieved within 10 s
- Characteristic curve ($Q=f(U)$); each set point has to be achieved between 10 s and 1 minute as defined by the system requirements. Coordination with the control times of the tap-changer of the feeding transformer is required.

12.2.4.3 Generation Connected to High-Voltage Systems ($U_n = 110\,kV$)

The Technical Guideline E VDE-AR-N 4120 "Technical conditions for connection of customer equipment to high-voltage systems" [69] generally applies to all kind of installations (generation plants, consumer facilities and mixed installations). The technical requirements for generation plants are only valid if the maximum received active power of the installation does not exceed 10% of the agreed active power of the customer facility.

Generation plants shall fulfill the technical requirements for operation as per Figure 12.8 under steady-state conditions as follows:

Frequency range:	$f=47.5\,Hz \ldots 51.5\,Hz$
Voltage range:	$U=93.5\,kV \ldots 127\,kV$
Frequency gradient:	$\Delta f/\Delta t < 0.005 \cdot f_n\,min^{-1}$
Voltage gradient:	$\Delta U/\Delta t < 0.05 \cdot U_n\,min^{-1}$

During normal operating conditions, generation plants shall participate in the static voltage control of the network by providing reactive power. Depending on the particular operating voltage at the PCC and following the frequency range of 47.5–51.5 Hz, generation units shall meet the requirements of the power factor as

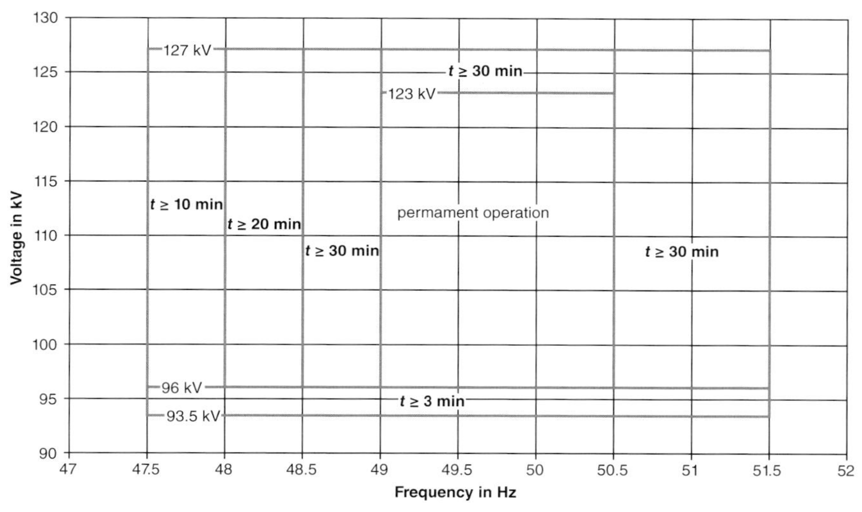

Figure 12.8 Frequency and voltage ranges for steady-state operation of generating plants as per E VDE-AR-N 4120 [69].

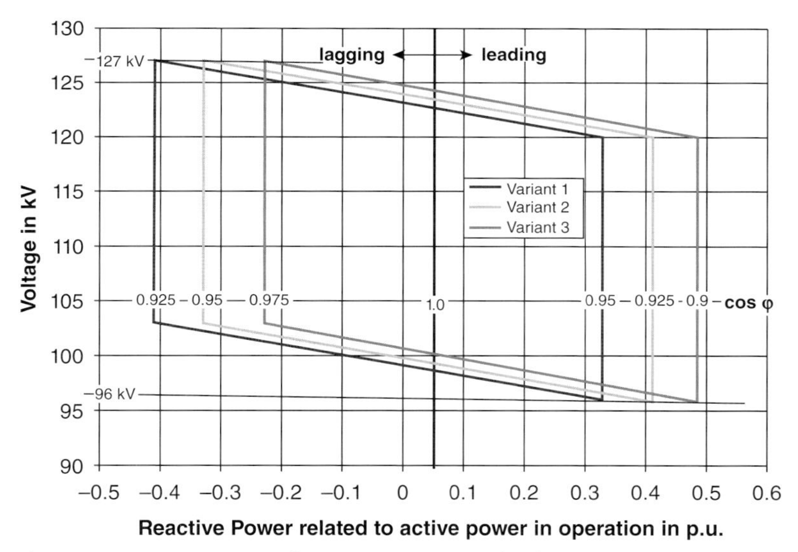

Figure 12.9 Operating ranges for reactive power supply of a generation plant at HV-systems; full-load conditions as per E VDE-AR-N 4120 [69].

outlined in Figure 12.9. Each set point has to be achieved within a maximum of 4 minutes.

For partial-load conditions, the generation plant shall be capable to be operated within the operating ranges presented in Figure 12.10. The ordinate is scaled in percent of the ratio of "instantaneous active power to operable installed active

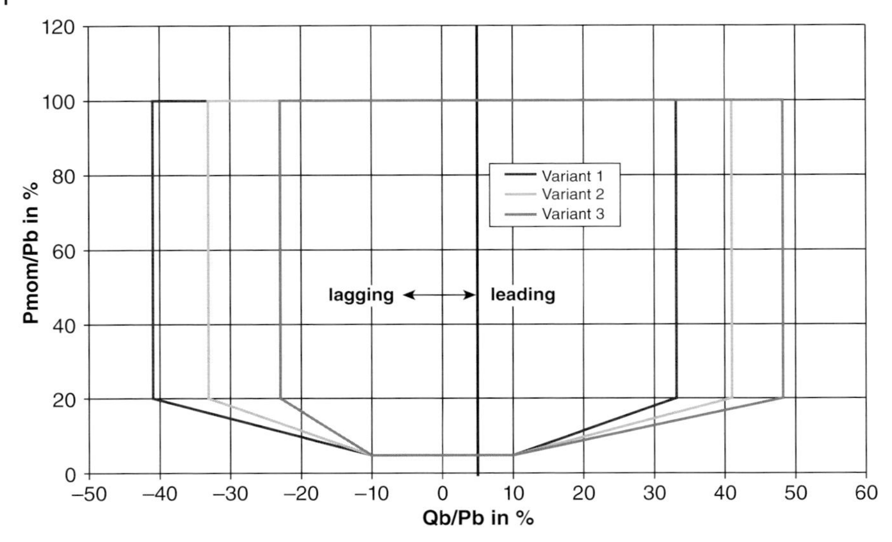

Figure 12.10 Operating ranges for reactive power supply of a generation plant at HV-systems; partial load conditions as per E VDE-AR-N 4120 [69].

power," the *x*-axis as a percentage of the ratio of "reactive power to be provided to operable installed active power."

Each set point has to be achieved within a maximum of 4 minutes. Choosing the required PQ-diagram for partial load conditions, reference Figure 12.10, it should be noted, that the diagrams for generation plants as per Variant 1, Variant 2, and Variant 3 are associated with the relevant diagrams for full-load conditions according to Figure 12.9. Depending on the network situation, the reactive power supply can prevail over the active power generation.

The reactive power has to be adjustable within several minutes. The reactive power set point resp. characteristic curve can be defined by

- Fixed value of the power factor cos φ
- Fixed determined reactive power Q
- Characteristic curve ($Q=f(P)$)
- Characteristic curve ($Q=f(U)$).

The specific requirements for set points, operating time, and so on, are outlined in detail in E VDE-AR-N 4120.

12.2.4.4 Generation Connected to Extra-High-Voltage Systems ($U_n \geq 110\,kV$)

The Transmission Code 2007 [66] and the new ENTSO Network Code [65] define general requirements for reactive power control relative to the generation plant. Special requirements are defined for generation the installations of which are covered under the German Renewable Energy Sources Act (EEG), such as

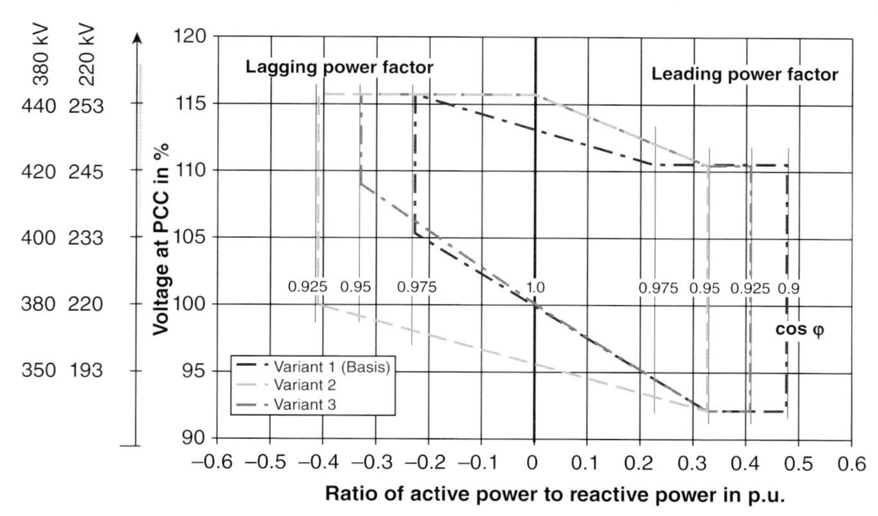

Figure 12.11 Operating ranges for reactive power supply of a generation plant at EHV-systems; full-load conditions [68].

photovoltaic and wind turbines. The VDN-Guide "Technical conditions for connection of renewable generating plants to high and extra-high voltage systems" [67] are to be applied. A new revised version of the technical requirements and conditions is in progress by VDE/FNN and is expected to be issued as a draft version in 2014 with the citation number E VDE-AR-N 4130 [68].

Depending on the actual operating voltage at the PCC, the generation plant has to fulfill the conditions corresponding to Figure 12.11 for full-load conditions under normal operating conditions and within the frequency range of 49.5–50.5 Hz. The voltage in percent refers to the respective nominal system voltage. In order to match different power system conditions, three different operation scenarios referred to as Variant 1 (Basis), Variant 2, and Variant 3 can be distinguished.

For partial-load conditions, the generation plant shall be capable to be operated within the operating ranges outlined in Figure 12.12. The ordinate is scaled in percent of the ratio between "instantaneous active power to operable installed active power," the *x*-axis as a percentage of the ration of "reactive power to be provided to operable installed active power." The operating ranges are applicable for different voltage ranges, stated as follows:

Variant 1 (basis):	U=400 kV . . . 420 kV, 233 kV . . . 245 kV
Variant 2:	U=409 kV . . . 420 kV, 239 kV . . . 245 kV
Variant 3:	U=380 kV . . . 420 kV, 220 kV . . . 245 kV

Choosing the required PQ-diagram for partial load conditions as shown in Figure 12.12, it should be noted that the graphs for generation plants of Variant 1, Variant 2, and Variant 3 are associated with the relevant diagrams for full-load

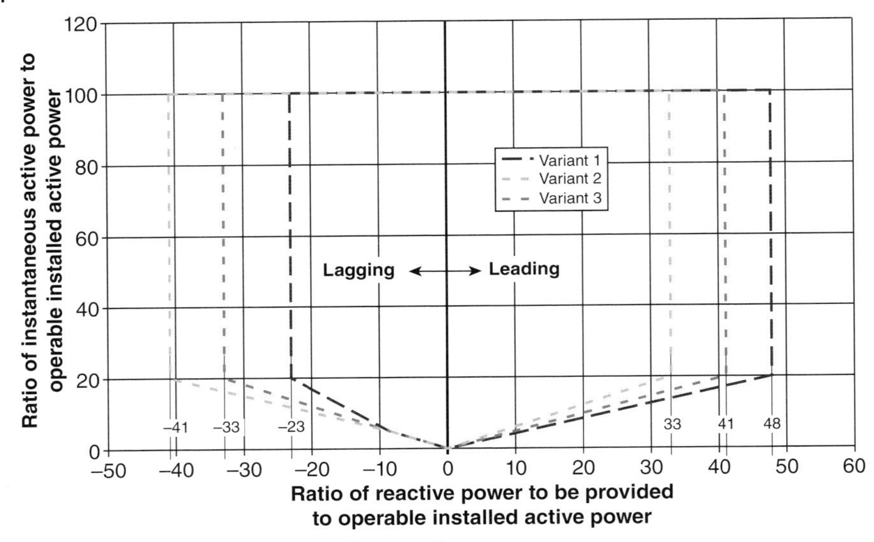

Figure 12.12 Operating ranges (PQ-diagram) for reactive power supply of a generation plant at EHV-systems; partial load conditions [68].

conditions as per Figure 12.11. Depending on the network situation, the reactive power supply can prevail over the active power generation.

The reactive power has to be adjustable within several minutes. The reactive power set point resp. characteristic curve can be defined by

- Fixed value of the power factor cos φ
- Fixed determined reactive power Q
- Characteristic curve (cos $\varphi = f(P)$); each set point has to be achieved within 10 s
- Characteristic curve ($Q = f(U)$), each set point has to be achieved between 10 s and 1 min as defined by the system requirements.

In Figure 12.13 some typical characteristic curves or settings for a generation plant are illustrated.

12.2.5 Frequency Control and Active Power Reduction

Under steady-state conditions the power system frequency is identical at all system locations. The control of the frequency is possible by changing the generated active power or the system load. Generation based on renewable energy sources can contribute significantly to the frequency control by means of reducing the generated power in case of increasing frequency. Most of the generation by renewable energy sources, such as photovoltaic and wind energy, is not capable to increase the generated active power in case of decreasing frequency.

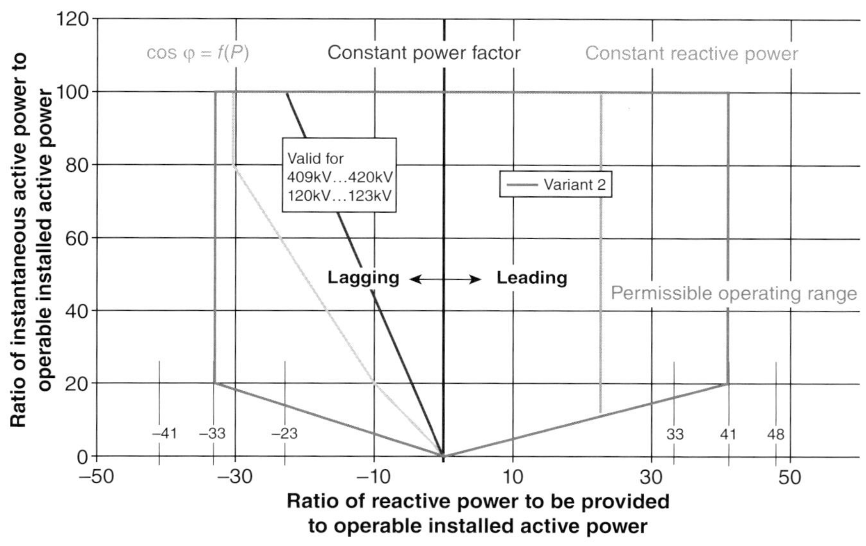

Figure 12.13 Example of characteristic curves for generation plant at EHV-system, partial-load condition.

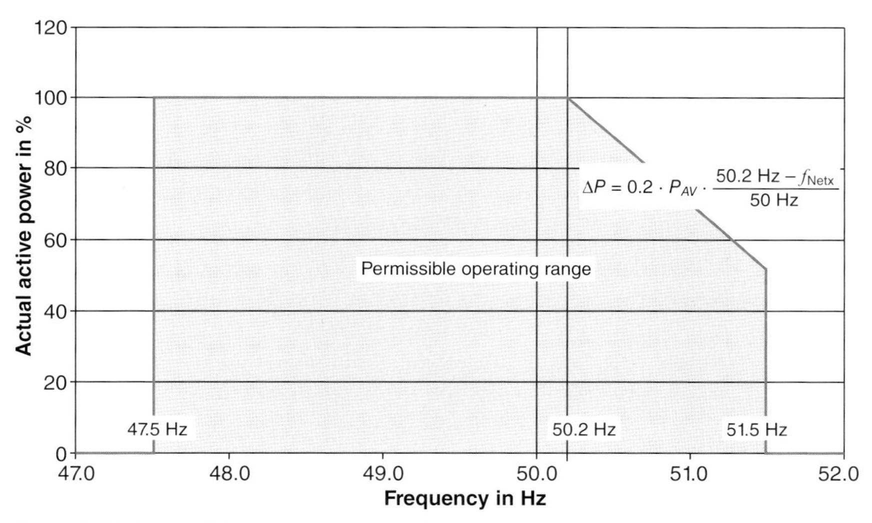

Figure 12.14 Permissible operating range, active power versus frequency of generation units connected to each voltage level.

Figure 12.14 indicates the required gradients for reduction of generated power in case of increasing frequency. If the frequency exceeds 50.02 Hz, active power has to be reduced applying a gradient of 40% of the available active power per hertz. The active power can be increased only, if the frequency is below 50.05 Hz again. In case the frequency exceeds 51.5 Hz, the generation unit has to be switched off within 100 ms; the same also applies if the frequency falls below 47.5 Hz. The permissible operating range as per Figure 12.14 extends to generation units at all voltage levels.

The reduction of the active power generation is also permitted under the condition that

- There is a danger of secure network operation
- There is a risk of overloading of equipment
- There is a risk of network separation (islanding)
- The dynamic or static stability is rated as critical
- The power system frequency rise is rated as dangerous
- Construction and maintenance works are to be carried out and must not be delayed
- Subnetworks need to be synchronized.

The reduction of the active power is intended to be made possible on any given new set point in steps under every possible operating condition. Target values of 100%, 60%, 30%, and 0% of the agreed active connection power are recommended. The power reduction must be implemented within 1 minute.

12.3 Fault-Ride-Through (FRT) Conditions and Dynamic Voltage Control

12.3.1 Types of Generation Units

The Transmission Code and the Technical Guidelines distinguish between two categories of generators within the generation plant:

- Type 1 generator refers to generation units with synchronous generator connected directly or via a step-up transformer (unit transformer) to the power system,
- Type 2 generator includes all other generation units, that is, those which are coupled via power electronic inverter with the power system.

12.3.2 Conditions for Generation Units of Type 1

In MV and HV systems, generation plants of Type 1 may not be separated from the system in case of symmetrical and asymmetrical system faults (short-circuits) with voltage drops of 100% (residual voltage at the PCC is equal to 0%) for a period of 150 ms after the fault occurrence, reference Figure 12.15. For short-duration voltage dips between 70% and 90%, the generation plant shall be capable to remain in operation for up to 1.5 s without shutdown. For periods of up to 3 seconds and residual voltage above the lower value of the permissible

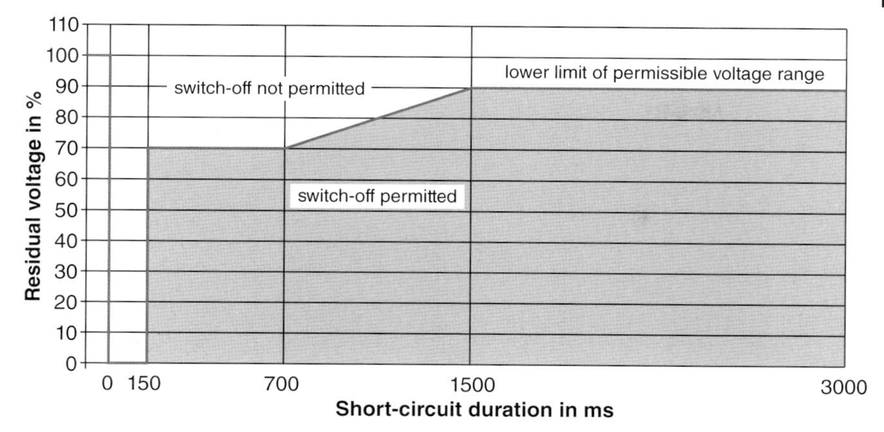

Figure 12.15 Fault-ride-through (FRT) conditions (residual voltage and fault-duration) for generation of Type 1, applicable to medium-voltage networks.

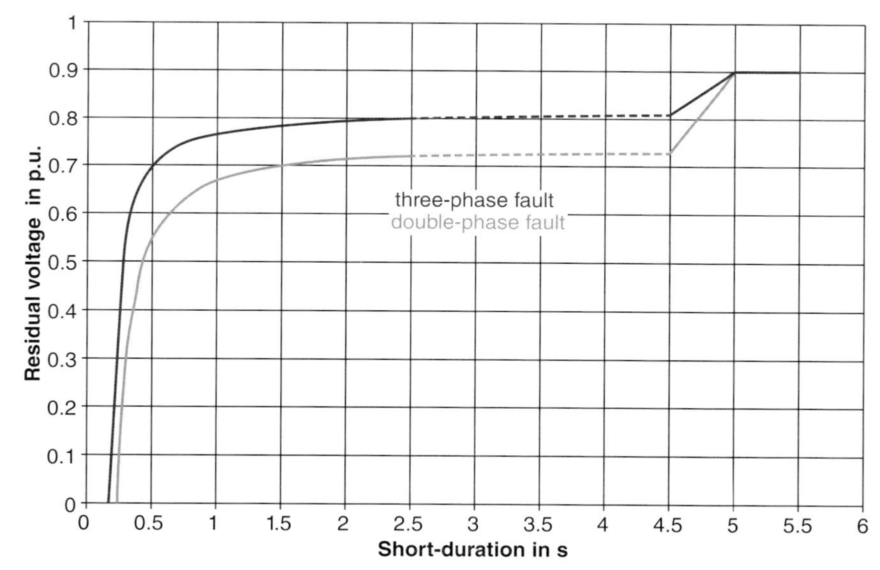

Figure 12.16 Fault-ride-through-curves for generation plants Type 1 as per E VDE-AR-N 4120, valid for HV and EHV networks.

voltage range (90% of rated voltage), the generation plant shall remain in operation.

Generation plants connected to HV systems (110-kV systems) shall be operated according to the fault-ride-through (FRT)-curves as per Figure 12.16. The FRT-curves represent the limiting curves of the voltage (values relative to the nominal voltage) at the PCC for the fault-duration and the residual voltage. Residual voltage above the FRT-curve must not lead to a separation of the generation plant from the power system. The lowest voltage value of the three phases shall be considered in case of a fault. For generation plants with rated active power $P_r > 10\,\text{MW}$ damping devices shall be provided against pole-swings.

12.3.3
Conditions for Units of Type 2

Generation plants of Type 2 connected to the MV system shall remain in operation in case of symmetrical and asymmetrical system faults (short-circuits) with voltage drops of 100% (residual voltage at the PCC equal to 0%) for a period of 150 ms after the fault occurrence as per Figure 12.17. For longer fault-durations different requirements are defined depending on the fault-duration and the level of the residual voltage at the PCC.

Between boundary line 1 and boundary line 2, reference Figure 12.17, a temporary disconnection of the generation plant from the system is allowed for short-circuit durations from 150 to 1500 ms and residual voltages between 30% and 90% (voltage dip 70% and 10%) if resynchronization will be achieved within 2 seconds. After resynchronization real power has to be increased with a gradient of at least 10% of the rated active power per second defined in the Transmission Code 2007.

In case of longer fault-durations and for larger voltage dips (below boundary 2 as per Figure 12.17), disconnection may be gradually allowed. A longer time for resynchronization and smaller gradients of the active power increase, after resynchronization, shall be permitted by agreement with the network operator. Below a residual voltage of 30% and for a fault-duration greater than 150 ms, a shutdown is generally permitted.

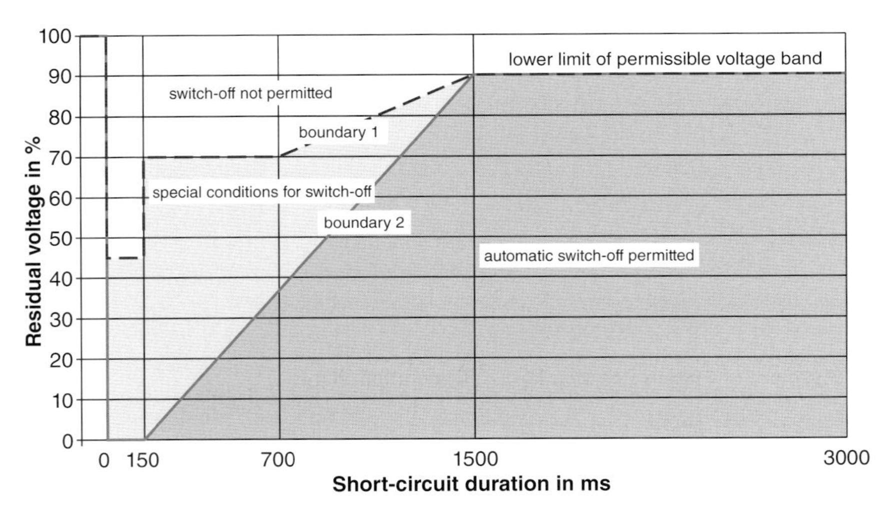

Figure 12.17 Fault-ride-through conditions (residual voltage and fault-duration) for generation plants of Type 2, valid for medium-voltage networks.

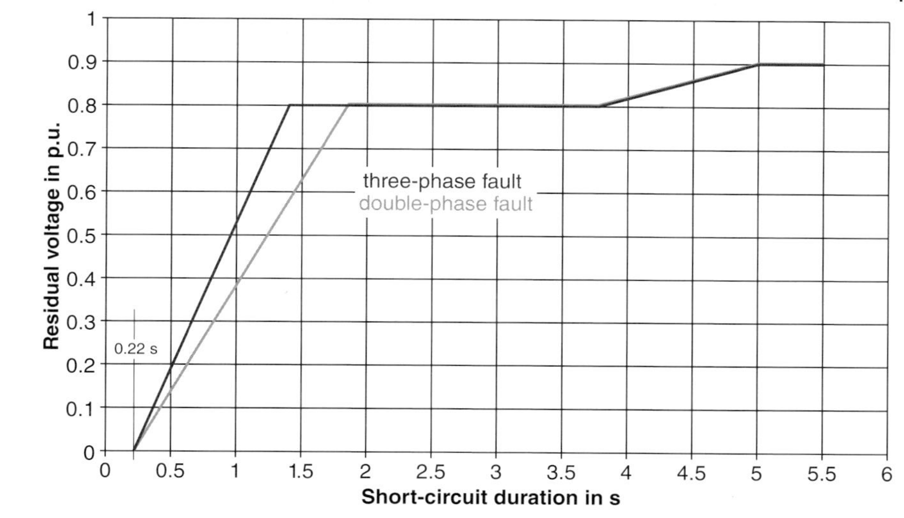

Figure 12.18 FRT-curves for generation plants Type 2 as per E VDE-AR-N 4120, valid for HV and EHV networks.

Generation plants connected to the HV system (110-kV system) shall be operated according to the FRT-curves as per Figure 12.18. The FRT-curves represent the limiting curves of the voltage (values relative to the nominal voltage) at the PCC for the fault-duration and the residual voltage. Residual voltages above the FRT-curve must not lead to a disconnection of the generation plant from the power system. The lowest voltage value of the three phases shall be considered in case of faults. For generation plants with rated active power $P_r > 10\,\text{MW}$, damping devices shall be provided against pole-swings.

Generation plants of Type 1 and Type 2 shall support the voltage by reactive power supply in case of operation during fault conditions. The amount of reactive power supply shall be within the permissible limits of the power factor as defined in the relevant technical guidelines, see Section 12.2.4. In case of asymmetrical short-circuits, the reactive power supply at the PCC shall be limited such, that the voltage of the non-faulted phases at the PCC does not exceed the value of 110% of the agreed supply voltage U_c.

12.4 Assessment of System Perturbations of Generation Plants

12.4.1 General

Generation plants must be operated in such a way that power system perturbations are kept to a bare minimum, for example, such that interference problems are avoided and power quality is maintained to the greatest possible extent. The assessment of power perturbations and voltage quality has to be carried out at the PCC (see Section 12.2). The parameters for the voltage quality are defined in DIN EN 50160 for all voltage levels.

Perturbations, affecting the power quality, occur due to different operation modes of generation plants and are discussed in the following chapters:

- Voltage increase
- Rapid voltage changes due to switching
- Flicker
- Harmonics and interharmonics
- Asymmetry and voltage unbalance
- Commutation dips
- Effects on carrier signals

Other types of network perturbations, such as frequency variations, transient and power-frequency surges and voltage dips, voltage swells (threshold voltage), and interruptions of the supply voltages, are not caused by generation, but can affect its operation. These phenomena are not discussed within this book. A detailed approach to the assessment of network effects, even in industrial systems, can be found in references [73] and [74].

12.4.2 Voltage Increase

The generation of active power, as well as capacitive reactive power (referred to as generation of reactive power), leads to an increase of the RMS-value of the voltage, whereas inductive reactive power (termed consumption of reactive power) leads to a decrease of the RMS-value of the voltage. On the basis of the equivalent circuit diagram illustrated in Figure 12.19, the Equation 12.3 for the voltage increase Δu are compiled.

The voltage increase Δu is calculated as per Equation 12.3

$$\Delta u = \frac{S_G}{S_{k,PCC}} \cdot \cos(\psi_{k,PCC} + \varphi) \tag{12.3a}$$

$$\Delta u = \frac{S_G}{U_N^2} \cdot (R_{k,PCC} \cdot \cos\varphi - X_{k,PCC} \cdot \sin\varphi) \tag{12.3b}$$

With:

S_G	apparent power of generation plant
$S_{k,PCC}$	short-circuit power at the PCC (see Section 12.2.2)
U_N	power system voltage without generation plant
φ	angle of power factor (leading power factor: $\varphi < 0$; lagging power factor: $\varphi > 0$)
$\psi_{k,PCC}$	phase-angle of power system ($\psi_{k,PCC} = \arctan(X_{k,PCC}/R_{k,PCC})$)

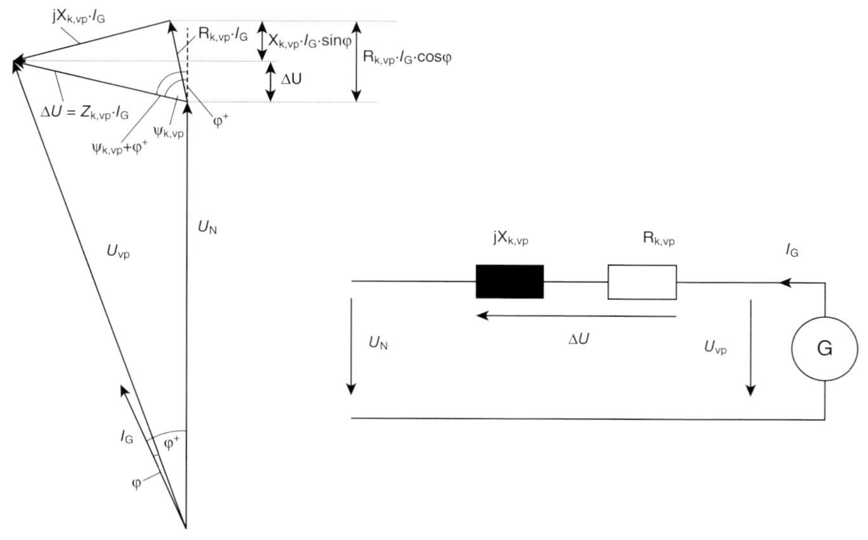

Figure 12.19 Equivalent circuit diagram of a generation plant connected to a power system. Generated active power and capacitive power (leading p.f.) are counted positive with positive values.

Regarding the different system voltage levels, the permissible values for voltage increase at the PCC are stated as follows:

LV system	$\Delta u_{PCC} \leq 3\%$; permissible voltage bandwidth to be fulfilled
MV system	$\Delta u_{PCC} \leq 2\%$; permissible voltage bandwidth to be fulfilled
LV and MV system	$\Delta u_{PCC} \leq 5\%$ for both systems; permissible voltage bandwidth to be fulfilled
HV system	$\Delta u_{PCC} \leq 1\%$; permissible voltage bandwidth to be fulfilled

Concerning the permissible voltage bandwidth, see Section 3.2.

It shall be noted that the effect of several generation plants integrated in a meshed power system and connected at different PCCs cannot be estimated as per Equation 12.3. In case of such a network configuration, load-flow calculations have to be carried out using a suitable planning program.

12.4.3
Rapid Voltage Change due to Switching Operations

The maximum voltage change due to switching operations (switch-on or switch-off) at the PCC is calculated applying the current switching factor $k_{\mathrm{I,max}}$ according to Equation 12.4.

$$k_{I,\max} = \frac{I_{an}}{I_{r,G}} \tag{12.4}$$

With:

I_{an} starting current (RMS-value) of generation unit
$I_{\text{r,G}}$ rated current of generation units

If the starting current is not known, the following values for the current switching factor shall be used:

$k_{I,\max} \leq 1.2$	generation with converter, exact values to be obtained from the generating unit certificate
$k_{I,\max} = 4$	asynchronous generators, accelerated to 95% . . . 105% of synchronous speed before switching
$k_{I,\max} = I_{\text{a}}/I_{\text{r,G}}$	asynchronous generators, accelerated to nominal speed directly by the power system
$k_{I,\max} = 8$	for all other types of equipment

An approximate calculation of the relative voltage change can be made using Equation 12.5

$$\Delta u_{\max,\text{PCC}} = k_{I,\max} \cdot \frac{S_{\text{r,G}}}{S_{k,\text{PCC}}} \tag{12.5}$$

$S_{\text{k,PCC}}$ short-circuit power at the PCC (see Section 12.2.2)
$S_{\text{r,G}}$ rated apparent power of the generation plant

In MV networks, the supply or operating voltage can be agreed between network operators and network users. The supply voltage therefore may differ from the nominal system voltage. Reference voltage is defined as the value of the agreed supply voltage U_{agr}.

The relative voltage change generally shall be limited to:

LV system	$\Delta u \leq 3\%$
MV system	$\Delta u \leq 2\%$
LV and MV system	$\Delta u \leq 5\%$ for both systems
HV and EHV system	$\Delta u \leq 0.5\%$ for switching of a single generation unit
	$\Delta u \leq 2\%$ for switching of a generation plant

As can be seen from Equation 12.5, the phase-angle $\psi_{\text{k,PCC}}$ of the power system is not included in the calculation; the voltage change actually is smaller. In order to reach to more realistic figures, the voltage change factor k_ψ is introduced, taking account of the system impedance angle and the effect of the voltage change by flicker. The voltage change factor k_ψ is defined as the maximal value of either the switching voltage change factor $k_{\psi,\text{U}}$ or the flicker-related current switching factor $k_{\psi,\text{F}}$ as per Equation 12.6

$$k_\psi = \text{MAX}\{k_{\psi,U}; k_{\psi,F}\} \tag{12.6a}$$

$$k_{\psi,U} = \left(\frac{\Delta U}{U} \bigg|_{\mathrm{Ref}} \right) * \frac{S_{k,\mathrm{Ref}}}{S_{\mathrm{r,G}}} \tag{12.6b}$$

The switching voltage change factor $k_{\psi,\mathrm{U}}$ is measured in the test field (index Ref) as reference value for different impedance angles in steps of $\Delta\psi = 5°$ in the range of $\psi = 0° \ldots 90°$.

The flicker-related current switching factor $k_{\psi,\mathrm{F}}$ is calculated based on the switching voltage change factor $k_{\psi,\mathrm{U}}$, which takes account of the short-time flicker $P_{\mathrm{st,Ref}}$, and the permissible emission value $P_{\mathrm{E,st}}$ as per IEC/TR 61000-3-7 of the single generation unit. The short-time flicker is measured during the switching voltage test as well.

$$k_{\psi,F} = \left(\frac{\Delta U}{U} \bigg|_{\mathrm{Ref}} \right) * \frac{S_{k,\mathrm{Ref}}}{S_{\mathrm{r,G}}} * \frac{P_{st,\mathrm{Ref}}}{P_{\mathrm{E},st}} \tag{12.6c}$$

The permissible emission value $P_{\mathrm{E,st}}$ is defined as:

LV system	actually not defined, long-time flicker $P_{\mathrm{E,lt}} \leq 0{,}46$
MV system	$P_{\mathrm{E,st}} \leq 0{,}585$
HV (110-kV) system	$P_{\mathrm{E,st}} \leq 0{,}5$
EHV system	actually not defined, long-time flicker $P_{\mathrm{E,lt}} \leq 0{,}37$

12.4.4 Flicker Caused by Switching

Voltage changes due to switching can cause flicker, that is, change of light intensity of incandescent lamps. The flicker-value P depends on the frequency of voltage changes within 10 minutes (short-term flicker) or 2 hours (long-term flicker) and on the level of voltage change. The flicker-value is calculated based on the duration of the after-effect t_{f} of one single event of voltage change as per Equation 12.7 (time duration t_{f} in seconds).

$$t_{\mathrm{f}} = 2{,}3 \cdot \sqrt[3.2]{R \cdot F \cdot d} \tag{12.7}$$

with:

R correction factor ($R = 0.76$ for switching frequencies $r \leq 0.1\ \mathrm{min}^{-1}$; $R = 1.0$ for higher switching frequencies)
F form factor ($F = 1$ for step voltage changes)
d voltage change in%

The short-time and long-time flicker-values P_{st} and P_{lt} are calculated by Equation 12.8.

$$P_{\text{st}} = \sqrt[3.2]{\frac{\sum t_{\text{f}}}{600\ \text{sec}}} \tag{12.8a}$$

$$P_{\text{lt}} = \sqrt[3.2]{\frac{\sum t_{\text{f}}}{7200\ \text{sec}}} \tag{12.8b}$$

For voltage change $\Delta U_{\text{PCC}} \approx 2\%$, a minimum time delay between two switching actions of $t_{\min} \geq 3$ minutes is required, for changes of voltage $\Delta U_{\text{PCC}} < 2\%$, the minimum time delay (time in minutes) as per Equation 12.9 is to be applied.

$$t_{\min} = 23 \cdot (100 \cdot \Delta u)^3 \tag{12.9}$$

According to DIN EN 61400-21 (VDE 0127-2), the long-time flicker-value P_{lt} caused by switching can be determined from the permissible number of switching operations as per Equation 12.10. For a single generating plant, it follows:

$$P_{\text{lt}} = 8 \cdot (N_{120})^{0.31} \cdot k_{\psi,\text{F}} \frac{S_{\text{r,G}}}{S_{k,\text{PCC}}} \tag{12.10a}$$

and in case of several generation plants (number N):

$$P_{\text{lt},N} = \frac{8}{S_{k,\text{PCC}}} \left(\sum_{i=1}^{N} N_{120,i} \cdot \left(k_{i,\psi,\text{F}} \cdot S_{\text{r,G}}\right)^{3.2} \right)^{0.31} \tag{12.10b}$$

with:

N_{120} maximal number of switching operations within 2 hours
$k_{\psi,\text{F}}$ flicker-related current switching factor as per Equation 12.6c
S_{rG} rated apparent power of the generation plant

The calculation of the short-time flicker-value P_{st} due to switching makes no sense, due to the limited possible numbers of switching actions within 10 minutes.

The permissible values for short-time and long-time flicker-values are given in Section 12.4.5.

12.4.5
Flicker at Normal Operating Conditions

Flicker from generation units are to be assessed in accordance with IEC 61000-3-3, IEC/TS 61000-3-5, and IEC 61000-3-11 in case of connection to the LV system. The compatibility of the connection to the power system is documented by a conformity certificate of the manufacturer or by an independent certification institute. The emission limit values as per the different parts of IEC 61000-3 must be adhered to. For the connection of generation units to the MV system, it is recommended to take IEC/TR 61000-3-7 as basis for the assessment. Flicker emission owing to generation is mainly caused by wind-energy units due to the changing of generated power.

The flicker-values are calculated as per Equation 12.11.

$$P_{\mathrm{lt}} = c\frac{S_{\mathrm{n,E}}}{S_{\mathrm{k,PCC}}} \tag{12.11}$$

The nominal power $S_{\mathrm{n,E}}$ of the wind energy unit is to be set equal to the nominal apparent power of the generator $S_{\mathrm{n,G}}$. The flicker-coefficient c actually is only defined for wind-power units and is calculated as per Equation 12.12.

$$c = P_{\mathrm{lt}}\frac{S_{\mathrm{k,PCC}}}{S_{\mathrm{n,G}} * \cos(\psi_{\mathrm{k,PCC}} + \varphi_{\mathrm{f}})} \tag{12.12}$$

with:

$S_{\mathrm{k,PCC}}$ short-circuit power at the PCC
$\psi_{\mathrm{k,PCC}}$ phase-angle of system impedance at the PCC

Load changes with the phase-angle φ_{f} cause the highest flicker-value (see Equation 12.13). The flicker-coefficient c and the impedance angle $\psi_{\mathrm{k,PCC}}$ are indicated in the test certificate of the wind-energy unit.

$$\varphi_{\mathrm{f}} = \arctan\left(\frac{\Delta Q}{\Delta P}\right) \tag{12.13}$$

The long-time flicker-value of a wind energy plant with one unit is calculated as per Equation 12.14.

$$P_{\mathrm{lt}} = c \cdot \frac{S_{\mathrm{n,E}}}{S_{\mathrm{k,PCC}}} \cdot |\cos(\psi_{\mathrm{k,PCC}} + \varphi_{\mathrm{f}})| \tag{12.14}$$

The total flicker-value resulting from different generation units of different power rating or type of generation shall be calculated according to Equation 12.15.

$$P_{\mathrm{lt,tot}} = \sqrt{\sum_{i=1}^{N} P_{\mathrm{lt},i}^{2}} \tag{12.15}$$

A typical value of the flicker-coefficient is $c < 40$. The flicker-coefficients are calculated from measurements according to DIN EN 61400-21 (VDE 0127-21) for the different average wind-speeds $v_{\mathrm{a}} = 6\,\mathrm{m/s}$, $7.5\,\mathrm{m/s}$, $8.5\,\mathrm{m/s}$, $10\,\mathrm{m/s}$, applying a Rayleigh distribution of the wind speed and different impedance angles of the power system $\psi_{\mathrm{k,Ref}} = 30°, 50°, 70°, 85°$. A typical example of flicker-coefficients is given in Table 12.5.

Impedance angles and average wind speed can be interpolated linearly from the values as per Table 12.5.

Table 12.5 Flicker-coefficient determined in accordance with DIN EN 61000-21 (VDE 0127-21), Double-fed induction generator, $S_{r,G} = 1.5\,MVA$.

	Impedance angle			
Average wind speed	**30°**	**50°**	**70°**	**85°**
6.0 m/s	14.6	15.2	16.4	16.6
7.5 m/s	15.0	15.7	16.8	17.2
8.5 m/s	15.0	15.7	17.1	17.4
10.0 m/s	15.1	15.7	17.1	17.6

Permissible values of the long-time flicker-value P_{lt} at the PCC are given as below.

LV system	$P_{lt} \leq 0.46$
MV system	$P_{lt} \leq 0.46$
110-kV system	$P_{lt} \leq 0.35$; short-time flicker-value also defined $P_{st} \leq 0.5$
EHV system	$P_{lt} \leq 0.37$

Short-time flicker-values are actually only defined for 110-kV systems.

12.4.6 Harmonic and Interharmonic Currents and Voltages

12.4.6.1 LV and MV System

The connection of generation units shall not cause any system disturbances in the higher frequency range (harmonics and interharmonics). Special care has to be taken of this fact in case of the connection of generation equipped with power electronic converters, such as used in photovoltaic and wind-energy installations. The manufacturer has to furnish proof by a compatibility or conformity certificate or by any independent certification agency that the harmonic emissions comply with the different parts of IEC 61000.

In case of connection to the LV system, the standards as per IEC 61000-3-2 and IEC 61000-3-12 are to be considered. For the connection to the MV system, as well as in case of missing compatibility certificates, the permissible harmonic and interharmonic currents $I_{\nu,per}$ and $I_{\mu,per}$ must be determined as per Equation 12.16,

$$I_{\nu,per} = i_{\nu,per} \cdot S_{k,PCC} \quad \text{harmonics} \tag{12.16a}$$

$$I_{\mu,per} = i_{\mu,per} \cdot S_{k,PCC} \quad \text{interharmonics} \tag{12.16b}$$

with the relative current $i_{\nu,per}$ resp. $i_{\mu,per}$ (in ampere per unit of short-circuit power at the PCC) according to Table 12.6.

Table 12.6 Permissible relative harmonic and interharmonic currents (emission limits) for connection of generation units to LV and MV systems. Emission limits for voltages not indicated can be determined by linear interpolation.

Harmonic order ν Interharmonic order μ	Relative current in A/MVA			
	LV system	10-kV system	20-kV system	30-kV system
3	3	–	–	–
5	1.5	0.058	0.029	0.019
7	1	0.082	0.041	0.027
9	0.7	–	–	–
11	1.3	0.052	0.026	0.017
13	1	0.038	0.019	0.013
17	0.55	0.022	0.011	0.007
19	0.45	0.018	0.009	0.006
23	0.3	0.012	0.006	0.004
25	0.25	0.01	0.005	0.003
$25 < \nu < 40$	$0.25*25/\nu$	$0.01 \cdot 25/\nu$	$0.005 \cdot 25/\nu$	$0.003 \cdot 25/\nu$
Even order	$1.5/\nu$	$0.06/\nu$	$0.03/\nu$	$0.02/\nu$
$\mu < 40$	$1.5/\mu$	$0.06/\mu$	$0.03/\mu$	$0.02/\mu$
$\mu > 40$ (bandwidth 200 Hz)	$4.5/\mu$	$0.18/\mu$	$0.09/\mu$	$0.06/\mu$

The limit values as per Table 12.6 are determined for typical system configurations having already prevailing harmonic voltages. In the planning stage, detailed harmonic studies and measurements are recommended, measurements also after commissioning, to verify the analyses, especially if there is not much known about the prevailing system conditions. If several generation units are connected to one location (PCC) or in a closely limited system area of the same voltage level, the permissible harmonic and interharmonic currents (emission limits) of each unit shall only be a part of the harmonic and interharmonic emission depending on the part of generation as against the total system generation as per Equation 12.17.

$$I_{\nu,\mathrm{per}} = i_{\nu,\mathrm{per}} \cdot S_{k,\mathrm{PCC}} \cdot \frac{S_A}{S_0} \qquad \text{harmonics} \tag{12.17a}$$

$$I_{\mu,\mathrm{per}} = i_{\mu,\mathrm{per}} \cdot S_{\mathrm{k,PCC}} \cdot \frac{S_A}{S_0} \qquad \text{interharmonics} \tag{12.17b}$$

With:

S_A rated power of the generation units

S_0 maximal generation power to be connected to one substation (PCC) or in a limited system area

Table 12.7 Permissible relative harmonic and interharmonic voltages in MV systems caused by all generation plants.

Harmonic order ν Interharmonic order μ	Permissible voltage in % related to $U_n/\sqrt{3}$
5	0.5
7	1.0
11	1.0
13	0.85
17	0.65
19	0.6
23	0.5
25	0.4
> 25 odd order	0.4
> 25 even order	0.1
$\mu < 40$	0.1
$\mu > 40$ (bandwidth 200 Hz)	0.3

The emission limits as per Table 12.6 are determined assuming an ohmic-inductive impedance at the PCC. In due consideration of the parallel resonances in MV systems (see Section 12.2.2.3), the harmonic and interharmonic voltages, caused by the emission of harmonic and interharmonic currents of all generation plants in the closely limited system area of the same voltage level, shall not exceed the limit values as per Table 12.7.

12.4.6.2 Generation Connected to HV and EHV System

For the connection to HV systems, the permissible harmonic and interharmonic currents shall be kept below the limit values as per Equation 12.18; likewise, this applies to interharmonic currents, with the relative currents (in ampere per unit of short-circuit power at the PCC) as per Table 12.8.

$$I_{\nu,\mathrm{per}} = i_{\nu,\mathrm{per}} \cdot S_{\mathrm{k,PCC}} \cdot \frac{S_A}{S_0} \qquad \text{für } \nu \le 13 \tag{12.18a}$$

$$I_{\nu,\mathrm{per}} = i_{\nu,\mathrm{per}} \cdot S_{\mathrm{k,PCC}} \cdot \sqrt{\frac{S_A}{S_0}} \qquad \text{für } \nu > 13 \tag{12.18b}$$

Harmonics and interharmonics of the order $\nu > 13$; $\mu > 13$ (frequencies above 650 Hz) need not be considered, if the maximal power of one generation unit is smaller than 1% of the system short-circuit power.

The emission limits as per Table 12.8 are determined assuming an ohmic-inductive impedance at the PCC. In order to take due account of the parallel resonances in HV and EHV systems (see Section 12.2.2.3), a resonance factor is introduced for the 110-kV system (HV system) as per Table 12.9 and Equation 12.19.

Table 12.8 Relative harmonic and interharmonic currents (emission limits) for the connection of generation plants to HV and EHV systems.

Harmonic order ν Interharmonic order μ	Relative current in A/GVA		
	HV system	EHV systems	
	110-kV system	220-kV system	380-kV system
5	2.6	1.3	0.74
7	3.75	1.9	1.1
11	2.4	1.2	0.68
13	1.6	0.8	0.46
17	0.92	0.46	0.26
19	0.7	0.35	0.2
23	0.46	0.23	0.13
25	0.32	0.16	0.09
>25 or even order	5.25/ν	2.6/ν	1.5/ν
μ < 40	5.25/μ	2.6/μ	1.5/μ
μ > 40 (bandwidth 200 Hz)	16/μ	8/μ	4.5/μ

Table 12.9 Resonance factor for the calculation of permissible emission limits in HV systems.

Harmonic order ν (also appl. to interharmonics μ)	110-kV system (HV system)	
	Mainly cables	Mainly overhead lines
$\nu < (\nu_{res} - 2)$	1	1
$(\nu_{res} - 2) \leq \nu \leq (\nu_{res} + 2)$	1.5 . . . 2.5	2 . . . 3
$\nu > (\nu_{res} + 2)$	1	1

$$I_{\nu,\mathrm{per}} = i_{\nu,\mathrm{per}} \cdot S_{\mathrm{k,PCC}} \cdot \frac{1}{k_\nu} \tag{12.19a}$$

$$I_{\mu,\mathrm{per}} = i_{\mu,\mathrm{per}} \cdot S_{\mathrm{k,PCC}} \cdot \frac{1}{k_\mu} \tag{12.19b}$$

In EHV systems (220 kV and 380 kV), the harmonic and interharmonic voltages caused by the emission of harmonic and interharmonic currents of all generation plants in closely limited system areas of the same voltage level shall not exceed the limit values as per Table 12.10.

The method and limit values explained earlier consider only the superposition of harmonics and interharmonics emitted from generation units at one PCC or

Table 12.10 Permissible relative harmonic and interharmonic voltages in EHV systems caused by all generation plants.

Harmonic order ν **Interharmonic order** μ	**Permissible voltage in % related to** $U_n/\sqrt{3}$
5	0.25
7	0.5
11	0.5
13	0.4
17	0.3
19	0.25
23	0.2
25	0.15
> 25	0.1
$\mu < 40$	0.1
$\mu > 40$ (bandwidth 200 Hz)	0.3

in a limited system area of the same voltage level. If a high amount of generation is connected to different substations of a power system, the simplified approach is no longer valid. Detailed investigations by means of harmonic studies and measurements are to be carried out in this case (Section 12.7).

12.4.6.3 Superposition of Harmonics and Interharmonics

Harmonic and interharmonic currents emitted by different generation units are superposed in the following way:

- Line-commutated converter (pulse number p):

The characteristic harmonic currents ($\nu = n \cdot p \pm 1$) and the noncharacteristic harmonic currents $\nu < 7$ are superposed in an arithmetic manner as per Equation 12.20a.

$$I_\nu = \sum_i I_{\nu,i} \tag{12.20a}$$

The noncharacteristic harmonic currents $\nu > 7$ are superposed in an euclidic manner as per Equation 12.20b.

$$I_\nu = \sqrt{\sum_i I_{\nu,i}^2} \tag{12.20b}$$

- Self-commutated converters:

The harmonic currents $\nu < 11$ are superposed as per Equation 12.20a. The harmonic currents $\nu > 11$ and the interharmonic currents are superposed as per Equation 12.20b.

12.4.7 Asymmetry and Voltage Unbalance

Single-phase connection of generation plants is permitted only in LV networks. The single-phase connection is permissible, if

- the asymmetrical power of all generating units $S_{uns,max}$ remains equal or below 4.6 kVA and
- the total power of all single-phase generation units of all three phases at one PCC remains equal or below 13.8 kVA.

Equipment with rated power $S_r > 13.8$ kVA are to be connected in a three-phase scheme. It should be ensured that the asymmetry of the voltage in the system remains below 2%.

12.4.8 Commutation Dips

Commutation dips occur only in case of line-commutated converters, which are rarely used in the field of renewable energy generation units. In the worst operating condition, the level of the commutation dip related to the peak value of the rated voltage shall not exceed the value of 5%. Details can be found in [73].

12.4.9 Effects on Ripple-Control and Line-Carrier Systems

Ripple-control and line-carrier systems are operated in the frequency range of f = 110 Hz . . . 2 kHz, the signal voltage is modulated depending on the method and frequency with a voltage level typically between 1% and 4%, sometimes up to 9%. The duration of the ripple-control telegram is between 6.6 seconds and a few minutes.

Relevant for the influence of ripple-control systems is the signal or noise level. The interharmonic voltages close to the ripple-control frequency shall be limited to 0.2%, the induced noise voltage of a single network user shall be limited to less than 0.1%.

If pulse-width modulated inverters, typically for wind energy plants and photovoltaic plants, are in operation in power systems, an influence on ripple-control signals is not expected if the frequency range between 70% and 130% of the chopping frequency or multiples thereof is avoided. If the ratio of short-circuit power $S_{k,PCC}$ at the PCC and rated power $S_{r,G}$ of the generation plant is above 1000, any influence on ripple-control systems can be neglected.

Effects on line-carrier signals are not expected in one but only case. Any reactive power compensation by capacitors in the connecting circuit will reduce the impedance at the PCC and might lead to a reduction of line-carrier signals (and ripple-control signals as well). Suitable countermeasures, such as blocking devices, have to be installed in this case.

13
Protection of Equipment and Power System Installations

13.1
Faults and Disturbances

Electrical equipment has to be designed and constructed in such a way as to withstand the foreseeable loading during its lifetime under normal and emergency conditions. Generally it is economically not meaningful and technically not realistic to design equipment for all loadings and disturbances. Among others factors, the following should be mentioned:

- Unforeseeable site and ambient conditions such as flooding of basements where cables are laid
- External influences such as mechanical damage by construction work
- Atmospheric influences such as lightning strokes in line conductors and structures
- Aging and loss of dielectric strength of non-self-healing insulation, for example, oil-impregnated paper
- Internal influences such as short-circuits due to insulation failure.

It is therefore necessary to install devices for the protection of equipment which limit the effects of unforeseeable faults and loading on the equipment and protect it against cascading damage. These protection devices must be capable of differentiating between normal and disturbed operating conditions and they must operate reliably to isolate the damaged or endangered equipment as soon as possible from the power supply. It is not the task of the protective devices to avoid errors and disturbances. This can be achieved only by careful planning of the power system, by thorough project engineering of the equipment and by appropriate operation. Protective devices are to fulfill the following four conditions ("Four S criteria"):

- *Selectivity:* Protective devices shall switch-off only that equipment affected by the system fault or impermissible loading condition, the nonfaulted equipment shall remain in operation.
- *Sensitivity:* Protective devices must be able to distinguish clearly between normal and impermissible operating conditions or faults. Permissible high loading of equipment during emergency operation and small short-circuit currents are to be handled in a different way.

Power System Engineering: Planning, Design and Operation of Power Systems and Equipment, Second Edition.
Jürgen Schlabbach and Karl-Heinz Rofalski.

- *Speed:* Protective devices are to switch-off the faulted equipment from the power supply as soon as possible in order to limit the effects of the short-circuit or impermissible loading.
- *Security:* Protective devices with all their associated components such as transducers, cable connections, wiring and trip circuits must operate safely and reliably. As faults in power systems are comparatively rare, protective devices must continue to be able to fulfill their function after many years of stand-by operation.

The design and selection of the protection devices must be economically appropriate in relation to the equipment to be protected in terms of its importance for a safe and reliable power supply to the consumers. In general the protection schemes to be designed should be less oriented on the different voltage levels and more oriented on the tasks the equipment and the system structure have to fulfill in the power system.

As an example, a 220 kV overhead line and a 400 kV overhead line may be of the same importance if they are serving the same purpose of supporting the countrywide transmission system. If the 220 kV line is being used only for the regional subtransmission system, its importance is more or less similar to that of a 132 kV or 110 kV overhead line. Furthermore, the amount and importance of load or the amount of energy not supplied during outages determine the importance of the equipment and hence the type and structure of the protection concept (see Figure 3.1).

Relevant aspects of security must be considered. In case of an earth fault in a MV system, the faulted equipment needs to be switched off immediately from the power supply by the protection. The resulting voltage increase of the nonfaulted phases and the increased voltage at neutrals and neutral conductors may endanger maintenance staff and must be supervised until the fault has been switched off.

If several protective devices are to be integrated into a complex protection concept, the individual tasks and operation modes of the protective devices have to be coordinated to ensure reliable operation in case of faults and to provide backup operation in case of failure of one protection device.

It is strongly recommended to collect all information concerning faults and outages by means of statistical methods. Fault types, outage duration, fault locations, type of the faulted equipment, voltage level and other information collected and assessed in a suitable way can be used as a basis for reliability-based maintenance of equipment.

13.2 Criteria for Operation of Protection Devices

Protection devices and concepts are to be designed with regard to the faults and disturbances of the equipment to be protected. A list of possible faults and the related criteria and protective functions are summarized in Table 13.1. Additionally the criterion of "time" has to be considered; this is relevant, for example, in the minimum fault-clearing time for the determination of short-circuit withstand capability or for setting the backup-time of remote protection devices.

Table 13.1 Criteria, faults and operation criteria for protective devices [75].

Criteria	Operation criteria	Fault, disturbance	Protection of equipment
Current	Overcurrent	Short-circuit Overload	Overcurrent protection of lines, transformers, motors and generators
	Differential current	Short-circuit	Differential protection of lines, transformers and generators
Voltage	Voltage increase	Load-shedding	Overvoltage protection of lines, transformers and generators
	Neutral voltage	Earth-fault	Earth-fault protection of lines, transformers and generators
	Voltage dip	Motor start-up Short-circuit	Undervoltage protection of motors and generators
Impedance	Low impedance	Short-circuit	High-impedance protection of busbars
	Ratio R/X	Short-circuit	Distance protection of lines, transformers and generators
Power	Direction of power flow	Short-circuit	Directional overcurrent protection of lines Reverse power of generators
	Active power in zero-sequence component	Earth-fault	Earth-fault protection of lines
Frequency	Change of frequency	Short-circuit Loss of load Increase of load	Protection of generators General task of system protection
	Under-/over-frequency	Loss of load Increase of load	Protection of generators General task of system protection
Phase-angle	Change of phase-angle	Short-circuit	Vector relay of generators
Harmonics and high frequencies		Power electronic Short-circuit Earth fault	Protection of capacitors Rush stabilization Earth-fault protection
Temperature	Increase or decrease of temperature	Overload Short-circuit	Overload protection of transformers and cables
Arc	Radiation	Short-circuit	Protection of switchgear cubicles and switchgear rooms
Pressure	High pressure	Short-circuit	Protection of switchgear cubicles and switchgear rooms
	Low pressure	Leakage	Protection of gas-insulated cables and switchgear
Speed	Change of oil-flow in transformers	Overload Short-circuit	Oil-stream protection of transformers (pressure switch or Buchholz-protection)
Volume	Decrease	Leakage	Protection of equipment with oil-insulation

13.3
General Structure of Protective Systems; Transducers

Protective systems consist of several components whose reliable and coordinated operation ensures the intended function of the total system. Figure 13.1 shows the principal structure of a protective system as an example of distance or impedance protection.

Distance protection requires measurement of current and voltage, normally through current and voltage transformers or transducers. For adjustment of the phase-angle or adaptation of the transformation ratio, auxiliary transducers are necessary sometimes. The connection of the secondary side of the current and voltage transformers to the protection device is carried out with cables, and in modern protection devices also with glass-fiber cables. For the operation of the protective device, an auxiliary power supply is needed, which must be supplied in a secure and reliable way, usually from a battery powered directly from the power system. Trip-circuits between the protection device and the circuit-breaker or load disconnecting switch consist of wiring (cables), auxiliary relays and circuits and trip-coil. Auxiliary power is also needed for the operation of the breaker or switch, usually available directly from the current transducer, from a separate AC voltage source or from a capacitor release unit. The trip signal thus does not activate the switchgear directly, but through a relay, which initiates the actual trip release by control of the trip-coil or by release of pneumatic, hydraulic or spring-coil control devices. This kind of trip-release is called secondary relay release, indicated in Figure 13.2a. Interface arrangements need to be implemented for

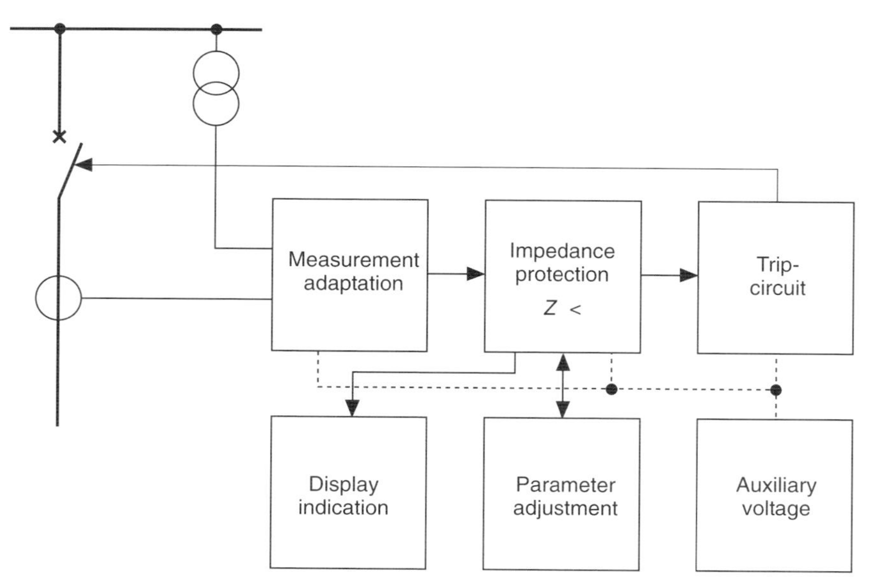

Figure 13.1 Principle of distance or impedance protection.

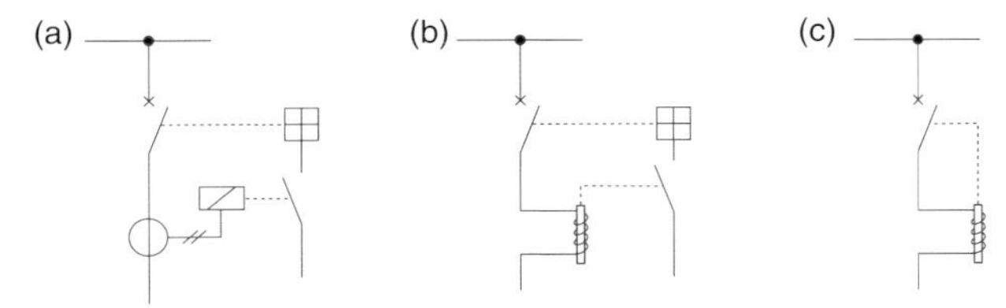

Figure 13.2 Different types of trip-release. (a) Secondary relay; (b) primary trip-relay; (c) primary relay.

indication of the operational status of the protection device, and these can be used for remote parameter setting and for the selection and remote transmission of internal information of the protective system, for example, for the clearance of disturbances.

Protective systems supervising the operation of circuit-breakers directly or by an auxiliary relay are called primary trip-relays or primary relays, as represented in Figure 13.2b and c. They are used today only in MV systems. Their disadvantages are the coupling between high-voltage circuit and secondary technology, the difficulty of maintenance checking during operation, the lack of adaptation of different currents and voltages to the desired measuring system, and increased thermal and dynamic stress on the protective device.

The arrangement, number and connection of current and voltage transformers are determined by the need for the protection. In the case of undirectional overcurrent protection, only current transformers are needed, whereas for directional overcurrent protection or impedance protection, additional voltage transformers are needed. For overcurrent protection in systems with isolated neutral, current transformers are needed in two phases only, as the single-phase earth fault cannot be detected by the overcurrent protection due to the small current; only two-phase and three-phase short-circuits will be detected by overcurrent protection. Three current transformers are required for protection of HV lines with single-phase autoreclosing as single-phase faults in any of the three phases need to be detected by the protection. If fault currents through earth are to be detected, current transformers are required in three phases in Holmgreen arrangement, or a separate current transformer is needed in the neutral. Differential protection schemes require current transformers at the sending and receiving ends of the equipment (line or transformer) to be protected. Figure 13.3 shows possible arrangements of current transformers.

Voltage transformers are needed in protection schemes if either the current direction, the power-flow, the impedance, the voltage or the power frequency is to be measured. For measurement of the line-to-earth voltage, line-to-earth voltage transformers are also necessary, whereas in medium-voltage systems two voltage transformers are used measuring line-to-line voltage (V-arrangement). The third line-to-line voltage can be determined by special arrangement of the secondary side or can be calculated from the two measured line-to-line voltages. The line-to-earth voltages can be determined correctly only in the case of equally loaded

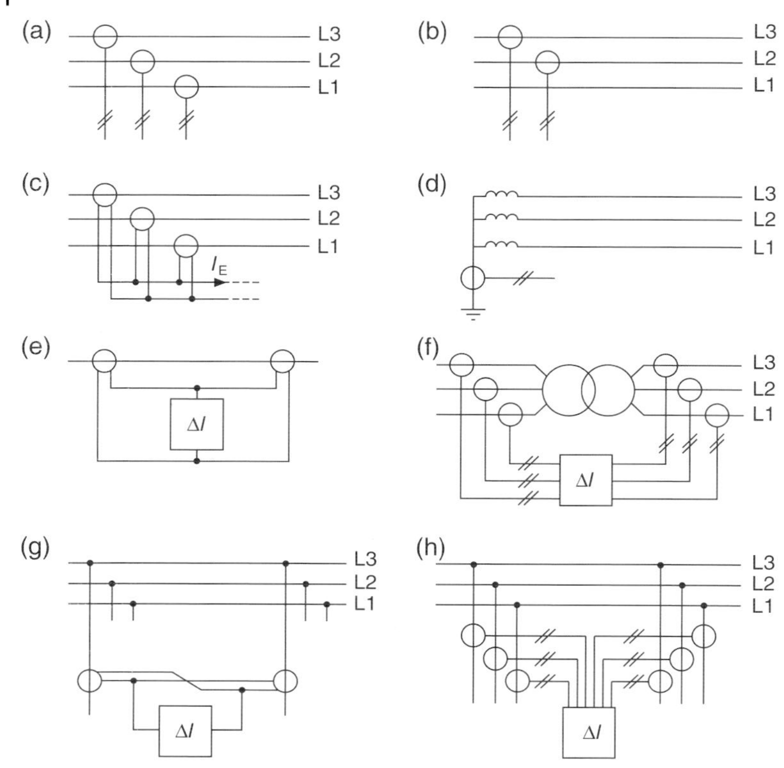

Figure 13.3 Arrangement of current transformers (switchgear not shown). (a) Current transformer in a HV system with low-impedance neutral earthing and single-pole autoreclosing; (b) current transformer in an MV system with ground-fault compensation; (c) Holmgreen-arrangement of current transformers; (d) current transformer in the neutral connection; (e) current transformer for longitudinal comparison protection; (f) current transformer for differential protection of a transformer (longitudinal comparison); (g) current transformer for transverse comparison protection; (h) current transformer for busbar protection (transverse comparison).

transformers. The voltage of the neutral can be measured by a separate voltage transformer provided in the neutral or can be obtained from the open secondary circuit of a delta-winding of three single-phase voltage transformers. Figure 13.4 outlines possible arrangements of voltage transformers.

13.4 Protection of Equipment

Selection of the protection scheme and the protection device for equipment depends on the types of faults, the type of power system and the handling of the

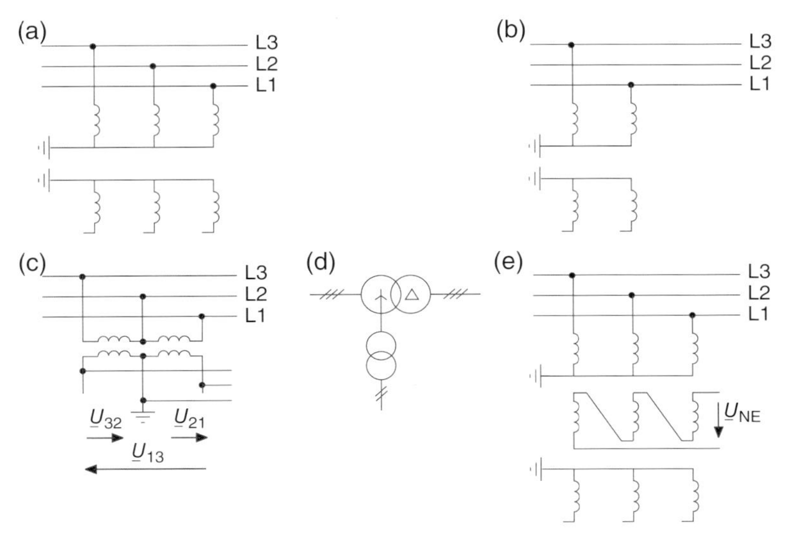

Figure 13.4 Typical arrangements of voltage transformers (switchgear not indicated). (a) Three single-phase voltage transformers in the HV system; (b) two single-phase voltage transformers in the MV system; (c) V-arrangement; (d) voltage transformer in the neutral point of a transformer; (e) single-phase voltage transformers with auxiliary winding in delta-connection.

neutral, the voltage level, the type of load to be supplied and the required selectivity. In setting the protection, the threshold value at which the operation of the protection device changes from stand-by modus into the working mode must be higher than the relapse value at which the protection device changes back into the stand-by mode. The relationship of the two values is called the threshold ratio, which for electromechanical relays is between 0.7 and 0.85, for electronic and digital devices between 0.9 and 0.95, and for special applications up to 0.99. The measured value during normal operation (for example, of current in case of an overcurrent relay) must be below the relapse value.

13.5 Protection of Lines (Overhead Lines and Cables)

13.5.1 General

Normally the protection of lines (overhead lines and cables) is implemented as short-circuit protection, which is not suitable for protection against overload. The selection of the protection device depends on the kind and the mode of operation of the power system (radial system, meshed system etc.). Table 13.2 indicates recommended protection for lines in HV transmission systems.

Table 13.2 Selection of protective measures for cables and overhead lines in HV-systems.

Protection task	Cables	Overhead lines	Remarks
Overcurrent protection	Necessary	Necessary	Directional and undirectional protection
Distance protection	Recommended in HV systems	Recommended in MV and HV systems	Necessary in meshed systems
Automatic reclosure	Not necessary	Recommended in MV and HV systems	Three-phase in MV systems Single-phase in HV systems
Earth-fault protection	Necessary	Necessary	In systems with isolated neutral or in case of resonance earthing
Differential protection	Necessary for short cables	Recommended in EHV and HV systems	Distance protection and overcurrent protection required additionally
Overload protection	Recommended	Recommended	Depending on the importance of the line

13.5.2
Overcurrent Protection

In power systems with isolated neutrals or with resonance earthing (earth-fault compensation) measurement of currents in two phases is adequate; however, one obtains better selectivity, as well as shorter operation time, if all three phase currents are measured in the case of double ground faults (line-to-line short-circuit with earth connection). Power systems with low-impedance earthing require the measurement of the three phase currents. Power systems with current-limiting neutral grounding (the single-phase short-circuit current is limited in this case to e.g. 2 kA) require an additional measurement of the earth-current by measurement of the total current sum or a current measurement in the neutral, in order to increase the sensitivity of the protection.

Radial power systems can easily be protected with independent maximum current time protection (UMZ) or with overcurrent relays having inverse time characteristic. The selectivity of the protection is obtained by progressive grading of the trip time in the individual substations such that the trip time increases toward the feeding substations as outlined in Figure 13.5a). Typical grading time is ~300 ms when electronic or digital protective devices and modern circuit-breakers are installed. The increase of the trip time can be avoided by reverse

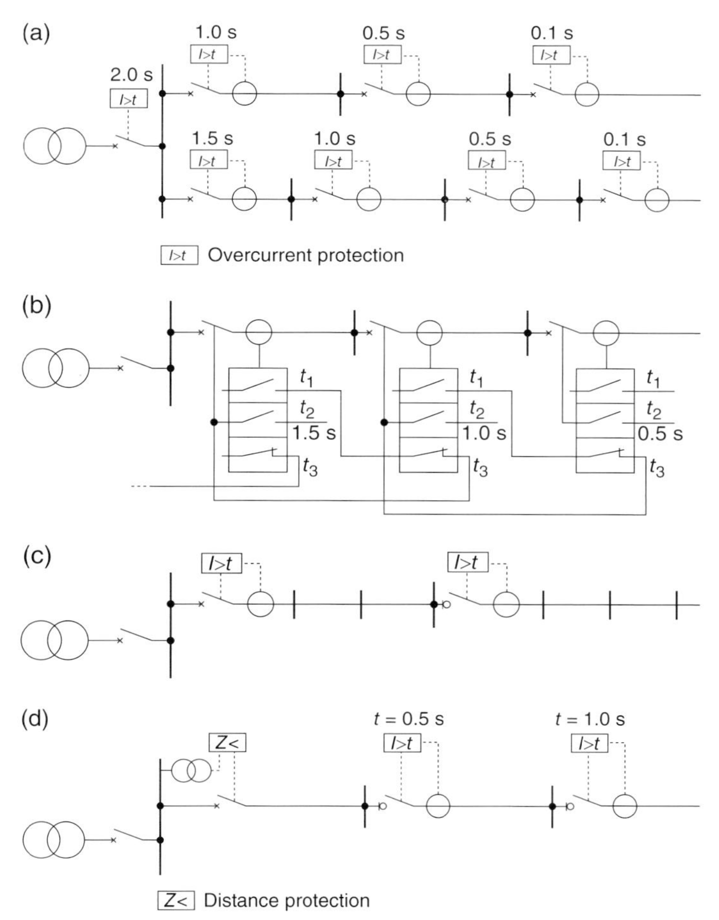

Figure 13.5 Overcurrent protection (UMZ protection) in radial networks. (a) Principal arrangement (circuit-breaker in each substation); (b) reverse interlocking scheme (t_1 = 200 ms; t_2, relay time; t_3, instantaneous tripping); (c) simple interval time connection; (d) graded interval time connection.

interlocking of the UMZ protection. This is achieved by interlocking of the tripping command of each circuit-breaker of each station with the status of the overcurrent relay in the next (downstream) substation (Figure 13.5b). Trip times in each substation can be realized in the range 300–500 ms. Reserve protection is given with both concepts by the upstream protective devices.

In radial and meshed systems in the medium-voltage range, the individual feeders are often equipped only with load-break switches and only the feeding transformer with a circuit-breaker (see Figure 13.5c). In case of a short-circuit on any feeder, the circuit-breaker opens first and afterwards the load-break switch related to the faulted feeder is operated. The circuit-breaker is reclosed after an interval of ~500 ms. Thus all faults are switched off selectively in the same interval. If several load-break switches are installed, one following another, impedance protection for the circuit-breaker and a time delay for the individual load-break switches are needed (see Figure 13.5d).

If lines are fed in the case of a short-circuit by two sides, as is possible in meshed systems or in systems with feeding from remote stations or in the case of parallel lines (see Chapter 5), a selective disconnection of the faulted line is only possible with the assessment of the direction of the energy flow. Voltage transformers are needed. If directional and unidirectional overcurrent devices are combined for the protection scheme, the number of devices can be reduced if a suitable grading of the trip time is realized, as represented in Figure 13.6.

Overcurrent protection of the UMZ type can also be used for the connection of the public system and industrial power systems having own generation, as outlined in Figure 13.7. In the case of faults in the public power system, the circuit-breaker is to be switched off immediately or with small delay, in order not to endanger reliable supply of the industrial system. In the case of faults in the industrial system, however, the circuit-breaker is to be operated with time delay, in order to ensure time for the disconnection of the fault by the protection device of the industrial system.

Overcurrent protection of the UMZ type has several disadvantages, such as increase of trip-time with increasing current, the lack of selectivity in meshed systems and the tendency to malfunction in power systems having multilateral infeed. UMZ protection therefore is a typical overcurrent protection used in urban MV systems, which are operated as ring-main or radial systems.

The method of protection with inverse time characteristic overcurrent relays is not dealt with in this book.

13.5.3
Distance (Impedance) Protection

Distance protection or impedance protection devices use current and voltage measurements at the location of the protective device to determine the location of the fault by assessment of the measured impedance. The measured impedance is compared with the impedance of the line, including existing current-limiting reactors or blocking reactors of carrier frequency signaling. If the measured impedance is below adjustable limit values, the associated circuit-breaker is released immediately or with a time delay. Each line section is switched off selectively by the next distance protection device in shortest time. The remote backup function is given by distance protection devices in the next line section; local backup is realized by other independent protection. Distance protection devices are used both for primary protection of lines and as backup protection of busbars and transformers. The basic scheme of distance protection is outlined in Figure 13.8.

Overcurrent excitation of the protection is normally sufficient. However, if the short-circuit current is in the same range as the normal operating current or if selective recognition of the faulted phase is desired, an underimpedance excitation is additionally necessary. Depending upon the type of distance protection, the

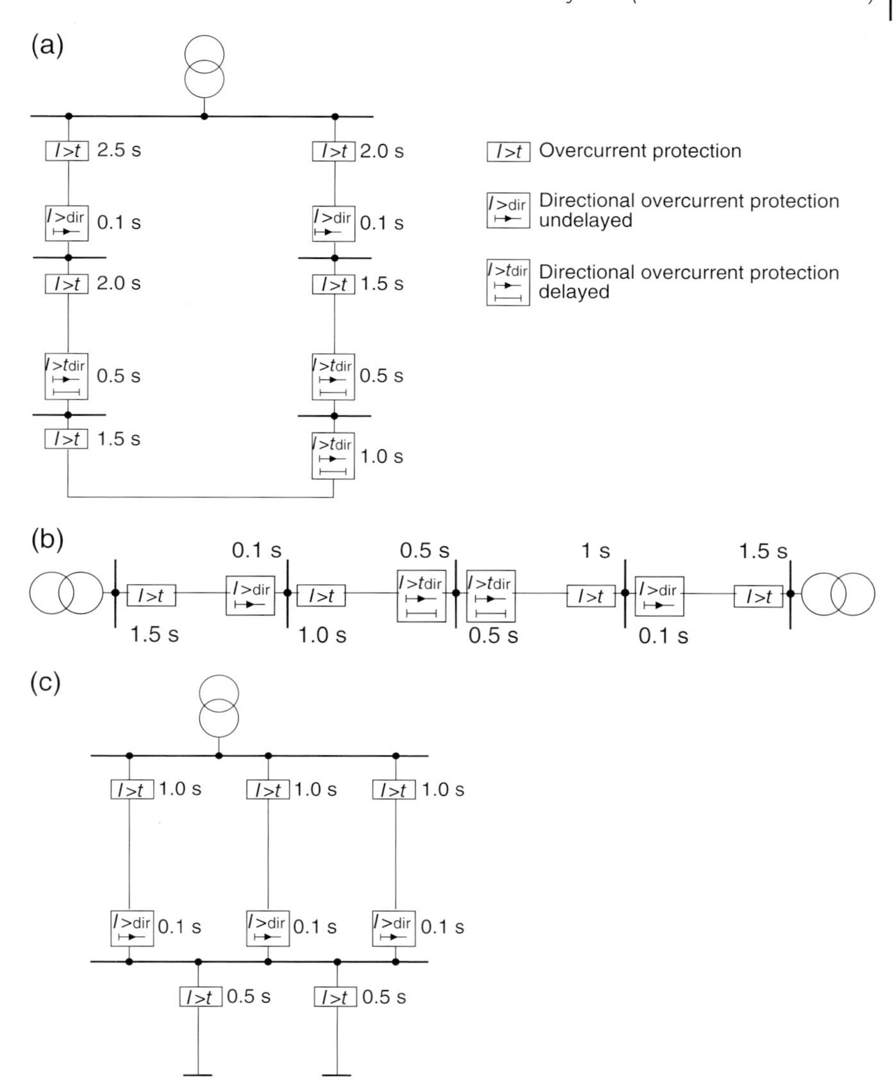

Figure 13.6 Directional overcurrent (UMZ) protection with circuit-breaker in each substation. (a) Ring-main system; (b) system fed from remote station; (c) parallel-operated lines.

measured currents and voltages are selected from a specially designed circuit. The actual protective function is then activated by an impedance measuring unit (the fault is within a certain distance of the location of the protection device) and a direction measurement unit (fault toward the busbar or toward the line). Parallel to the selection and measurement, a timer is activated, so that by combination of the distance measurement and the decision of direction for each line section, a time-dependent release of the associated circuit-breaker can be made. If the fault

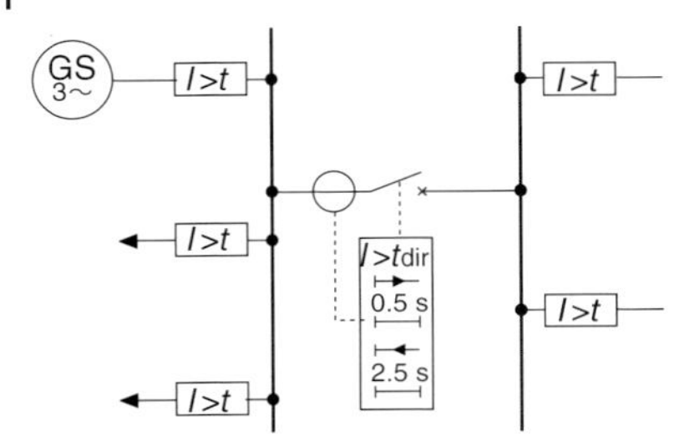

Figure 13.7 Directional overcurrent (UMZ) protection in the connection of public and industrial power system.

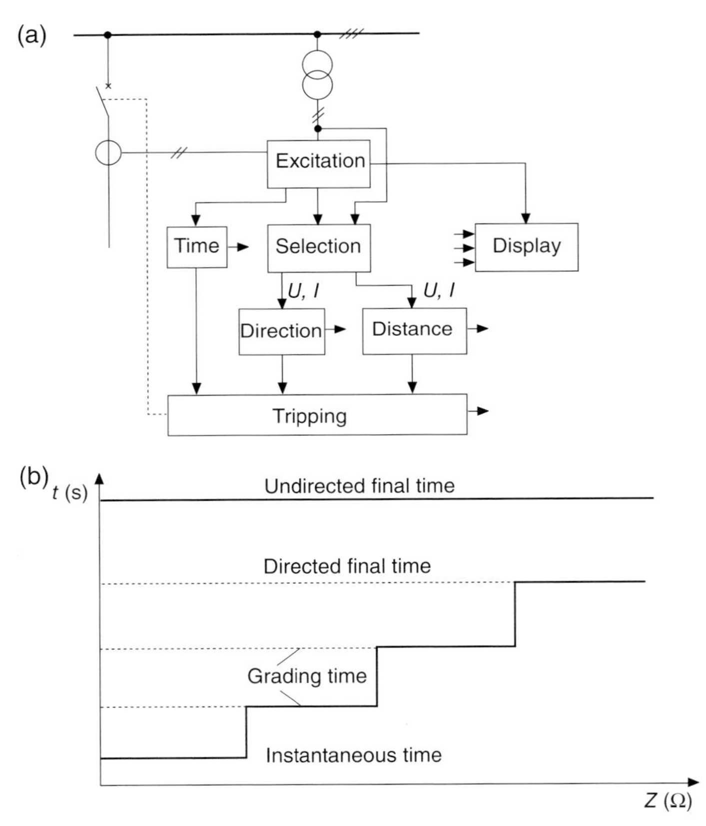

Figure 13.8 Basic scheme of distance protection. (a) Block diagram; (b) grading of operating times.

persists, because of failure of the distance protection or of the associated circuit-breaker, the other distance protection devices installed in the power system remain in the activated condition. The fault is then switched off by circuit-breakers in other substations with an adjustable time delay in the so-called grading time. Pos-

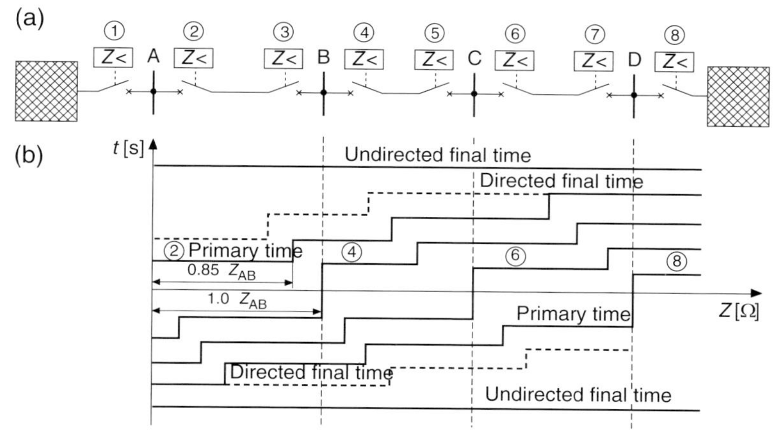

Figure 13.9 Progressive time grading of distance protection. (a) Circuit diagram; (b) impedance-time characteristic.

sible grading times depend on the operating time of the protective device, the minimal clearing time of the circuit-breaker, the quenching time of the arc in case of faults in air and on the necessary threshold value. In HV systems grading times are 300–500 ms; in MV systems and in systems having different protection and switchgear the grading time has to be set to a higher value than in protection schemes with identical protection devices and circuit-breakers. Several grading times as well as directional and unidirectional final or end times, during which the fault is switched off, are adjustable independently of the fault direction. Figure 13.9 outlines the principle of progressive grading time for distance protection devices.

The number of current and voltage transformers for distance protection depends upon the type of the neutral treatment and the type of autoreclosure. In HV transmission systems one should generally use distance protection devices with three measuring systems: thus three voltage and three current transformers are necessary.

The determination of the setting values of the distance protection devices for single-phase faults (short-circuits) is difficult because the zero-sequence impedances depend on other circuits (cables, pipes, overhead lines) installed in parallel to the line concerned, on the grounding conditions and also on the soil characteristics, see Chapter 14 for details. The zero-sequence impedances should be measured during commissioning of HV transmission lines. In order to allow adjustment of the parameters of the protection, these measurements should be repeated if cable and pipe construction work is carried out in the proximity of the lines.

Measuring errors, inaccuracies in the line data and very frequent disturbances of the measured currents and voltages reduce the accuracy of the protection devices and have to be taken into account when determining the parameter setting of the relay. Parameters for the primary time of the distance protection are set to cover approximately 80–90% of the line section; the remaining length of the section is protected by the first grading time only. If the total line section is to be protected in primary time, then interlocking of the distance protection devices at both ends of the line section has to be implemented. The circuit-breaker at the opposite end of the line section is then released in primary time in each case if one of the two distance protection devices operates in primary time. Thus an almost simultaneous opening of the circuit-breakers is possible, an important condition for successful autoreclosing in HV transmission systems.

In MV systems overlap switching in combination with autoreclosure can be used, which is much simpler to implement. The primary time is set to 130–150% of the line length. During the recovery time of the autoreclosing sequence the setting is put back to 80–90% of the length in order to enable selective disconnection by the remote protection in case of unsuccessful autoreclosing.

Circuits in urban HV systems have a comparatively short length, causing problems in selective switch-off in primary time due to measuring errors. If direction comparison of the status of the distance protection at both ends of the line section is introduced, the direction measurement of the two protection devices can be compared. The operating range is extended beyond the line ends and corresponds almost to the operating range of the overlap switching. The fault is switched off only if the direction measurements of both protective devices indicate the same fault location or direction. The reserve protection is here given by the distance protection device itself.

Distance protection is not applicable if the fault location cannot be determined from the comparison of measured current and voltage, which is the case with T-offs of lines, as outlined in Figure 13.10a. The current at the location of the distance protection device in case of faults on the T-off branch depends on the relationship of the currents in the substations A and B; in each case, however, the measured impedance is higher than the actual impedance. This error cannot

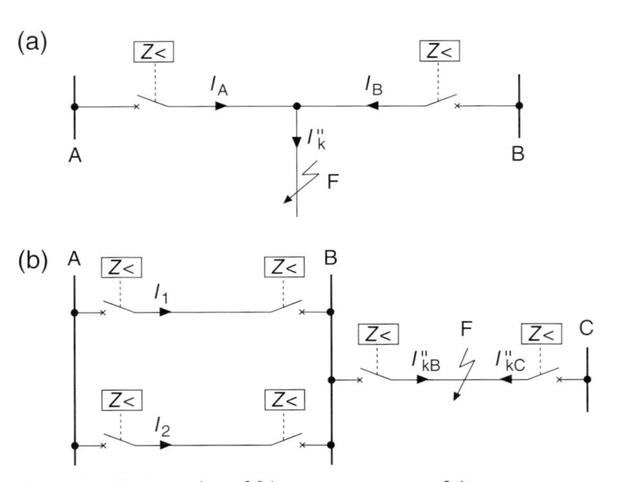

Figure 13.10 Examples of false measurement of distance protection devices. (a) T-off line; (b) double-circuit line.

be considered in the setting of the protection, since the relationship of the currents can change during operation. An improvement can only be reached by signal transmission of the currents and the correction of the measured current by means of suitable algorithms.

Protection setting of lines fed by double-circuit lines (as outlined in Figure 13.10b) have to take into account that only one circuit of the double-circuit line is in operation. For the fault case considered, the protection in substation A measures the distance as too long (equivalent to the impedance). The grading time has to be set to a higher value in this case.

13.5.4 Differential Protection of Lines

Differential protection can be installed in cases where UMZ type distance protection is not operating selectively or if very short operating times are required. Different types of differential protection are used. In general the currents measured at both line ends are compared. Information is exchanged by auxiliary cables or by fiber-optic cables. If the instantaneous values of the currents are compared, the amplitude and the phase-angle of the currents have to be measured, while for the phase-comparison protection, phase-angle measurement of the current zero crossing is sufficient. Digital differential protection devices additionally assess the gradient of current change. Thus the sensitivity of the protection for faults with high impedance can be improved and the dependence on the accuracy of the current measurement can be reduced. Differential protection devices offer no reserve protection in themselves and can be used only as primary protection. The section between the location of the current transformers (typically between circuit-breakers and line; see Figure 6.5) and the busbar remains unprotected. To cover this gap, distance protection or UMZ protection should be installed, which can also be used as reserve protection for the differential protection of the line.

13.5.5 Ground-Fault Protection

Ground-fault protection is used in MV and HV transmission systems (for example with nominal voltages 10 kV to 132 kV) having ground-fault compensation or isolated neutral. Since ground faults (single-phase to earth) do not cause short-circuit currents, an immediate disconnection of the fault is not necessary. The main task of ground fault protection therefore is the detection and signaling of the fault.

In general the detection of a ground fault is realized by means of the delta-winding of a potential transformer or by measurement of the voltage in the neutral (see Figure 13.4). The setting of the protection must be done in such a way that the displacement voltage in the neutral under normal operating conditions does not lead to a ground-fault signaling [76]. If transient ground faults are not to be signaled, the signal emission has to be delayed. The direction of the ground fault with regard to the location of the protection device is determined from the earth-fault current and the displacement voltage by determination of the active energy flow direction by watt-metric devices. By assessing the indications of several ground fault protection devices, the fault location can be found.

In power systems with isolated neutral, the reactive power is measured by means of a "sin φ" measurement. In case of an earth fault, the current in the faulted feeder is the superposition of the total earth fault current and the branch current of the faulted feeder. All other feeders carry branch fault currents only. Since the directions of the current of the faulted feeder and the current of the feeders without fault have a phase displacement of 180°, the ground fault direction can be definitively detected. In a radial system, the faulted branch can easily be detected if the earth fault current is significantly larger as the fault currents in the feeders.

In power systems with resonance earthing (ground fault compensation) the direction of the active power is measured by the "cos φ" measurement method since the superposition of the capacitive earth fault current with the inductive current from the resonance coil does not permit clear assessment of the reactive component. The evaluation of the active component from the resonance coil and those parts supplied from the system feeders (relatively evenly distributed) always results in the highest value for the ground-faulted feeder. Since the ohmic part of the ground fault residual current amounts to only a small percentage of the earth fault current [68, 69], assessment of the ohmic currents is sometimes difficult. In this case, the ohmic part of the residual current is to be increased by an additional resistance in the neutral. As the assessment of the earth fault direction is more important than the measurement of the actual current in the case of ground fault, current transformers with small error in phase-angle measurement must be used, preferably slip-over type. Since only steady-state ground faults are captured, an economical arrangement can be used with only one ground-fault protective device at the busbar, which is switched successively to each feeder.

In systems with resonance earthing, only the power-frequency component of the ground-fault current is compensated. The fifth harmonic, which is normally the harmonic with the highest voltage in power systems, can be used for determination of the ground fault direction. The resonance coil has a high impedance for 250 Hz (five times the impedance for 50 Hz) and the capacitances of the feeders have a lower impedance than at 50 Hz (one-fifth of the impedance for 50 Hz). For 250 Hz the system can therefore be seen as one with isolated neutral. The ground fault direction is determined with the "sin φ" circuit.

With the occurence of the ground fault, a transient swing (current pulse) is initiated in the frequency range of some hundrets of hertz in cable systems and up to 2 kHz in overhead line systems. The phase-angle of the transient traveling wave resulting from the ignition oscillation with regard to the displacement voltage can be evaluated for the direction decision. The phase-angle of the current pulse refers at all measuring locations to the ground-fault location. Each feeder has to be equipped with a protective device, since only the unique occurrence of the ground-fault ignition is evaluated in this case. Further methods for ground-fault protection and ground-fault detection, such as pulse detection method, temporary grounding and current injection, are not dealt with in this book; details are outlined in references [75] and [76].

13.6
Protection of Transformers

13.6.1
General

Transformers are one of the most important items in power systems, having long repair times, non-self-healing insulation with voltage stress from internal and outside overvoltages, and large thermal time constants in case of overloading. The protection scheme of transformers must therefore take into account the special importance of the transformers for a reliable supply of the consumer load. Besides the measurements of current and voltage, other parameters such as oil temperature, flow rate of oil and gas concentration of the insulating oil are measured for protection purposes. Table 13.3 indicates the recommended protective measures for transformers of different power rating.

13.6.2
Differential Protection

Differential protection of transformers is arranged as longitudinal comparison protection as outlined in Figure 13.3e and f. The connection of the protection device to the current transformers must be done in such a way that comparison of the currents can be carried out with respect to r.m.s. value and phase-angle. Protection schemes of transformers with vector groups different from 0° additionally require intermediate transformers to correct the differences in phase-angle between the HV and LV sides, as outlined in Figure 13.11. With three-winding transformers (see Figure 7.2) all windings have to be included in the differential protection scheme. With transformers having vector group YNynd, which are not grounded permanently, intermediate transformers likewise have to be planned, in order to ensure selectivity and avoid malfunction in the case of single-phase short-circuits by measuring the partial short-circuit current through the neutral.

Inrush-currents occurring during energizing of transformers act as differential current for the differential protection. To avoid malfunction of the differential protection during energizing, rush-stabilization of the differential protection is needed, which analyses the 100 Hz-component in the inrush-current.

Differential protection is meaningful only if all windings of the transformer are connected to the power system by means of circuit-breakers. Differential protection of transformers therefore is scarcely used for transformers with rated power up to 2 MVA, generally only with larger ratings in the range above 10 MW. The differential protection represents a primary protection and offers no reserve protection.

Table 13.3 Selection of protection measures for transformers.

Protective measure	LV-transformers	HV-transformers	
	<2 MVA	<10 MVA	≥10 MVA
Overcurrent protection	Necessary, if fuses are not installed	Recommended	Necessary
Distance protection	Not necessary	Not necessary	Strongly recommended
Differential protection	Not necessary	Necessary in some cases	Necessary
Thermal protection	Not necessary	Not necessary	Recommended
Buchholz protection	Necessary in some cases	Necessary	Necessary
Earth-fault protection	Not necessary	Usual outside Europe	
Protection of cooling system	Necessary	Necessary	Necessary

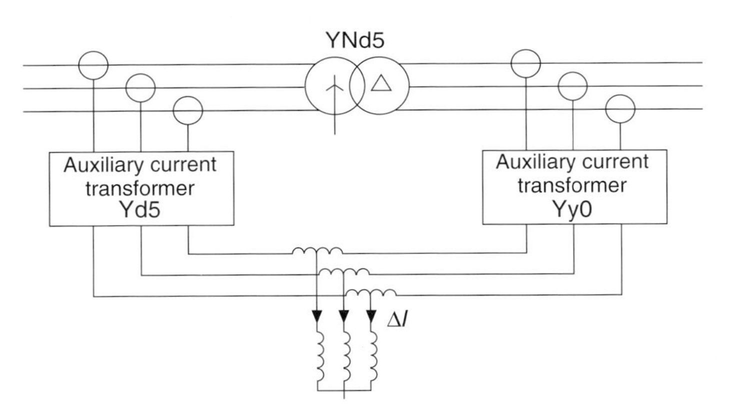

Figure 13.11 Arrangement of transformer differential protection.

13.6.3
Overcurrent Protection, Distance Protection, Ground-Fault Protection

Overcurrent protection is generally used for the protection of transformers if protection is not realized by fuses, as prevails with distribution transformers. Overcurrent protection is a backup protection for the differential protection and also a backup protection for the overcurrent protection of the line feeders fed from the transformer and the associated busbar. The operation time can become quite high due to the coordination of the grading time with the overcurrent protection of the lines; see also Section 13.5.2.

Monitoring of the temperatures of the oil and the transformer tank can likewise be regarded as overcurrent protection. Usually the temperature measurement does not release the circuit-breakers but is only signalized, since other causes than short-circuits can lead to the rise in temperature, which can then be reduced, for example, by suitable countermeasures such as reducing the transformer load.

If high grading time and as a result long time delays in operation of the overcurrent protection are to be avoided, distance protection must be selected as the backup protection concept, which depending on the type of fault can operate on the busbar side and on the transformer side at different times, thus also in primary time.

A separate ground-fault (earth-fault) protection is not used for transformers in German power systems. In other countries, ground-fault protection is applied for transformers with rated power above 10 MVA. As criteria for earth faults, one can evaluate the displacement voltage in the neutral and the earth fault current. Due to the existing unbalance of the power system and the displacement voltage arising thereby, it is difficult to achieve reliable operation of the protection, especially for faults near the transformer neutral, as the remaining voltage and thus the earth-fault current in case of faults at these locations becomes very small. This part of the winding can only be protected by additional measures, for example, by switching on a resistance to increase the fault current.

13.6.4
Buchholz Protection

An important primary protection for transformers is the Buchholz protection, a mechanical relay, used for oil-immersed transformers with rated power of 630 kVA and above. Buchholz protection is also used for auxiliary transformers in substations and power stations, for installations in explosive environments and for transformers in mining installations. Buchholz protection is released by the oil current-flow between the transformer tank and the oil expansion tank or radiators as well as by the oil level. It also works as a warning device in case of slow oil loss, for example by leakage, or during gasification in case of faults with low fault currents. With sudden oil loss or with strong gas bubble formation or large oil flow rates, for example following faults with high currents, the Buchholz protection releases the circuit-breakers on the HV and LV sides of the transformer. If an onload tap-changer is installed with separate oil-insulated housing, a separate Buchholz-protection is needed for the tap-changer. The time delay of the Buchholz protection is in the range 20–60 ms. The Buchholz protection responds only to faults inside the transformer tank and represents no backup protection for other protective devices.

13.7 Protection of Busbars

The protection of busbars requires special care because of the special importance of the equipment for a reliable power supply. The loss of a busbar following a busbar fault can result in subsequent loss of lines and transformers connected to the busbar.

13.7.1 Current Criteria for Busbar Protection

Simple busbar arrangements with one infeed and outgoing feeders without the possibility of backfeed can be protected using overcurrent (UMZ) protection devices. In the case of a busbar fault, the excitation of the UMZ protection in the feeding transformer is active and all other protection devices are inactive; all circuit-breakers are tripped in primary time. In the case of a fault on an outgoing feeder, the UMZ protection devices are all in active status; the busbar protection is blocked, since the faulted line is to be switched off by the assigned protection of the faulted feeder in primary time. The UMZ protection of the feeding transformer works as backup protection. This type of busbar protection does not represent an independent protective device of the equipment; the operating time is comparatively long at ~100 ms. Its advantage lies in the simple and inexpensive realization.

Single busbars with several infeeds can also be protected with a high-impedance protection. The secondary currents of the current transformers of all feeders are connected in parallel to a high-impedance resistance, the voltage at the resistance serving as tripping criterion. Since the currents are added with the correct phase-angle, busbar faults lead to high voltages and feeder faults lead to low voltages. Operating time is in the range up to 60 ms.

If independent busbar protection with short operation time is needed, busbar differential protection is to be used. This scheme is adequate for protection of busbars with arbitrary numbers of feeders. The switching status of the substation – that is, which feeders are connected to each busbar – is modeled by an auxiliary circuit (switching image), for example, by connection of auxiliary contacts of the circuit-breakers and isolating switches into the protective device. False representation of the switching image can result in inaccurate operation of the busbar protection in case of normal operation or can result in unnecessary operation in case of faults on the feeders due to the "preloading" of the protection. Monitoring of the current sum of all feeders is an appropriate countermeasure. Special considerations are necessary if the busbar protection operates only on those circuit-breakers which carry partial short-circuit currents in the direction to the busbar or when transformers with isolated neutral, for example, in power station feeders, are connected [78]. Operation time of the protection is in the range up to 20 ms. With the combination of differential protection and high-impedance protection, the operating time can be reduced to less than 10 ms.

A further scheme for busbar protection is based on the comparison of the phase-angles of the currents of all feeders with the phase-angle of the total current sum. A switching image is also necessary for this scheme.

13.7.2 Impedance Criteria for Busbar Protection

If lines are protected by distance protection, the busbar protection is already realized as backup protection in the grading time of the distance protection (see Figure 13.9). The operating time is accordingly high. If shorter operating time is desired, the protection scheme can be extended by assessing the direction measurement of the distance protection of all feeders connected to the busbar. However, it should be noted that the accuracy of the direction measurement is not always reliable in case of faults close to the busbar. Furthermore, this scheme is not seen as an independent protection of the equipment busbar but represents an auxiliary function of the line protection. The concept should be applied therefore only in MV installations or for HV busbars of low importance.

13.8 Protection of Other Equipment

Besides the protection of lines, transformers and busbars, other equipment such as capacitor banks, resonance and earthing reactors, short-circuit limiting devices, FACTS, motors and generators have to be equipped with protective devices as well. Considerations of security, sensitivity, selectivity and the importance of the equipment must lead to an appropriate design and to selection of a protection concept. Further information on these aspects can be found in the technical literature [75, 78, 79].

13.9 Reference List of IEC-Symbols and ANSI-Code-Numbers

The symbols for protective devices according to IEC 60617 (DIN EN 60617) should be used. However the use of ANSI code numbers for protective devices is quite common. A reference list correlating symbols for protection according to IEC EN 60617-7 and ANSI (IEEE C37.2) is given in Table 13.4.

Table 13.4 Symbols for protection equipment according to IEC 60617 and ANSI code numbers IEEE C37.2 (selection).

ANSI-code-number	IEC-symbol	Protection equipment, auxiliary equipment
2		Time-delay
21	$Z<$	Minimum impedance relay, distance protection relay
25AR	$\Delta U<$, $\Delta f<$, $\Delta\varphi<$	Voltage- and synchro-check for autoreclosure
26	$\Theta>$	Overtemperature relay
27	$U<$	Undervoltage relay
30		Annunciator relay
32	$P>$	Directional power relay
32R	$-P$	Reverse power relay
50	$I>$, $I>>$	Instantaneous overcurrent relay
50N	$I_E>$	Neutral instantaneous overcurrent relay
51	$I>t$	Time overcurrent relay
51/27R		Voltage restraint time overcurrent relay
51/27C		Voltage controlled time overcurrent relay
59	$U>$	Overvoltage relay
63		Pressure switch, Buchholz protection
64RF		Restricted earth-fault protection
67	$I>_{\text{dir}}$	Directional relay
87	$\Delta I>$	Differential protection relay
87B	$I_d>$	Bus differential relay
87H		High-impedance differential relay
87T	$I_d>$	Transformer differential relay

14
Overvoltages and Insulation Coordination

14.1
General; Definitions

In the context of this book the topic of insulation coordination is dealt with for three-phase AC systems with nominal voltages $U_n > 1\,kV$ in accordance with IEC 60071. The insulation of equipment must be designed in such a way as to withstand all foreseeable voltage stresses or to limit the damage in insulation to the faulted equipment. The voltage withstand capability of the insulation depends on shape, duration, height and frequency of the overvoltages that arise and on the insulation material. Insulation can be solid insulation (e.g. XLPE, PVC), liquid insulation (e.g. oil), gas insulation (e.g. air, SF_6) or combinations of different kinds of materials (e.g. paper–oil insulation). The voltage withstand capability is generally greater with shorter duration of the voltage stress. Voltages and overvoltages are classified according to their shape and duration (see also Table 14.1):

- **Continuous power frequency voltage** with constant r.m.s. value, which may achieve at the most the values specified in IEC 60038, see Section 3.3.1.
- **Temporary overvoltage**, which can occur in case of short-circuits (in power systems with low-impedance neutral earthing) or ground faults (in power systems with resonance earthing or with isolated neutral). The duration can vary from some seconds (short-circuits) up to some hours (ground faults, resonances). The frequency of the temporary overvoltages can also be below or above the nominal frequency of the power system.
- **Transient overvoltages** with duration of a few milliseconds, immediately followed by temporary overvoltages. Normally the transient overvoltage occurs as highly damped overvoltage.
- **Slow-front overvoltage**, resulting from switching of capacitive and inductive currents, from long lines or as high-frequency AC voltages in the case of ground faults or short-circuits. Slow-front overvoltages are normally unidirectional, with a rise time to peak between 20 μs and 5 ms and a tail duration below 20 ms.
- **Fast-front overvoltages** occur due to lightning strokes into phase conductors or due to backward flashover following lightning strokes in towers or earth conductors or can result from switching actions. Fast-front overvoltages are

Power System Engineering: Planning, Design and Operation of Power Systems and Equipment, Second Edition.
Jürgen Schlabbach and Karl-Heinz Rofalski.

Table 14.1 Classes, shape and parameters of voltages and overvoltages as per IEC 60071-1.

Class	Low-frequency		Transient	
	Continuous	Temporary	Slow-front	Fast-front
Voltage shape				
Frequency range, time duration	$f = 50\,\text{Hz}$ or $60\,\text{Hz}$ $T_t \geq 3.600\,\text{s}$	$10\,\text{Hz} < f < 500\,\text{Hz}$ $3.600\,\text{s} \geq T_t \geq 30\,\text{ms}$	$20\,\mu\text{s} < T_p \leq 5.000\,\mu\text{s}$ $T_2 \leq 20\,\text{ms}$	$0.1\,\mu\text{s} < T_1 \leq 20\,\mu\text{s}$ $T_2 \leq 300\,\mu\text{s}$
Standardized voltage shape	$f = 50\,\text{Hz}$ or $60\,\text{Hz}$ T_t^a	$48\,\text{Hz} \leq f \leq 62\,\text{Hz}$ $T_t = 60\,\text{ms}$	$T_P = 250\,\mu\text{s}$ $T_2 = 2500\,\mu\text{s}$	$T_1 = 1.2\,\mu\text{s}$ $T_2 = 50\,\mu\text{s}$
Standardized voltage test procedure	a	Short-duration power-frequency test	Switching impulse test	Lightning impulse test

a To be defined by the responsible equipment committees for standardization.
T_p and T_1, time to peak or front time; T_2, time to half-value of tail; T_t, total time duration.

normally unidirectional; the time to peak is in the range 0.1–20 μs; the tail duration is below 300 μs.

- **Very-fast-front overvoltages** result from operation of disconnecting switches within gas-insulated switchgear. They are normally unidirectional; the time to peak is below 0.2 μs; the total duration is less than 3 μs. In most cases very-fast-front overvoltages have superimposed oscillations with frequency between 30 kHz and 100 MHz.

The terms explained below are used for insulation coordination:

External insulation is defined as the air between contacts or life parts. The surfaces of solid insulation in contact with air, subject to dielectric stress or to the effects of atmospheric conditions such as pollution and humidity, also belong to external insulation. If external insulation is operated inside closed rooms, the insulation is called weather-protected otherwise it is called non-weather-protected. **Internal insulation** is defined as the solid, gaseous or liquid parts of insulation or any combination of these parts protected from the effects of atmospheric conditions, for example, the paper–oil insulation (oil-impregnated insulation) of a transformer.

Self-restoring insulation completely recovers its insulating properties after a disruptive discharge in the insulation. Atmospheric or pressurized air, nitrogen and SF_6 are self-restoring insulating materials. **Non-self-restoring** insulation loses its insulating properties completely or partly after a disruptive discharge. Synthetic materials such as XLPE or paper–oil insulation are of the non-self-restoring type.

Withstand voltage is the value of a testing voltage applied during the voltage withstand test, during which a defined number of disruptive discharges are tolerated. For the definition of the **conventional assumed withstand voltage** the number of disruptive discharges is equal to zero; this value is indicated in IEC 60071 for non-self-restoring insulation, and the withstand probability is equal to 100%. The **statistical withstand voltage** allows for a specified number of disruptive discharges during the test, which is equal to a defined withstand probability. The withstand probability for self-restoring insulation defined in IEC 60071 is 90%. The **required withstand voltage** is the testing voltage the insulation must resist during a standard withstand voltage test. The **coordination withstand voltage** indicates the value of the withstand voltage which fulfills the selected criteria in each voltage class.

The **standard withstand voltage** is the standard voltage value which is used in a standard withstand voltage test. The standard withstand voltage is the rated value of the insulation for one or more required withstand voltages. The **standard short-duration power-frequency voltage** is a sinusoidal voltage with frequency between 48 and 62 Hz in 50 Hz systems with total duration of 60 s. Standard voltage shapes are also the **standard switching impulse** voltage, an impulse voltage with time to peak of 250 μs and a time to half-value of the tail (see Figure 14.1 for explanation) of 2500 μs, and the **standard lightning impulse** voltage, an impulse voltage with time to peak of 1.2 μs and a time to half-value of the tail of 50 μs.

The **rated insulation level** is a set of standard withstand voltages indicating the dielectric strength of the insulation. The **standard insulation level** is a rated insulation level, the standard withstand voltages are associated with the highest voltage of equipment, which are given as recommended values in IEC 60071-1. The **highest voltage for equipment** is the highest phase-to-phase r.m.s.-voltage the equipment is designed for with respect to the insulation and other characteristics related to this voltage value in the relevant standards.

14.2 Procedure of Insulation Coordination

Insulation coordination procedure is carried out according to IEC 60071. Part 1 contains definitions and requirements and describes fundamentals, Part 2 describes the detailed procedure of insulation coordination in the form of an application guideline. The insulation levels are specified separately for range I ($1\,\text{kV} < U_m \leq 245\,\text{kV}$) and range II ($U_m > 245\,\text{kV}$). The procedure described is divided into four stages.

1. On the basis of calculations and analyses of the expected overvoltages, **representative overvoltages** and/or representative **voltages** are determined (see Section 14.3), which cause the same dielectric effects on the insulation with respect to insulation coordination as the expected overvoltages or voltages during system operation. The representative overvoltage has a standardized voltage shape as used in the withstand voltage tests and can deviate in amplitude from the expected overvoltage. The determination of the representative overvoltage is to be accomplished independently of the voltage level (ranges I and II).

2. In the second step the **coordination withstand voltages** (Section 14.4) for all defined voltage classes are determined. Equipment must withstand at least the withstand voltage specified by the coordination withstand voltage during the total lifetime, taking account of the aging of the insulation. The definition of the coordination withstand voltage considers a certain probability of a disruptive discharge in the insulation. The definition of the coordination withstand voltage is to be accomplished independently of the voltage level (ranges I and II).
3. In the third step the **required withstand voltage** for each class of voltage (see Section 14.4) under standard test conditions is determined. Since the conditions at the installation site can deviate from the standard test conditions, the coordination withstand voltage can deviate from the required withstand voltage. The defined withstand voltage is to be accomplished independently of the voltage level (ranges I and II).
4. In the last step the necessary **rated insulation level** (Section 14.5) is selected. In range I ($1\,\text{kV} < U_m \leq 245\,\text{kV}$) combinations of standard short-duration power-frequency voltage and the standard lightning impulse voltage can be selected. Within range II ($U_m > 245\,\text{kV}$) combinations of the standard switching impulse voltage and the standard lightning impulse voltage can be selected. In both ranges one of the possible voltage shapes is disregarded, that is, the standard switching impulse voltage within range I and the standard short-duration power-frequency voltage within range II. This voltage is to be covered by a suitable selection of insulation levels for the two remaining voltage shapes. In this step of insulation coordination the withstand voltages are determined for the line-to-earth insulation, the line-to-line insulation and the longitudinal insulation, for example, between the open switching poles of a circuit-breaker or disconnecting switch.

Overvoltages can by calculated using suitable computer programs, such as EMTP or NETOMAG. The mathematical basis of the calculation methods, the kind of modeling of equipment and system parameters and the required details of calculation are not dealt within this book.

14.3 Determination of the Representative Overvoltages

14.3.1 Continuous Power-Frequency Voltage and Temporary Overvoltages

The determination of the continuous power-frequency voltage is based on the highest voltage for equipment U_m, specified in IEC 60071-1 and IEC 60038. For solidly earthed power systems (system with low-impedance earthing) the maximal permissible earth fault factor $k = 0.8\sqrt{3}$ can be taken to estimate the highest value for the continuous power-frequency voltage. It is recommended to calculate the earth fault factor k from the results of a short-circuit study. The earth fault factor is defined as the ratio of maximal line-to-earth voltage of the non-faulted phases to the voltage prior to fault, normally the nominal system voltage divided by $\sqrt{3}$ as defined in Equation 14.1.

$$k = \frac{U_{\mathrm{LEmax}}}{U_{\mathrm{n}}/\sqrt{3}} \tag{14.1}$$

In power systems with resonance earthing or with isolated neutral, the earth fault factor is $k = \sqrt{3}$.

Load-shedding can likewise lead to voltage increase (overvoltage) with power-frequency. If the system consists of relatively short lines and if the short-circuit power is comparatively high, detailed analysis of the voltage increase due to load-shedding can be neglected, whereas the increase factor u, related to U_{m}, the highest voltage of equipment, can be estimated according to IEC 60071-2 from Equation 14.2.

$$u < \frac{U}{\sqrt{2} \cdot U_{\mathrm{m}}/\sqrt{3}} \tag{14.2}$$

In case of doubt, the voltage increase due to load-shedding should be calculated by means of an appropriate program, in particular if long transmission lines, HV cables or low short-circuit power are characteristic of the system. In the case of load-shedding, a subsequent single-phase short-circuit can occur due to the voltage increase. Particular attention should be paid to this combination in the case of highly polluted insulation. Based on the considerations mentioned, the representative power-frequency overvoltage U_{rp} is calculated as in Equation 14.3.

$$U_{\mathrm{rp}} = k \cdot u \cdot \frac{U_{\mathrm{m}}}{\sqrt{3}} \tag{14.3}$$

Generally the highest overvoltages occur in the following cases:

- For the line-to-earth insulation as a result of earth faults or load-shedding
- For the line-to-line insulation in case of load-shedding
- For the longitudinal insulation for incorrect synchronization in the case of switching in phase opposition.

14.3.2 Slow-Front Overvoltages

Determination of representative slow-front overvoltages necessitates the calculation of the frequency distribution of the expected overvoltages. The frequency distribution considers the statistical dispersion of the three poles of the circuit-breaker while opening and closing. The switching simulations have to be carried out for a sufficient number of cases (at least 50), with the dispersion of the poles (first to last pole) and the total time lag fixed due to the technical characteristics of the circuit-breaker. Possible switching actions in power systems are to be considered:

- Switch-on and switch-off of lines
- Single-phase short-circuit with single-phase successful or unsuccessful autoreclosing
- Single-phase short-circuit with three-phase successful or unsuccessful autoreclosing
- Three-phase short-circuit with three-phase successful or unsuccessful autoreclosing
- Switching of capacitive or inductive currents
- Load-shedding.

In principle, two different methods are possible for the calculation. For phase-related analysis the maximal overvoltage value of each phase or between the phases is assumed for the frequency distribution; each simulation of switching results in three overvoltage values. This frequency distribution is then to be used for the determination of the insulation coordination for all voltages under consideration (line-to-earth, line-to-line and longitudinal insulation). During the case-related analysis the maximal overvoltage value of all three phases or of the three phase-to-phase voltages for each case is considered for the frequency distribution; each simulation of switching results in one overvoltage value. This frequency distribution is to be used for the associated insulation coordination separately for line-to-earth, line-to-line and longitudinal insulation.

Ferranti effect, power system resonances and the type of system feeding, among other things, determine the type of switching that will lead to the highest slow-front overvoltage. In power systems with resonance earthing or with isolated neutral, special attention must be dedicated to the overvoltages that occur in case of earth faults. Equipment operated in no-load operation mode always receives the highest slow-front overvoltages at the open-circuit end. In power systems with solidly earthed neutral, only switching actions are of any importance for the determination of slow-front overvoltages. Three-phase switching, especially three-phase autoreclosing, is significant for the line-to-line insulation. For the longitudinal insulation, only wrong synchronization must considered, since as in synchronous power systems the longitudinal insulation is stressed less than the line-to-earth insulation.

The statistical evaluation is based on a modified Weibull or Gaussian distribution, which is described by the 50%-value u_{e50}, the 2%-value u_{e2}, the standard deviation and the abort-value u_{et}, which is given by the 2%-value depending upon the type of analysis applied according to Equations 14.4. The maximal value of the representative overvoltage is equal to the abort-value u_{et} of the distribution.

$$u_{et} = 1.25 \cdot u_{e2} - 0.25 \quad \text{for phase-related analysis} \tag{14.4a}$$

$$u_{et} = 1.13 \cdot u_{e2} - 0.13 \quad \text{for case-related analysis} \tag{14.4b}$$

If surge arresters are installed, then the protection level of the arrester is used as the representative overvoltage.

14.3.3 Fast-Front Overvoltages

14.3.3.1 General

Mainly lightning overvoltages are to be considered as fast-front overvoltages. In rare cases, for example, if lightning overvoltages are to be restricted to a limit value below three times the line-to-earth voltage or in cable systems, where lightning overvoltages are not possible, fast-front overvoltages can also originate in substations through switching operations. Lightning overvoltages as a cause of fast-front overvoltages can originate from direct lightning strokes to the phase conductor or by back-flashover after strokes to the earth conductor or the tower or they can be induced by lightning strokes in the proximity of the line. Induced lightning overvoltages can reach amplitudes up to ~400 kV and need only be considered in power systems with voltages up to approximately 60 kV. Back-flashovers are more likely in power systems in range I ($1\,kV < U_m \leq 245\,kV$) than for those in range II ($U_m > 245\,kV$) and can usually be neglected for power systems with operating voltage above 500 kV.

The amplitude of the lightning overvoltages depends, among other things, on the arrangement of earth and phase conductor (lightning protection angle φ, see Figure 14.5), the flashover characteristic of the insulators due to pollution of the surface, the grounding conditions of the towers (lightning earthing resistance in case of lightning strokes), the type and radius of the phase- and earth-conductors (corona-losses), the switchgear design, the capacities of voltage transformers, coupling devices for power line carrier signals, joints, the number of lines connected to the substation, the distances between the individual parts of the substation and the timing of the lightning stroke with regard to the instantaneous value of the power-frequency voltage.

14.3.3.2 Simplified Approach

The amplitude of the representative lightning overvoltage U_{rp} can be determined using a simplified approach in accordance with IEC 60071-2, appendix F using Equation 14.5.

$$U_{rp} = U_{pl} + \frac{A}{n} \cdot \frac{l}{l_{Sp} + l_t} \tag{14.5}$$

with

U_{pl} = protection level of the arrester for lightning strokes
A = factor as in Table 14.2
n = number of lines connected to the substation
l = distance between arrester and equipment to be protected
l_{Sp} = span of the overhead line
l_t = length of overhead line with outage rate equal to the return rate: $l_t = R_t/R_{km}$
R_t = adopted overvoltage return rate per year
R_{km} = overhead line outage rate per year corresponding to the first kilometer in front of the substation.

Table 14.2 Factor *A* and corona damping *K* of different types of overhead lines.

Type of line	Factor A (kV)	Corona damping K (10^{-6} µs kV^{-1}m^{-1})
MV systems (distribution lines)		
Earthed cross-arms (low flashover voltage to earth)	900	–
Wood-pole lines (high flashover voltage to earth)	2700	–
HV-systems (transmission lines)		
Single conductor	4500	1.5
Double conductor bundle	7000	1.0
Bundle of four conductors	11000	0.6
Bundle of six or eight conductors	17000	0.4

The simplified approach must not be used if transformers are connected, as the capacities of the windings cause higher overvoltages than determined with this simplified approach. An intensive overvoltage study with detailed analysis must be accomplished in this case, which requires experience in the details and extent of modelization, apart from a substantial expenditure of time above all, in order to be able to make statistically reliable assessment of the amplitude and frequency of lightning overvoltages. Lightning currents must be considered in term of amplitude and frequency as well as the probability of strokes into phase-conductors, tower or earth-conductor.

14.3.3.3 Detailed Calculation; Parameters of Lightning Current

The detailed calculation of the overvoltages caused by lightning strokes is in principle very difficult because

- The range of variation of the lightning current parameters, such as initial slope or front duration time, amplitude, energy, charge and total duration is very large.
- Negative, positive and multiple strokes have to be distinguished
- The stroke location depends on the parameters of the lightning current.
- The effective grounding impedance of the overhead line towers depends on the lightning current.
- The corona-losses of the overhead line influences the shape of the overvoltage.

See Figure 14.1 as well as references [80] and [81].

In reference [82] a procedure of three steps is suggested as a basis for international standardization. In the first step the limit distance is determined within which lightning strokes can lead to critical, that is, impermissible overvoltages. In the second step the probability of lightning strokes within the limit distance is defined. If necessary the overvoltage as the basis of the insulation coordination is determined in the third step.

Due to the corona-losses of the overhead line, all lightning strokes within a distance larger than the limit distance are damped in such a way that they do not

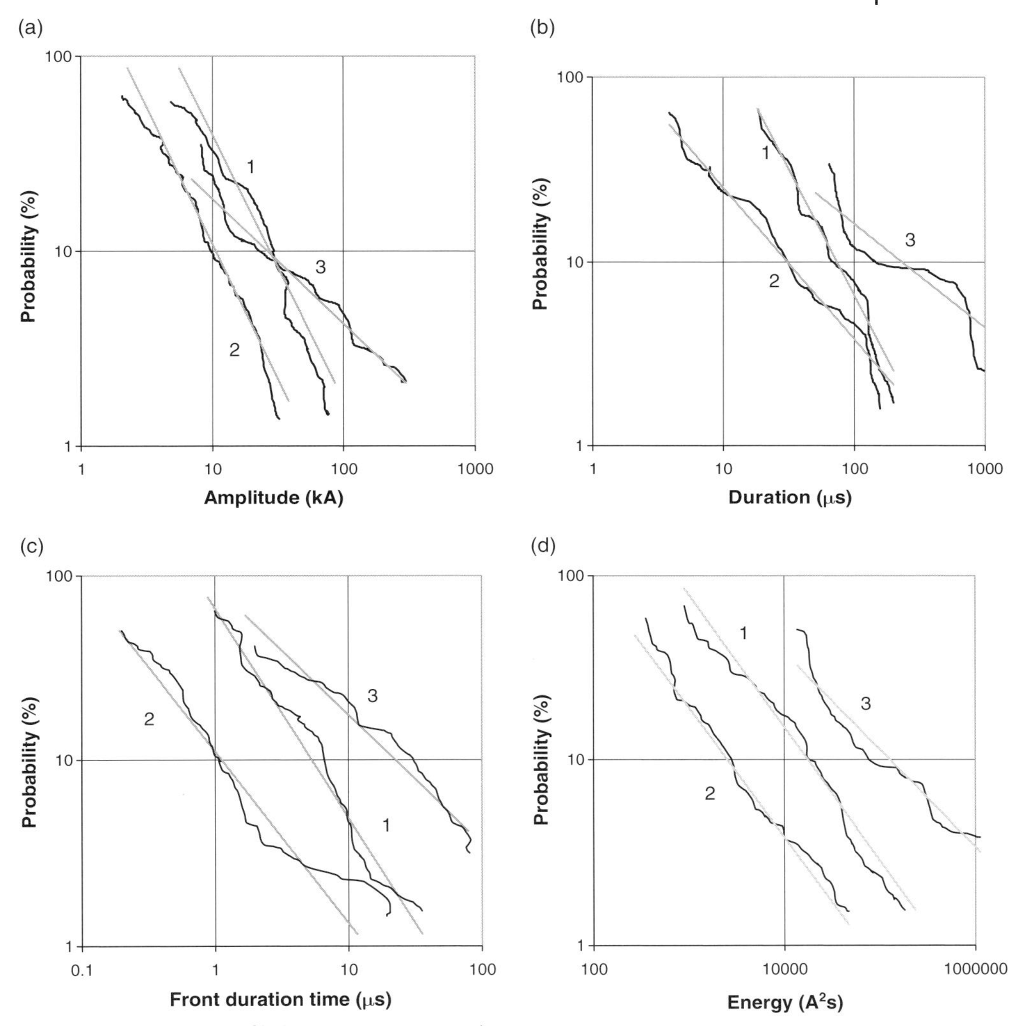

Figure 14.1 Parameters of lightning currents according to reference [80]. **1**, First stroke, negative polarity; **2**, subsequent stroke, negative polarity; **3**, positive polarity. (a) Amplitude; (b) total duration; (c) front duration; (d) energy.

cause impermissible overvoltages in the substation. The limit distance X_P is determined either by the lightning impulse protection level of arresters according to Equation 14.6 or by the self-protection effect of the substation. This self-protection effect is due to the change of the surge impedance, for example, by a cable connection between overhead line and switchgear, by capacitors or due to other connected overhead lines.

$$X_P = \frac{2 \cdot \tau}{n \cdot K \cdot (U_{cw} - U_{pl})} \quad (14.6)$$

with

τ = maximal traveling time in the substation between the nearest arrester and the equipment to be protected
n = number of overhead lines connected to the substation
K = factor for corona-damping according to Table 14.2
U_{cw} = coordination lightning impulse withstand voltage
U_{pl} = lightning impulse protection level of the arrester.

If the overvoltage amplitude is limited to a value below the coordination withstand voltage, the self-protection effect operates. The number of lines n necessary for self-protection of the substation can be determined from Equation 14.7.

$$n \geq 4 \cdot \frac{U_{50}^{-}}{U} - 1 \quad (14.7)$$

with

U_{50}^{-} = 50%-lighning impulse flashover voltage of the line insulation (negative polarity)
U = overvoltage amplitude under consideration.

In a gas-insulated substation (GIS) or with cable connections, the self-protection effect operates if Equation 14.8 is fulfilled.

$$U > \frac{6 \cdot Z_S}{Z_S + Z_L} \cdot U_{50}^{-} \quad (14.8)$$

with

Z_S = surge impedance of GIS or cable
Z_L = surge impedance of overhead line.

It should be noted in this respect that the reflection of the traveling wave at any location inside the switchgear may not reach the stroke location within the total time duration of the lightning stroke. One span distance for earth conductor strokes or a distance including two towers for backward flashover are to be considered as minimum limit distance.

In substations without cables, the self-protection effect is present only if the front slope of the incoming overvoltage is low due to corona damping. This can be assumed if the limit distance X_P is large compared with six times the longest distance l_{max} within the substation (Equation 14.9):

$$X_P >> 6 \cdot l_{max} \quad (14.9)$$

The numbers of lightning strokes, direct strokes and back-flashover strokes within the limit distance are calculated from the geometrical data of the overhead line. If the number of estimated strokes is smaller than the accepted failure rate, the insulation of the substation is sufficient.

The number of strokes N into the line is calculated using Equation 14.10,

$$N = (2 \cdot r + w) \cdot N_g \tag{14.10}$$

where N_g is the number of lightning strokes toward earth (isoceraunic level), w is the distance between the earth conductors or between the outer phase conductors, and r is the equivalent impact radius as given by Equation 14.11.

$$r = 16.3 \cdot (h_E - 0.667 \cdot d_M)^{0.61} \tag{14.11}$$

The term in parenthesis describes the average effective height of the earth conductor with the height h_E at the tower and the average sag d_M of the overhead line at mid span.

14.3.3.4 Direct Strokes to the Phase Conductor

Depending on the arrangement of earth and phase conductors, lightning strokes with current amplitudes I defined by the equivalent impact radius r [83] in accordance with Equation 14.12 can reach the phase conductor.

$$r = I^{0.64} \cdot h_E^{(0.66+0.0002 \cdot I)} \tag{14.12}$$

For towers up to 100 m high and lightning currents below 30 kA, Equation 14.12 can be simplified, see Equation 14.13. The error amounts to less than 5%.

$$r = 0.84 \cdot h_E^{0.6} \cdot I^{0.74} \tag{14.13}$$

with h_E the conductor height at the tower.

The probability $F(I)$ of lightning strokes into the phase conductor (failure rate) is calculated using Equation 14.14 from the failure rate R_p within the limit distance.

$$F(I) = F(I_{max}) + \frac{R_t}{R_p} \tag{14.14a}$$

$$R_p = \frac{R_{sf}}{F(I_{cr}) - F(I_{max})} \tag{14.14b}$$

with

R_t = adopted overvoltage return rate
R_{sf} = outage rate of overhead line insulation

$F(I_{max})$ = probability of maximal lightning amplitude I_{max} into the phase conductor as per Equation 14.15

$F(I_{cr})$ = probability of lightning amplitude I_{cr}, resulting in an overvoltage equal to the 50%-lightning impulse flashover voltage of the overhead line.

The probability of a lightning current with amplitude I is given by Equation 14.15.

$$F(I) = \left[1 + \left(\frac{I}{31}\right)^{2.6}\right]^{-1} \quad (14.15)$$

The lightning amplitude I_{cr} that causes an overvoltage equal to the 50%-lightning impulse flashover voltage is given by Equation 14.16.

$$I_{cr} = 2 \cdot \frac{U_{50}}{Z_L} \quad (14.16)$$

with

U_{50} = 50%-lightning impulse flashover voltage of the line
Z_L = surge impedance of the overhead line.

The steepness S of the front is calculated from Equation 14.17.

$$S = \frac{4}{K \cdot X_p} \quad (14.17)$$

with the corona damping K as given in Table 14.2 and the limit distance X_p. The half-time of the tail (the decaying part of the overvoltage after the maximum) is to be less than 140 μs. If the maximal amplitude of the lightning current is below the current I_{cr}, then the line is perfectly protected.

The representative overvoltage for all locations within the substation is calculated using a suitable program with the incoming transient overvoltage determined by the method described above.

For calculation of the probability for back-flashover, the number of lightning strokes into tower and earth conductor is calculated using Equation 14.18.

$$F(I) = \frac{R_t}{R_{sf}} \quad (14.18)$$

with

R_t = adopted overvoltage return rate
R_{sf} = impact rate into the overhead line with the limit distance.

The lightning current through the tower and the earthing of the tower causes a time-dependent voltage which also depends on the current shape. If the area

Table 14.3 Specific earth resistance and specific earth conductivity according to reference [84].

Kind of soil	ρ (Ω m)	$1/\rho$ (μS m^{-1})
Swampy soil, marly clay	30	333
Clay	50	200
Porous limestone, sandstone, loam, agricultural soil	100	100
Quartz, firm limestone, wet sand	200	50
Quartz, wet gravel	500	20
Granite, gneiss, dry sand and gravel	1000	10
Granite, gneiss, stony soil	3000	3.33

of the effective earthing resistance is limited to a radius of 30 m around the tower, then the time-dependence can be neglected. The earthing resistance R_{hc} is calculated using Equation 14.19 from the low-current resistance R_{lc} and the current I_g as per Equation 14.20, taking account of the ionization of the ground.

$$R_{hc} = \frac{R_{lc}}{\sqrt{1 + \frac{I}{I_g}}} \tag{14.19}$$

$$I_g = \frac{E_0 \cdot \rho}{2\pi \cdot R_{lc}^2} \tag{14.20}$$

with the specific ground resistance ρ as in Table 14.3 and the electric field strength of the ground ionization E_0 to be limited to a value $E_0 = 400\,\text{kV m}^{-1}$. With consideration of the coupling factor c_f between earth and phase conductor ($c_f \approx 0.15$ in case of one earth conductor; $c_f \approx 0.35$ in case of two earth conductors) the incoming overvoltage is calculated from Equation 14.21.

$$U_I = \frac{1 - c_f}{\sqrt{1 + \frac{I}{I_g}}} \cdot R_{lc} \cdot I \tag{14.21}$$

The steepness of the front S and time to half-value of tail τ are given by Equations 14.22.

$$S = \frac{4}{K \cdot X_p} \tag{14.22a}$$

$$\tau = \frac{Z_E}{R_{lc}} \cdot \frac{I_{Sp}}{c} \tag{14.22b}$$

with

Z_E = surge impedance of soil ($Z_E \approx 500\,\Omega$ for one earth conductor; $Z_E \approx 270\,\Omega$ for two earth conductors)
l_{Sp} = width of span
c = speed of light
R_{lc} = low-current earthing resistance
K = corona-damping
X_P = limit distance.

The modeling for transient wave analysis with computer programs includes a voltage source with the voltage shape determined by the method explained above and an internal impedance equal to the tower earthing resistance R_{lc}, connected with the substation by a conductor of length $X_p/4$, that is, one-quarter of the limit distance. If the amplitude of the incoming overvoltage is more than 1.6 times the 50%-lightning impulse flashover voltage of the line or if the effective earthing resistance of the tower has a larger radius than 30 m, detailed investigations have to be undertaken.

14.4 Determination of the Coordination Withstand Voltage and the Required Withstand Voltage

The next step of insulation coordination procedure covers the determination of the coordination withstand voltages as reference value for the lowest value of the withstand voltage of the insulation. Two methods, the deterministic or the statistical procedures, can be used. Since the statistical procedure is based on event probabilities, described by the probability density function of the overvoltages and the probability of disruptive discharge of the insulation, but sufficient data are rarely available, the deterministic procedure is to be preferred; this is described below in more detail.

The coordination withstand voltage for AC operating voltage of the line-to-earth insulation is equal to the highest value of the line-to-line voltage divided by $\sqrt{3}$, to be taken for the total lifetime of the equipment. This value is thus equal to the maximal representative voltage. For temporary power frequency overvoltages, the coordination short-time withstand voltage is equal to the representative overvoltage. The influence of pollution of the external insulation is not considered in the insulation coordination procedure but is used in the selection of the insulators. Pollution classes are defined in IEC 60071-2, Table 1; see also Section 9.3.

The coordination withstand voltage for slow-front overvoltages is determined from the representative overvoltages by multiplication with the deterministic coordination factor K_{dc}, which accounts for inaccuracies in the data and models in the determination of the withstand voltage and the representative overvoltage. The deterministic coordination factor depends on the relationship of the switching impulse protection level U_{ps} of the arrester to the 2%-value of the line-to-earth

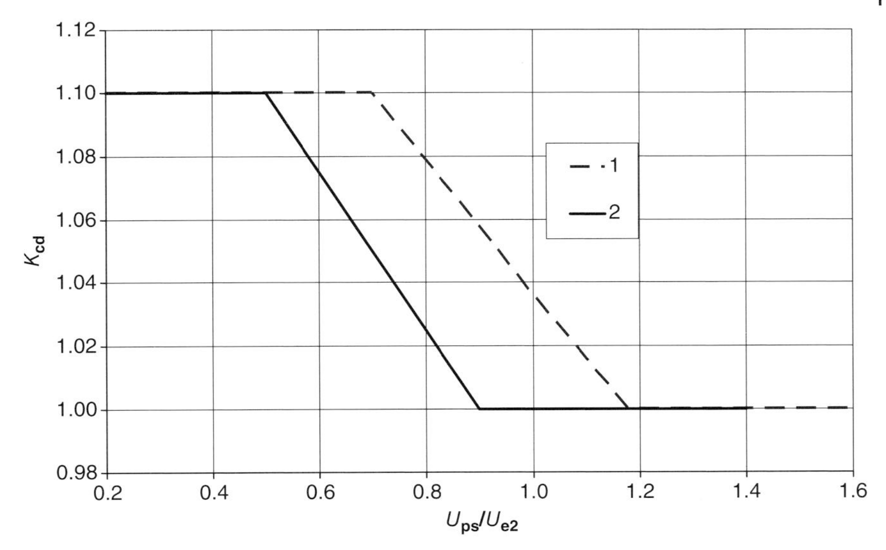

Figure 14.2 Deterministic coordination factor. **1**, for line-to-earth coordination withstand voltage (also for longitudinal insulation); **2**, for protection level of line-to-line coordination withstand voltage.

overvoltage U_{e2}; values are outlined in Figure 14.2. For equipment without arrester protection the maximal overvoltage is equal to the abort value of the distribution u_{et} and the deterministic coordination factor is $K_{cd} = 1$.

The coordination withstand voltage for fast-front overvoltages is equal to the maximal value of the overvoltage. The deterministic coordination factor is $K_{cd} = 1$. The same conditions and assumptions apply as with slow-front overvoltages.

The required withstand voltage is calculated from the coordination withstand voltage with a correction factor K_a accounting for atmospheric conditions deviating from the standard conditions for type testing and with a safety factor K_S that takes account of differences in the insulation characteristic between operating conditions and conditions during the standard withstand voltage test.

The correction factor K_a for atmospheric conditions is to be used only for external insulation and is defined in Equation 14.23 as a function of the geographic height H above sea level.

$$K_a = e^{m(H/8150)} \tag{14.23}$$

with factor m according to Table 14.4 and Figure 14.3.

The safety factor K_S considers differences in the equipment assembly, variations in product quality, aging of the insulation during the lifetime and so on. If safety factors are not fixed, the values $K_S = 1.15$ (internal insulation) and $K_S = 1.05$

Table 14.4 Factor m for the calculation of atmospheric correction factor K_a.

Voltage	Factor m
Coordination withstand voltage	
Lightning impulse voltage	1.0
Short-time withstand voltage (air and clean insulators)	1.0
Short-time withstand voltage (polluted insulators and anti-fog insulators)	0.8
Short-time withstand voltage (polluted insulators and normal insulators)	0.5
Switching impulse voltage	See Figure 14.3

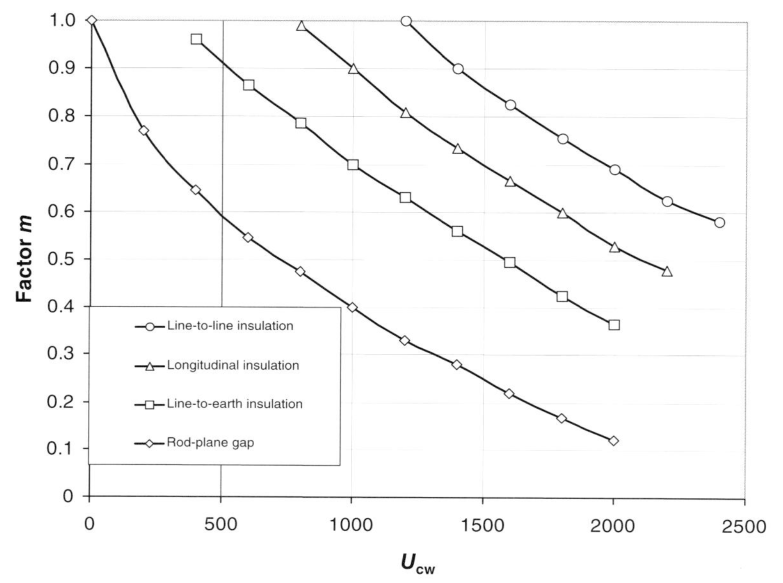

Figure 14.3 Factor m for the calculation of the atmospheric correction factor in case of switching impulse voltage.

(external insulation) should be used. Higher safety factors can be specified for gas-insulated switchgears in range II ($U_m > 245$ kV).

14.5 Selection of the Rated Voltage

IEC 60071-1 specifies the standard test voltages for which standard insulation levels and rated voltages and the highest voltage for equipment U_m are assigned. For range I ($1\,\text{kV} < U_m \leq 245\,\text{kV}$) standard short-duration power-frequency with-

Table 14.5 Standard insulation levels (selection) in range I (1 kV < $U_m \leq 245$ kV).

Highest voltage for equipment U_m (kV)	Standard short-duration power-frequency withstand voltage (kV)	Standard lightning impulse withstand voltage (kV)
12	28	60
		75
		95
24	50	95
		125
		145
36	70	145
		170
123	(185)	450
	230	550
145	(185)	(450)
	230	550
	275	650
170	(230)	(550)
	275	650
	325	750
245	(275)	(650)
	(325)	(750)
	360	850
	395	950
	460	1050

stand voltages and standard lightning impulse withstand voltages are specified; see Table 14.5. These limit values are to cover the required switching impulse line-to-earth and line-to-line withstand voltages and the required withstand voltage of the longitudinal insulation. For range II ($U_m > 245$ kV) the standard lightning impulse withstand voltages and the standard switching impulse withstand voltages are defined; see Table 14.6. It should be noted that the highest voltage for equipment U_m is stated as the r.m.s. value, all other voltages are peak values. If the voltages indicated in parentheses as per Table 14.5 are not sufficient, additional withstand voltage tests are required in order to prove the necessary line-to-line withstand voltage.

The standard switching impulse withstand voltage is to cover the power-frequency withstand voltage also, and if no other values are fixed, the required short-time AC-withstand voltage as well. The required withstand voltages are to be modified with test conversion factors according to Tables 14.7 and 14.8 into those voltages defined for the standard voltages. The test conversion factors include the factor $1/\sqrt{2}$ for the conversion of peak values (switching impulse withstand voltage, lightning impulse withstand voltage) into r.m.s. values (short-time AC-withstand voltage).

Table 14.6 Standard insulation levels (selection) in range II (U_m > 245 kV).

Highest voltage for equipment U_m (kV)	Standard switching impulse withstand voltage			Standard lightning impulse withstand voltage (kV)
	Line-to-earth (kV)	Ratio line-to-line to line-to-earth (kV)	Longitudinal insulation[a] (kV)	
300	750	1.5	750	859 950
	850	1.5	750	950 1050
362	850	1.5	850	950 1050
	950	1.5	850	1050 1175
420	850	1.6	850	1050 1175
	950	1.5	950	1175 1300
	1050	1.5	950	1300 1425
525	950	1.7	950	1175 1300
	1050	1.6	950	1300 1425
	1175	1.5	950	1425 1550

a Value of impulse voltage in combined test.

Table 14.7 Test conversion factors in range I (1 kV < $U_m \leq$ 245 kV) for conversion of required switching impulse voltage into short-time standard AC-withstand voltages and standard lightning impulse withstand voltage (U_{rw} = required standard switching impulse voltage in kV).

Type of insulation	Test conversion factor	
	Short-time AC-withstand voltage	Lightning impulse withstand voltage
External insulation		
Air, clean and dry insulators		
Line-to-earth	$0.6 + U_{rw}/8500$	$1.05 + U_{rw}/6000$
Line-to-line	$0.6 + U_{rw}/12700$	$1.05 + U_{rw}/9000$
Clean and wet insulators	0.6	1.3
Internal insulation		
GIS	0.7	1.25
Impregnated insulation	0.5	1.1
Solid insulation	0.5	1.0

Table 14.8 Test conversion factors in range II ($U_m > 245\,kV$) for conversion of required short-time AC-withstand voltage into standard switching impulse withstand voltage.

Insulation	Switching impulse withstand voltage
External insulation	
Air, clean and dry insulators	1.4
Clean and wet insulators	1.7
Internal insulation	
GIS	1.6
Impregnated insulation	2.3
Solid insulation	2.0

14.6 Application Example

Figure 14.4 indicates the arrangement of a 380/115 kV-switchyard (GIS) with two interbus-transformers. The 380 kV-switchgear is connected by SF_6-ducts to the transformers as well as to the attached 380 kV overhead line (double-circuit line). Lattice steel towers with vertical arrangement of the phase conductors and two earth conductors are installed. Figure 14.5 indicates the 380 kV tower outlines with dimensions.

The insulation coordination procedure is carried out separately for the different voltage classes, that is, power-frequency voltage, and switching and lightning overvoltages. The equipment was modeled for the analysis according to the requirements of the computer program EMTP/ATP [85] and [86] in various ways, see also IEC/TR 60071-4.

The power-frequency voltages and temporary overvoltages are indicated in Table 14.9. The highest voltage for equipment U_m are defined according to IEC 60038 and IEC 60071-1. The 380 kV and the 115 kV systems are operated with low-impedance earthing. The results of short-circuit current analysis indicate an earth-fault factor k according to Table 14.9. Both power systems are characterized by high short-circuit power and comparatively short line lengths. Overvoltages resulting from load-shedding can therefore be neglected. The values for the representative overvoltage are $U_{rp} = 87.8\,kV$ for the 115 kV system and $U_{rp} = 340.5\,kV$ for the 380 kV system.

The representative slow-front overvoltages U_e are calculated with EMTP/ATP [86]. The results are outlined in Table 14.10.

The isoceraunic level in the area is $N_g = 10$ strokes/(km·year). With the dimensions of the overhead tower according to Figure 14.5, the number of strokes into the overhead line is $N = 0.37$ strokes/(km·year). The overhead line has a surge impedance of $Z_L = 317\,\Omega$, the 50%-lightning impulse flashover voltage is $U_{50} = 1425\,kV$. Thus the maximal permissible lightning current into the phase conductor

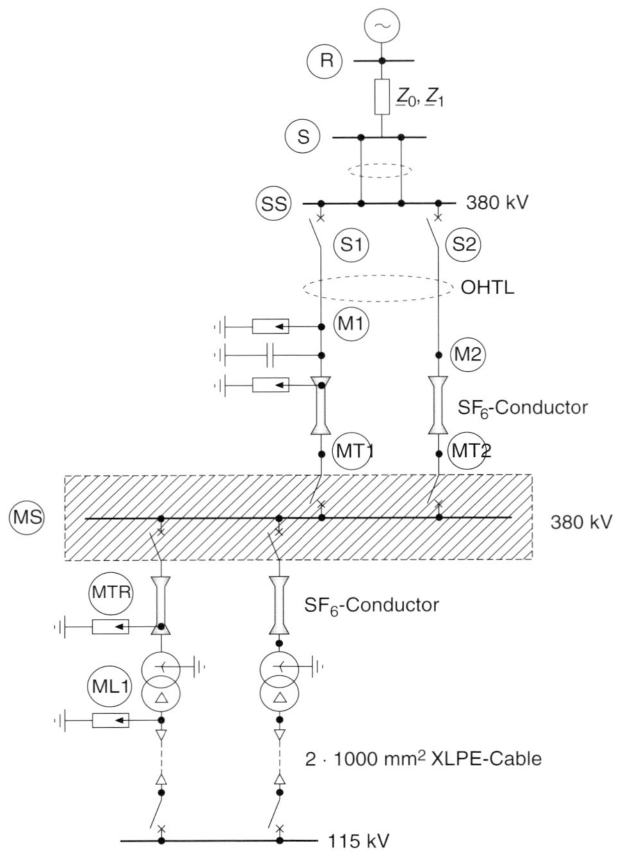

Figure 14.4 Diagram of the 380/115 kV switchyard under investigation.

to cope with the self-protection effect is $I_C = 8.99\,\text{kA}$. Based on the actual arrangement of the overhead tower, the maximal lightning current that can strike the phase conductor is $I_{max} = 9.2\,\text{kA}$. Further analysis is therefore carried on with a lightning current amplitude of 9.2 kA.

The probabilities of lightning currents of $I_c = 8.99\,\text{kA}$ and $I_{max} = 9.2\,\text{kA}$ are $F_c = 0.962$ and $F_{max} = 0.959$. Thus the failure rate of the earth conductor protection is $N_{SF} = 0.001$ strokes/(km · year). The maximal voltage amplitude caused by lightning strokes is $U_l = 1458.2\,\text{kV}$. The transient overvoltage is damped by the corona effect. The limit distance is to be defined for the calculation of the front-time . The longest traveling time inside the switchgear is $\tau = 0.178\,\mu\text{s}$ (distance $l = 48\,\text{m}$). The limit distance therefore is $X_p = 1249\,\text{m}$. Thus the slope of the front of the incoming overvoltage will be $S = 2.669\,\text{kV}\,\mu\text{s}^{-1}$.

The calculation of the back-flashover is based on a tower earthing resistance of $R_{lc} = 10\,\Omega$ in the range of the limit distance, which is approximately three spans long. The probability for currents that can cause back-flashover is $F = 0.57$; thus the number of back-flashovers is $N = 0.0021$ events/(km · year). Based on the

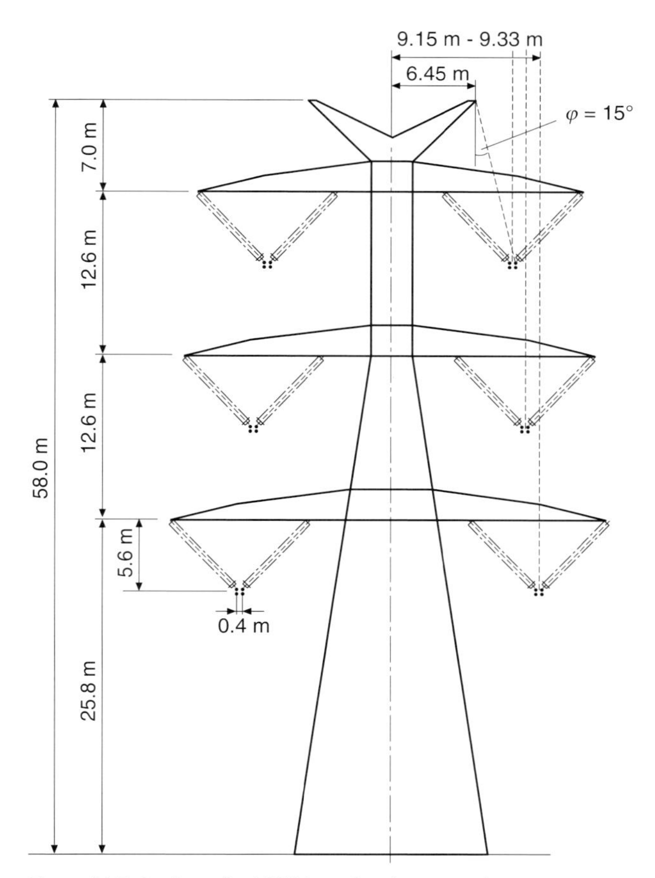

Figure 14.5 Outline of a 380 kV overhead tower with dimensioning (φ: lightning protection angle).

Table 14.9 Power-frequency voltages and temporary overvoltages.

System level	Voltages; factors			
	U_m (kV)	k	u	U_{rp} (kV)
115 kV	123	<1.03	<1.2	87.8
380 kV	420	<1.17	<1.2	340.5

frequency distribution of the lightning current amplitudes, the considered current for the back-flashover is $I < 300\,\text{kA}$. Thus the overvoltage resulting from the lightning current is $U_l < 1687.5\,\text{kV}$ with a slope of $S = 2.669\,\text{kV}\,\mu\text{s}^{-1}$ and time to half-value of tail of $T_2 = 66\,\mu\text{s}$.

The incoming overvoltage is modeled for calculations involving larger amplitudes, which arise either with direct strokes into the phase-conductor or with

Table 14.10 Calculation results of representative slow-front overvoltages U_e.

	Equipment 380 kV				Equipment 115 kV
	Line-to-earth voltage		Line-to-line voltage		Voltage line-to-earth
	Internal insulation	External insulation	Internal insulation	External insulation	
U_e (kV)	737.3	823	–	–	253.1
Factor u_e	2.15	2.4	–	–	2.52
2%-value U_{e2} (kV)	726.9	797.2	1163.0	1275.5	232
Factor u_{e2}	2.12	2.32	3.39	3.72	2.31
Abort-value u_{et}	2.27	2.5	3.7	4.07	2.48

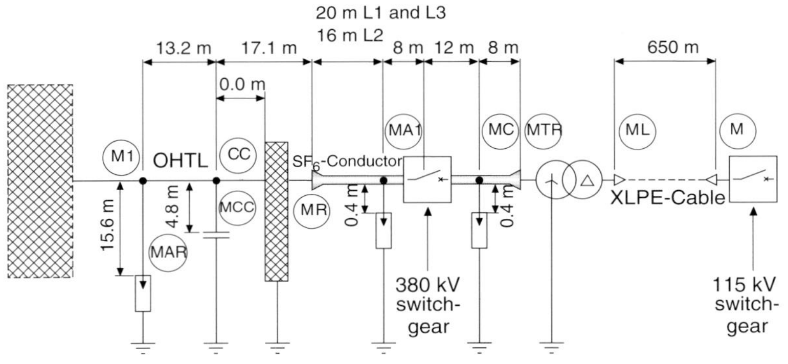

Figure 14.6 Representation of 380/115 kV switchgear for the calculation of fast-front overvoltages; data for dimensioning.

back-flashover. In this case the overvoltage caused by back-flashover will be U_l = 1687.5 kV. For the calculation of the overvoltages with lightning strokes (fast-front overvoltages), the 380/115 kV switchgear was modeled in accordance with Figure 14.6.

Different calculations for direct strokes into the phase conductor and for back-flashover were carried out. The results are outlined in Table 14.11 for the 380 kV installations and in Table 14.12 for the 115 kV installations. As can be seen from the tables (indicated in bold print) the highest overvoltages arise for direct strokes into the phase conductor.

The representative overvoltages of the 380 kV installations are specified in Table 14.13.

The coordination withstand voltages U_{cw} are as in Table 14.14 for internal insulation and Table 14.15 for external insulation.

The standard voltages for internal insulation are adequate, as the standard switching impulse voltage (1050 kV) covers both the switching impulse withstand

Table 14.11 Results of calculation of fast-front overvoltages of a 380 kV installation as in Figure 14.6; all data in kV.

Case	External insulation							
	Isolating switch closed at line-side				Isolating switch opened at line-side			
	M1	MAR	CC	MR	M1	MAR	CC	MR
Direct stroke	844.0	804.2	**891.7**	825.4	1026.5	897.3	**1164.5**	963.4
Back-flashover	825.4	800.1	855.4	801.6	885.6	834.5	952.1	854.0
	Internal insulation							
	Isolating switch closed at line-side				**Isolating switch opened at line-side**			
	MCC	**MA1**	**MC**	**MTR**	**MCC**	**MA1**	**MC**	**MTR**
Direct stroke	**945.7**	807.0	852.2	944.9	**1219.2**	856.7	857.9	–
Back-flashover	885.1	779.1	804.1	845.6	1000.6	811.5	812.2	–

Table 14.12 Result of calculation of fast-front overvoltages of a 115 kV installations as in Figure 14.6; all data in kV.

Case	Internal insulation	
	ML	M
Direct stroke	90.7	**98.4**
Back-flashover	65.8	68.0

Table 14.13 Representative overvoltages of 380 kV installations as in Figure 14.6.

Voltage category	Symbol	Value (kV)	Comment
Nominal system voltage	U_n	400	r.m.s., line-to-line
Highest voltage for equipment	U_m	420	r.m.s., line-to-line
Temporary overvoltage	U_{rp}	340.5	r.m.s., line-to-earth
Slow-front overvoltage (internal insulation)	U_{e2}	726.9	Peak value, line-to-earth, case-related
Slow-front overvoltage (external insulation)	U_{e2}	797.2	Peak value, line-to-earth, case-related
Fast-front overvoltage (internal insulation)	U_{el}	1219.2	Peak value
Fast-front overvoltage (external insulation)	U_{el}	1164.5	Peak value

Pollution class IV (see chapter 9.4.1), maximum height above sea level 100 m.

Table 14.14 Voltages to be used for internal insulation coordination of 380 kV installations.

Type of overvoltage; factor	Power-frequency	Slow-front	Fast-front
Coordination withstand voltage U_{cw}	340.5 kV	745.2–828 kV	1219.2 kV
Safety factor K_S	1.15	1.15	1.15
Required withstand voltage U_{rw}	391.6 kV	857.0–952.2 kV	1402.1 kV
Test conversion factor (GIS)	1.6		
Test conversion factor (impregnated insulation)	2.3		
Switching impulse withstand voltage (GIS)	626.6 kV		
Switching impulse withstand voltage (impregnated insulation)	900.7 kV		
Standard voltage	Standard switching impulse voltage 1050 kV	Standard lightning impulse voltage 1425 kV	

Table 14.15 Voltages to be used for external insulation coordination of 380 kV installations.

Type of overvoltage; factor	Power-frequency	Slow-front	Fast-front
Coordination withstand voltage U_{cw}	340.5 kV	932.7 kV	1164.5 kV
Atmosperic correction factor K_a	1.0	1.0	1.0
Safety factor K_S	1.05	1.05	1.05
Required withstand voltage U_{rw}	357.5 kV	979.3 kV	1222.7 kV
Test conversion factor	1.7		
Switching impulse withstand voltage	607.8 kV		
Standard voltage	Standard switching impulse voltage 1050 kV	Standard lightning impulse voltage 1425 kV	

voltage (900.7 kV) and the required slow-front withstand overvoltage (857.0–952.2 kV). The standard lightning impulse voltage (1425 kV) is larger than the required withstand voltage for fast-front overvoltages (1402.1 kV).

The selected standard voltages for external insulation are sufficient, as the switching impulse voltage (1050 kV) covers both the switching impulse withstand voltage (607.8 kV) and the required withstand voltage for slow-front overvoltages (979.3 kV). The standard lightning impulse withstand voltage (1425 kV) is larger than the required withstand voltage for fast-front overvoltages (1222.7 kV).

15
Influence of Neutral Earthing on Single-Phase Short-Circuit Currents

15.1
General

Currents and voltages in case of short-circuits with earth connection (e.g. single-phase short-circuits) depend on the positive-sequence and zero-sequence impedances Z_1 and Z_0. If the ratio of zero-sequence to positive-sequence impedance is $k = Z_0/Z_1$ the voltages in the non-faulted phases and the single-phase short-circuit current are calculated according to Equation 15.1.

$$|\underline{U}_{L2}| = |\underline{U}_{L3}| = E_1 \cdot \sqrt{3} \cdot \frac{\sqrt{k^2 + k + 1}}{2 + k} \quad (15.1a)$$

$$I''_{k1} = \frac{E_1}{Z_1} \cdot \frac{3}{2 + k} \quad (15.1b)$$

If the voltage E_1 is set to $E = U_n/\sqrt{3}$, similar to the equivalent voltage at the short-circuit location, then

$$|\underline{U}_{L2}| = |\underline{U}_{L3}| = U_n \cdot \frac{\sqrt{k^2 + k + 1}}{2 + k} \quad (15.1c)$$

$$I''_{k1} = \frac{U_n}{\sqrt{3} \cdot Z_1} \cdot \frac{3}{2 + k} \quad (15.1d)$$

The impedances in the positive-sequence (and negative-sequence) component are determined only by the network topology. The single-phase short-circuit current and the voltages of the non-faulted phases can be changed only by changing the ratio of positive-sequence to zero-sequence impedance, that is, by changing the handling of transformer neutrals.

The type of neutral earthing determines the impedance Z_0 of the zero-sequence component and has a dominating influence on the short-circuit current through earth, that is, I''_{k1} in case of single-phase short-circuits and I''_{k2EE} in case of two-phase short-circuit with earth connection. In order to change the zero-sequence

Power System Engineering: Planning, Design and Operation of Power Systems and Equipment, Second Edition.
Jürgen Schlabbach and Karl-Heinz Rofalski.

impedance of the system, it is possible to earth any number of neutrals – that is, none, a few or all transformer neutrals – leading to the highest zero-sequence impedance (no neutral earthed) or to the lowest zero-sequence impedance (all neutrals earthed). The system is characterized less by the number of neutrals to be earthed than by the value of the single-phase short-circuit current and by the voltages in the non-faulted phases.

The different types of neutral handling in power systems (high-voltage systems only) are outlined in Table 15.1.

15.2
Power System with Low-Impedance Earthing

Low-impedance earthing is applied worldwide in medium-voltage and high-voltage systems with nominal voltages above 10 kV. Power systems having nominal voltages $U_n \geq 132\,\text{kV}$ are generally operated with low-impedance earthing. In order to realize a power system with low-impedance earthing, it is not necessary that the neutrals of all transformers are earthed, but the criterion should be fulfilled that the earth-fault factor (ratio of the line-to-earth voltages of the non-faulted phases to the line-to-earth voltage prior to the fault) remains below $k \leq 0.8\sqrt{3}$. The disadvantage of earthing all neutrals is seen in an increased single-phase short-circuit current, sometimes exceeding the three-phase short-circuit current. The neutral of unit transformers in power stations should not be earthed at all, as the single-phase short-circuit current will then depend on the generation dispatch. As the contribution of one unit transformer is in the range of some kiloamps, the influence on the single-phase short-circuit currents is significant.

Based on Figure 15.1 and assuming a far-from-generator short-circuit with positive-sequence impedance equal to negative-sequence impedance, $\underline{Z}_1 = \underline{Z}_2$, the single-phase short-circuit current is calculated by

$$\underline{I}''_{k1} = \frac{c \cdot \sqrt{3} \cdot U_n}{2 \cdot \underline{Z}_1 + \underline{Z}_0} \tag{15.1}$$

with voltage factor c according to Table 11.1. If the single-phase short-circuit current is related to the three-phase short-circuit current as

$$\underline{I}''_{k3} = \frac{c \cdot U_n}{\sqrt{3} \cdot \underline{Z}_1} \tag{15.2}$$

it follows that

$$\frac{\underline{I}''_{k1}}{\underline{I}''_{k3}} = \frac{3 \cdot \underline{Z}_1}{2 \cdot \underline{Z}_1 + \underline{Z}_0} \tag{15.3}$$

The relation of the ratio of single-phase to three-phase short-circuit current with the Z_0/Z_1 for different values of phase-angle difference $(\gamma_1 - \gamma_0)$ of the impedances

Table 15.1 Characteristics of different types of neutral handling in power systems.

	Isolated neutral	Low-impedance earthing	Earthing with current limitation	Resonance earthing
Single-phase fault current (short-circuit current)	Capacitive earth-fault current $\underline{I}_{CE} = j\omega\sqrt{3}C_E U_n$	Single-phase (earth-fault) short-circuit current $\underline{I}''_{k1} = \frac{\omega\sqrt{3}U_n}{(2\underline{Z}_1 + \underline{Z}_0)}$	Single-phase (earth-fault) short-circuit current $\underline{I}''_{k1} = \frac{\omega\sqrt{3}U_n}{(2\underline{Z}_1 + \underline{Z}_0)}$	Residual earth-fault current $\underline{I}_{Res} = j\omega\sqrt{3}C_E U_n \cdot (\delta_0 + j\nu)$
Increase of voltages at non-faulted phases	Present $U_{0max}/U_n \approx 0.6$	No increase $U_{0max}/U_n < 0.3$–0.45	No increase $U_{0max}/U_n \approx 0.45$–$0.6$	Present $U_{0max}/U_n \approx 0.6$
Earth-fault factor δ	$\approx\sqrt{3}$	<1.38	1.38 . . . $\sqrt{3}$	$\approx\sqrt{3}$ to $1.1\times\sqrt{3}$
Ratio of impedances Z_0/Z_1	Generally high	2–4	>4	$\rightarrow$ infinity
Extinguishing of fault arc	Self-extinguishing (see Figure 15.7)	Not self-extinguishing	Self-extinguishing in rare cases	Self-extinguishing (see Figure 15.11)
Repetition of faults	Double earth-fault Re-ignition of earth-fault	None	None	Double earth-fault
Voltage at earthing electrode U_E	$U_E \leq 125\,V$	$U_E > 125\,V$ permitted	$U_E > 125\,V$ permitted	$U_E \leq 125\,V$
Touch voltage U_B	$U_B \leq 65\,V$	see VDE 0141	see VDE 0141	$U_B \leq 65\,V$

U_n = nominal system voltage.
U_{0max} = maximal voltage in the zero-sequence system, that is, at neutral of transformer.
ω = angular velocity of the power system.
C_E = line-to-earth capacitance of the power system.
$\underline{Z}_0$, $\underline{Z}_1$ = zero-sequence and positive-sequence impedance of the system, respectively.

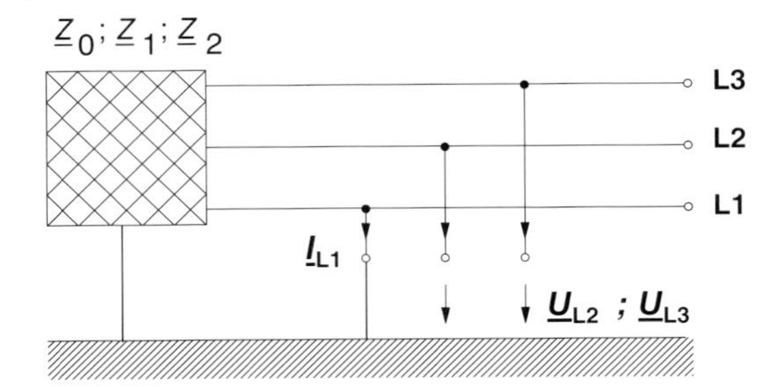

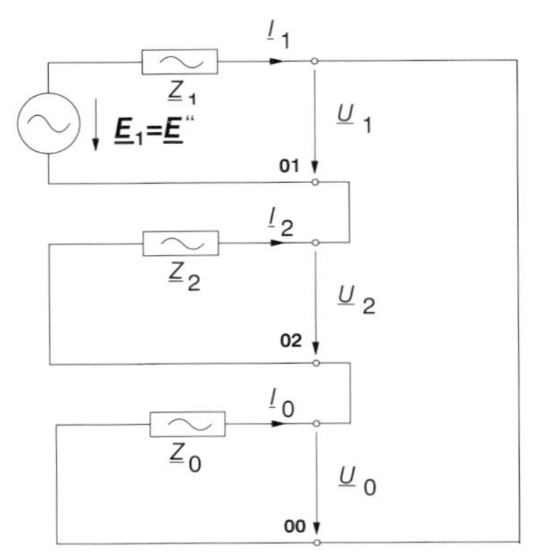

Figure 15.1 Equivalent circuit diagram of a single-phase short-circuit (system with low-impedance earthing). (a) Equivalent circuit diagram in the three-phase system; (b) equivalent circuit diagram in the system of symmetrical components.

as parameter is outlined in Figure 15.2. The phase angles γ_1 and γ_0 are defined by the arctan-functions $\gamma_1 = \arctan(X_1/R_1)$ in the positive-sequence component and $\gamma_0 = \arctan(X_0/R_0)$ in the zero-sequence component.

The voltages (power-frequency voltage) of the non-faulted phases L2 and L3 are defined by Equation 15.4

$$\underline{U}_{L2} = \underline{E}_1 \cdot \frac{\underline{Z}_0 \cdot (\underline{a}^2 - 1) + \underline{Z}_2 \cdot (\underline{a}^2 - \underline{a})}{\underline{Z}_0 + \underline{Z}_1 + \underline{Z}_2} \tag{15.4a}$$

$$\underline{U}_{L3} = \underline{E}_1 \cdot \frac{\underline{Z}_0 \cdot (\underline{a} - 1) + \underline{Z}_2 \cdot (\underline{a} - \underline{a}^2)}{\underline{Z}_0 + \underline{Z}_1 + \underline{Z}_2} \tag{15.4b}$$

Equations 15.4a and b can be simplified if $\underline{Z}_1 = \underline{Z}_2$ is assumed and by taking account of the meaning of a and a^2 according to Equations 15.5 to obtain Equations 15.4c and d

$$\underline{a} = e^{j120^\circ} = -0.5 + j0.5 \cdot \sqrt{3} \tag{15.5a}$$

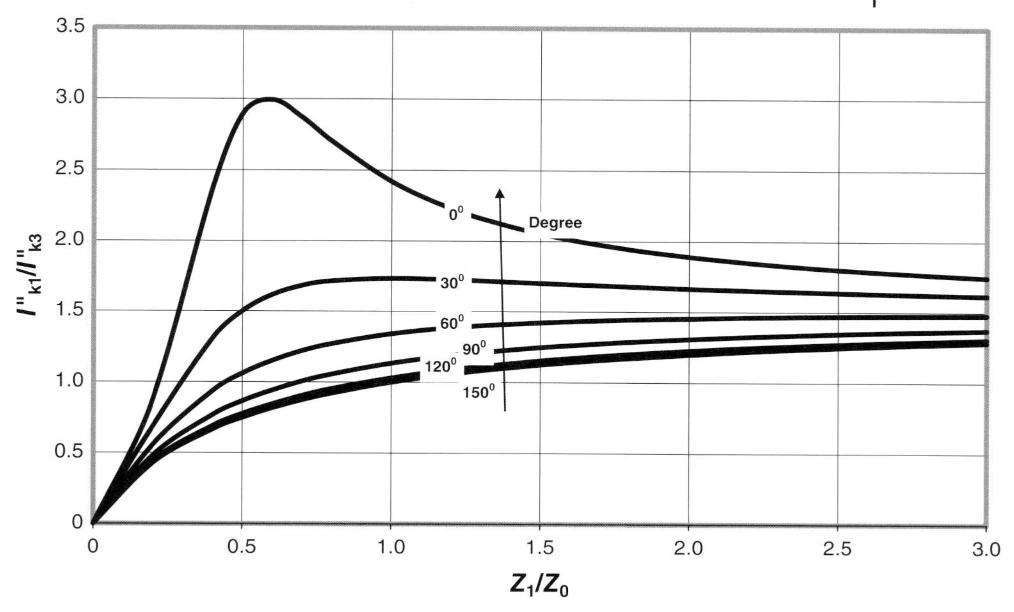

Figure 15.2 Ratio of single-phase to three-phase short-circuit current in relation to Z_1/Z_0 and $(\gamma_1 - \gamma_0)$.

$$\underline{a}^2 = e^{j240^\circ} = -0.5 - j0.5 \cdot \sqrt{3} \tag{15.5b}$$

$$\underline{U}_{L2} = -0.5\sqrt{3} \cdot \underline{E}_1 \cdot \frac{\sqrt{3}}{1 + (2\underline{Z}_1/\underline{Z}_0) + j} \tag{15.4c}$$

$$\underline{U}_{L3} = -0.5\sqrt{3} \cdot \underline{E}_1 \cdot \frac{\sqrt{3}}{1 + (2\underline{Z}_1/\underline{Z}_0) - j} \tag{15.4d}$$

Relating the voltages $\underline{U}_{L2}$ and $\underline{U}_{L3}$ to the voltage $\underline{E}_1$ the earth-fault factors δ_{L2} and δ_{L3} of the phases L2 and L3 are obtained:

$$\delta_{L2} = \left| -0.5 \frac{3}{1 + (2\underline{Z}_1/\underline{Z}_0) + j} \right| \tag{15.6a}$$

$$\delta_{L3} = \left| -0.5 \cdot \frac{3}{1 + (2\underline{Z}_1/\underline{Z}_0) - j} \right| \tag{15.6b}$$

which are different from each other depending on the impedances and the phase-angle. The effect of the earthing can be described by the earth-fault factor δ according to VDE 0141 and is defined to be the maximum of the earth-fault factors δ_{L2} and δ_{L3}:

$$\delta = \mathrm{MAX}\{\delta_{L2}; \delta_{L3}\} = \frac{U_{\mathrm{LEmax}}}{U/\sqrt{3}} \tag{15.7}$$

with

U_{LEmax} = highest value of the power-frequency voltage line-to-earth of the non-faulted phases in case of a short-circuit with earth connection
U = voltage between phases prior to fault.

Power systems having an earth-fault factor $\delta < 1.4$ are defined as having low-impedance earthing. It should be noted that the single-phase short-circuit currents should be below the permissible limits, which are defined by the breaking capability of circuit-breakers, by the short-circuit withstand capability of switchgear, installations and equipment and by other criteria such as earthing voltage, induced voltages, and so on.

Figure 15.3 indicates the interdependence between the earth-fault factors δ_{L2} and δ_{L3} and the ratio Z_1/Z_0 and the difference of impedance angles $(\gamma_1 - \gamma_0)$. An impedance angle above 90° is only possible in case of a capacitive impedance of the zero-sequence component but not in systems with low-impedance earthing.

Figure 15.4 presents the earth-fault factor δ in relation to X_0/X_1 with the parameter R_0/X_0 when the impedance angle in the positive-sequence component remains constant. The earth-fault factor δ remains below 1.4 if $X_0/X_1 \leq 5$ can be achieved and if R_0/X_0 is kept below 0.2 (alternatively $X_0/X_1 \leq 4$ and $R_0/X_0 < 0.3$).

An impedance ratio X_0/X_1 = 2–4 can easily be achieved if the relation of zero-sequence to positive-sequence impedances of equipment is

$X_0/X_1 \approx 4$ Parallel double-circuit overhead lines
$X_0/X_1 \approx 3$ Single-circuit overhead lines and HV transformers Yy(d)
$X_0/X_1 \approx 0.3$ Unit-transformers Yd in power stations (normally not to be earthed).

which is the case with most of the equipment in power systems.

15.3 Power System Having Earthing with Current Limitation

Earthing with current limitation can be seen in some cases as a special case of low-impedance earthing, provided the earth-fault factor is below 1.4. Earthing with current limitation is applied in urban power systems having rated voltage $U_n \leq$ 20 kV. Some applications are known in systems with nominal voltage up to 132 kV see Section 11.4.3.

The criterion for the design of the earthing conditions is the value of the single-phase short-circuit current, which can be limited to some kiloamps (1 or 2 kA) in medium-voltage systems or to some tens of kiloamps in high voltage systems, for example, below the three-phase short-circuit current. To realize the scheme of

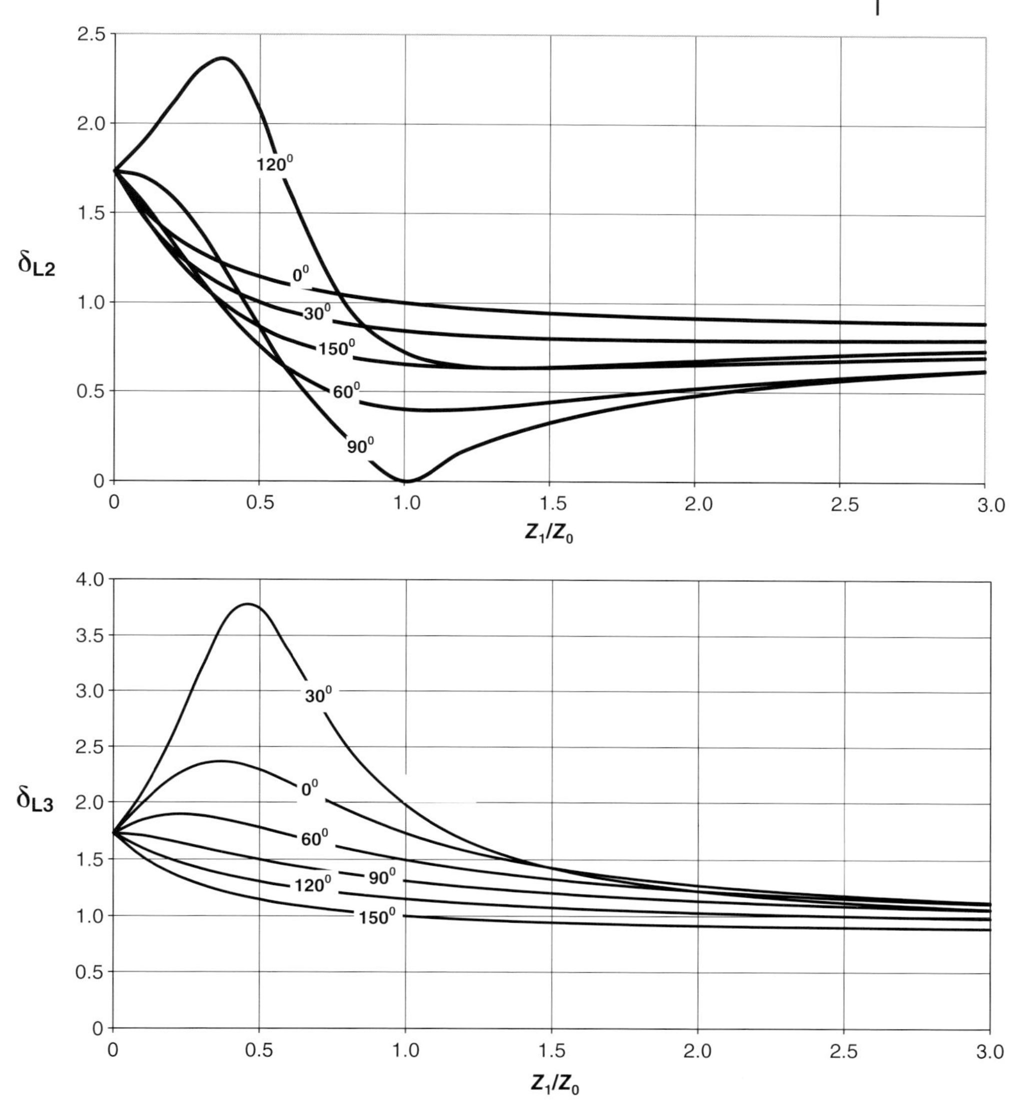

Figure 15.3 Earth-fault factors in relation to Z_1/Z_0 and $(\gamma_1 - \gamma_0)$. (a) Earth fault factor δ_{L2}; (b) earth fault factor δ_{L3}.

earthing with current limitation, the neutrals of some or all transformers are earthed through reactances or resistances to such a value that the condition for the single-phase short-circuit current is fulfilled. As a disadvantage it should be noted that the earth-fault factor might exceed the value of $\delta = 1.4$; this seems to be acceptable in medium voltage systems with nominal voltages $U_n = 10$–$20\,\text{kV}$. In high-voltage systems with $U_n = 110$–$132\,\text{kV}$ the advantages and disadvantages have to be analyzed in more detail.

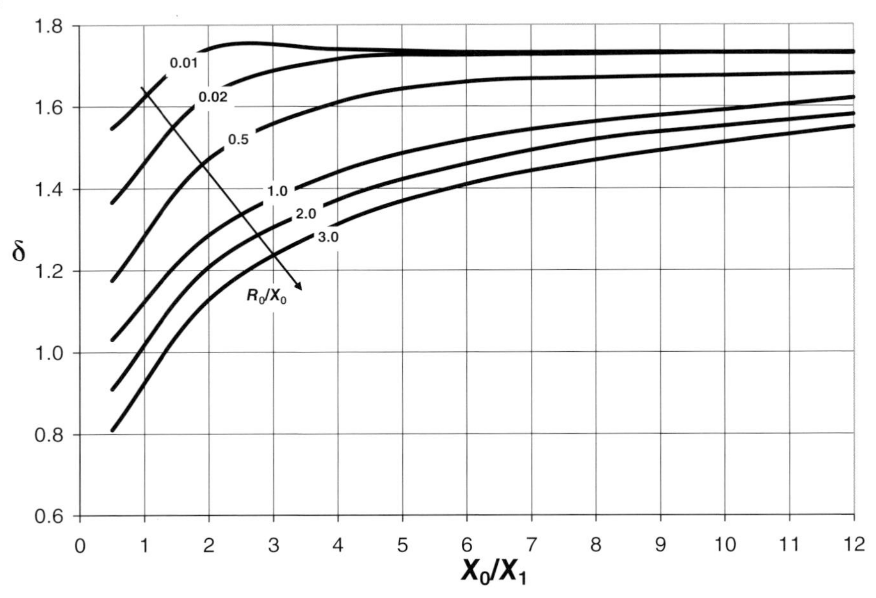

Figure 15.4 Earth-fault factor δ in relation to X_0/X_1 for different ratios R_0/X_0 and $R_1/X_1 = 0.01$.

In the estimation of the required value of the earthing impedance, the determination of the zero-sequence impedance is based on Figure 15.5, indicating the ratio I''_{k1}/I''_{k3} as well as the earth-fault factor δ in relation to X_0/X_1. As an example, a medium-voltage system with $U_n = 10\,\text{kV}$ having an initial three-phase short-circuit power $S''_{k3} = 100 - 250\,\text{MVA}$ ($I''_{k3} = 5.8 - 14.4\,\text{kA}$) is considered. As for the limitation of single-phase short-circuit current to $I''_{k1} = 2\,\text{kA}$, an impedance ratio $X_0/X_1 = 6.7$–19.6 is required. The earth-fault factor in this case will be $\delta = 1.44$–1.61. By this means, the system is no longer a system with low-impedance earthing.

15.4
Power System with Isolated Neutral

The operation of power systems with isolated neutrals is applicable to systems with nominal voltages up to 60 kV, but the main application is seen in power station auxiliary installations and industrial power systems with voltages up to 10 kV. In public supply systems, isolated neutrals are not very common.

The analysis of a single-phase earth-fault is based on Figure 15.6.

Contrary to power systems with low-impedance earthing or earthing with current limitation the capacitances phase-to-earth (capacitances in the zero-sequence component) cannot be neglected in power systems with isolated neutral as can be seen from Figure 15.6. To determine the respective parameters of the equipment, no-

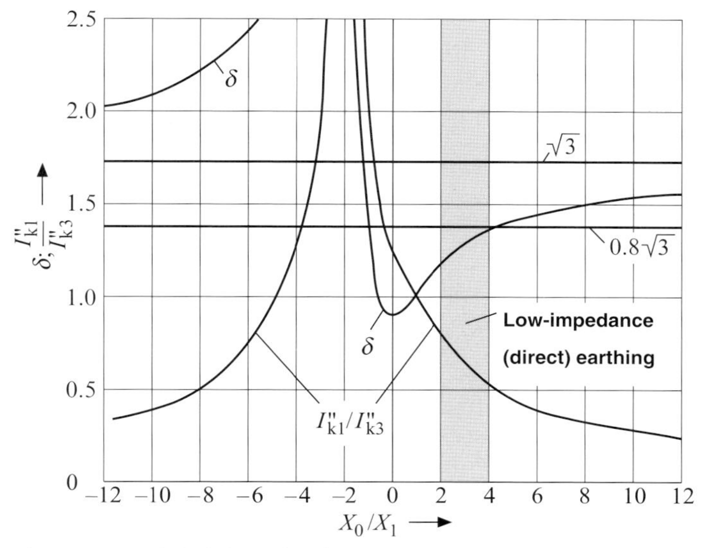

Figure 15.5 Earth-fault factor δ and ratio I''_{k1}/I''_{k3} in relation to on X_0/X_1.

load measurements are necessary. The single-phase earth-fault current in general is calculated by Equation 15.8

$$I''_{k1} = \frac{c \cdot \sqrt{3} \cdot U_n}{|2 \cdot \underline{Z}_1 + \underline{Z}_0|} \tag{15.8}$$

with

U_n = nominal system voltage
c = voltage factor according to Table 11.1
$\underline{Z}_1$, $\underline{Z}_0$ = impedance of the positive-sequence and zero-sequence component, respectively.

The zero-sequence impedance Z_0 is determined by the capacitance phase-to-earth C_E and is significantly higher than the positive-sequence impedance Z_1. The single-phase earth-fault current is determined through the capacitive component by Equation 15.9

$$\underline{I}_{L1} = \underline{I}_{CE} = j\omega \cdot C_E \cdot \sqrt{3} \cdot U_n \tag{15.9}$$

and is called capacitive earth-fault current I_{CE}. As the capacitive earth-fault current is considerably lower than a typical short-circuit current – in most of the cases even lower than the normal operating current – the single-phase fault in a system with isolated neutral is called earth-fault instead of short-circuit. The earth-fault current increases with increasing phase-to-earth capacitance and because of this with

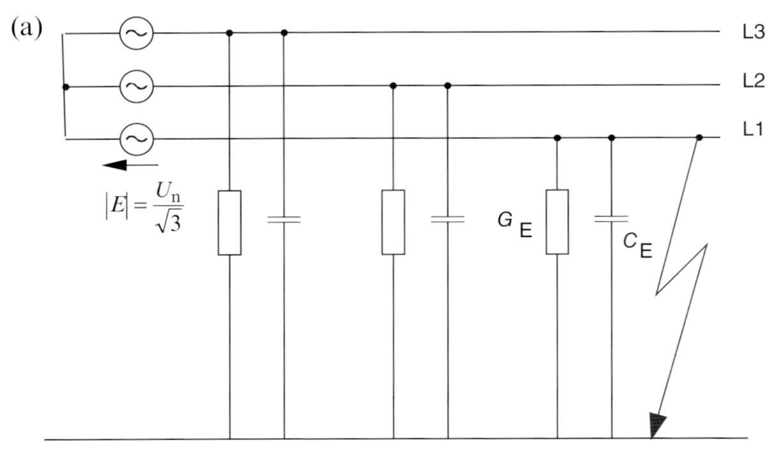

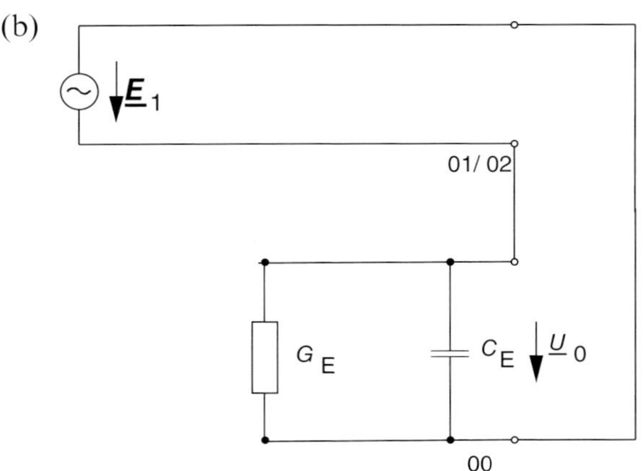

Figure 15.6 Power system with isolated neutral with single-phase earth-fault. (a) Equivalent circuit diagram in the three-phase system; (b) equivalent circuit diagram in the system of symmetrical components.

increasing line length, as can be seen from Equation 15.8. Small capacitive currents, for faults in self-restoring insulation (e.g. air), can be extinguished by themselves if they remain below some tens of amps depending on the voltage level. Figure 15.7 indicates the limits of self-extinguishing capacitive currents I_{CE} according to VDE 0228-2.

The voltages (line-to-earth) of the non-faulted phases, in case of an earth-fault, are increasing to the amount of the line-to-line voltage, as can be seen from Figure 15.8. Prior to fault the voltage potential of earth (E) and neutral (N) are identical, the line-to-earth voltages are symmetrical as well as the line-to-line voltages. During the earth-fault, the voltage of the faulted phase (L1) is identical to the voltage of the earth (E). The voltage potential of the neutral (N) is given by defini-

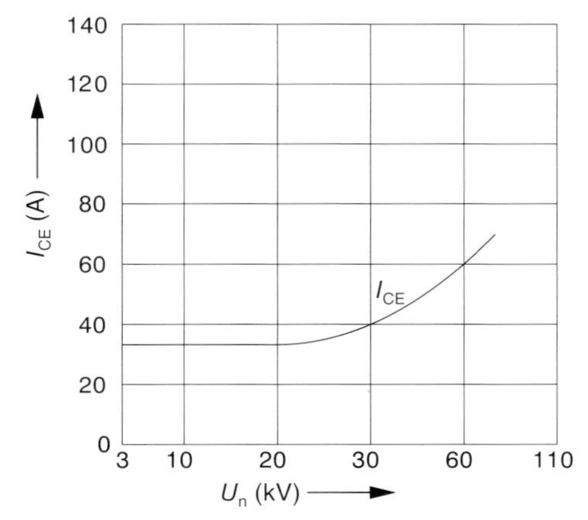

Figure 15.7 Limit of self-extinguishing capacitive currents in air according to VDE 0228-2:1987.

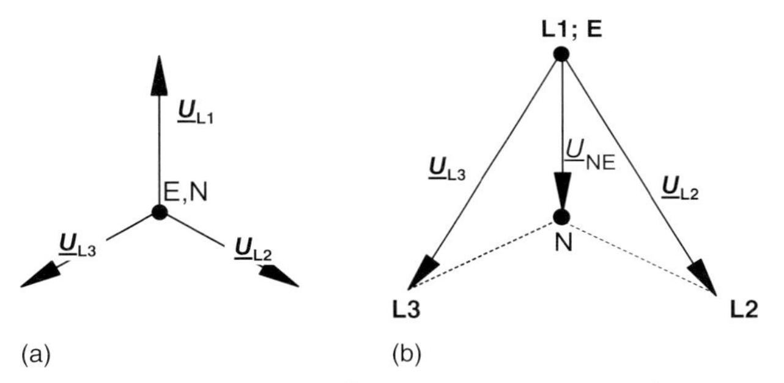

Figure 15.8 Vector diagram of voltages, power system with isolated neutral. (a) Prior to fault; (b) during earth-fault.

tion as the mean value of the three phases L1, L2 and L3, which is not changed by the earth-fault. A voltage displacement $\underline{U}_{NE}$ between neutral and earth equal the line-to-earth voltage originates from the earth-fault. The voltage displacement is equal to the voltage $\underline{U}_0$ of the zero-sequence component. As the impedance of the zero-sequence component is significantly higher than the impedances of the positive- and negative-sequence component, the displacement voltage is identical to the voltage at the transformer neutral. The voltages of the non-faulted phases are increased, but the three phase-to-phase voltages remain symmetrical as outlined in Figure 15.8b.

The capacitive earth-fault current and the recovery voltage at the fault location have a phase displacement of nearly 90°. At the instant of the maximum of the recovery voltage or shortly after it, a re-ignition of the fault arc is possible and probable. The time courses of the phase-to-earth voltages u_{L1}, u_{L2} and u_{L3} and of the displacement voltage u_{NE} as well as the earth-fault current i_{CE} are outlined in

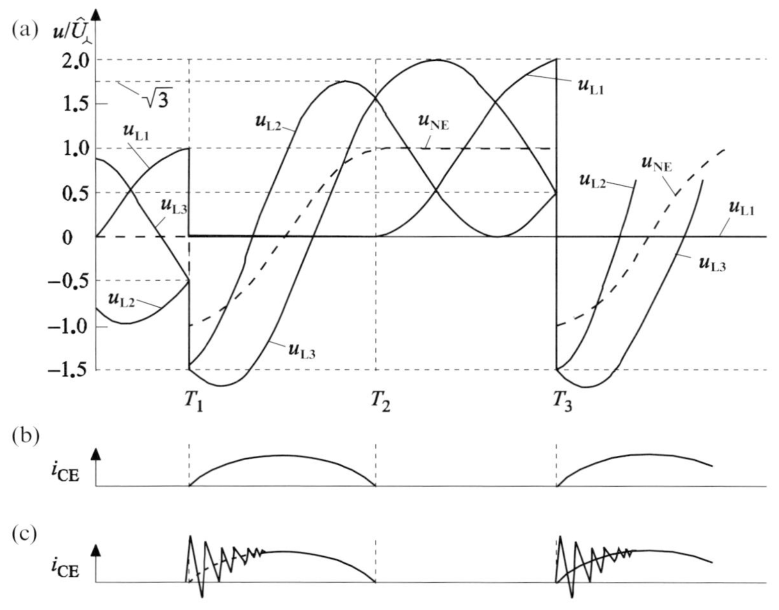

Figure 15.9 Time course of line-to-earth voltages, displacement voltage and earth-fault current. System with isolated neutral, earth-fault in phase L1.

Figure 15.9, indicating the time prior to, during and after the occurrence of the earth-fault.

The earth-fault occurs at time instant t_1, phase L1 having the maximal voltage. The line-to-earth voltage of the non-faulted phases L2 and L3 increase to the value of the line-to-line voltage. The displacement voltage u_{NE} increases from a very low value, ideally zero, to the line-to-earth voltage. The transient frequency can be calculated by Equation 15.10.

$$f \approx \frac{1}{2 \cdot \pi \cdot \sqrt{3 \cdot L_1 \cdot C_0}} \tag{15.10}$$

with

L_1 = inductance of the positive-sequence component
C_0 = capacitance of the zero-sequence component.

The earth-fault arc is extinguished at time t_2 approximately 10 ms after ignition of the earth-fault; the current i_{CE} has its zero-crossing, whereas the displacement voltage has nearly reached its peak value. The three line-to-earth voltages u_{L1}, u_{L2} and u_{L3} are symmetrical to each other, but with a displacement determined by the displacement voltage at the time of arc extinguishing; that is, the displacement voltage is equal to the peak value of the line-to-earth voltage. Approximately 10 ms after the extinguishing of the arc, the line-to-earth voltage of phase L1 reaches the new peak value $U_{L1} = 2\sqrt{2}U_{\mathrm{n}}$.

This voltage may cause a re-ignition of the earth-fault due to the very high voltage stress. This re-ignition takes place at time instant t_3 with the line-to-earth voltage of phase L1 having its peak value. The voltages of the non-faulted phases again are increasing, this time starting from a higher value and reaching the peak value nearly to $U = \sqrt{3}\sqrt{2}U_{\mathrm{n}}$.

As well as the power-frequency overvoltage in the case of an earth-fault, the transient overvoltage with frequency according to Equation 15.9 has to be considered. The overvoltage factor k_{LE} taking account of both types of overvoltages is given by the maximal peak voltage related to the peak value of line-to-earth voltage according to Equation 15.11.

$$k_{\mathrm{LE}} = \frac{u_{\mathrm{p}}}{\sqrt{2} \cdot \dfrac{U}{\sqrt{3}}} \tag{15.11}$$

with

u_{p} = maximal peak voltage during the earth fault
U = line-to-earth voltage (power-frequency).

In theory, the overvoltage factor, after multiple re-ignition of the earth-fault, can reach $k_{\mathrm{LE}} = 3.5$. Due to the system damping, the overvoltage factor will be below $k_{\mathrm{LE}} < 3$ in most cases.

15.5 Power System with Resonance Earthing (Petersen Coil)

15.5.1 General

Power systems with resonance earthing are widely in operation in Central European countries. Statistics on the German power system [87] indicate that 87% of the MV systems, having nominal voltages U_{n} = 10–30 kV, and nearly 80% of 110 kV-systems are operated with resonance earthing (criterion: total line length). Some MV systems are operated with a combined scheme of resonance earthing (normal operating conditions) and low-impedance earthing in case of earth-fault. Resonance earthing is therefore the predominant type of system earthing in Germany for power systems with voltages from 10 kV up to 110 kV. In other countries, such as India, South Africa and China, power systems with resonance earthing have gained increasing importance during recent decades, but they are still not so common in practice as systems with low-impedance earthing.

Resonance earthing is realized by earthing of one or several neutrals of transformers through reactances (Petersen coils) – normally adjustable – which

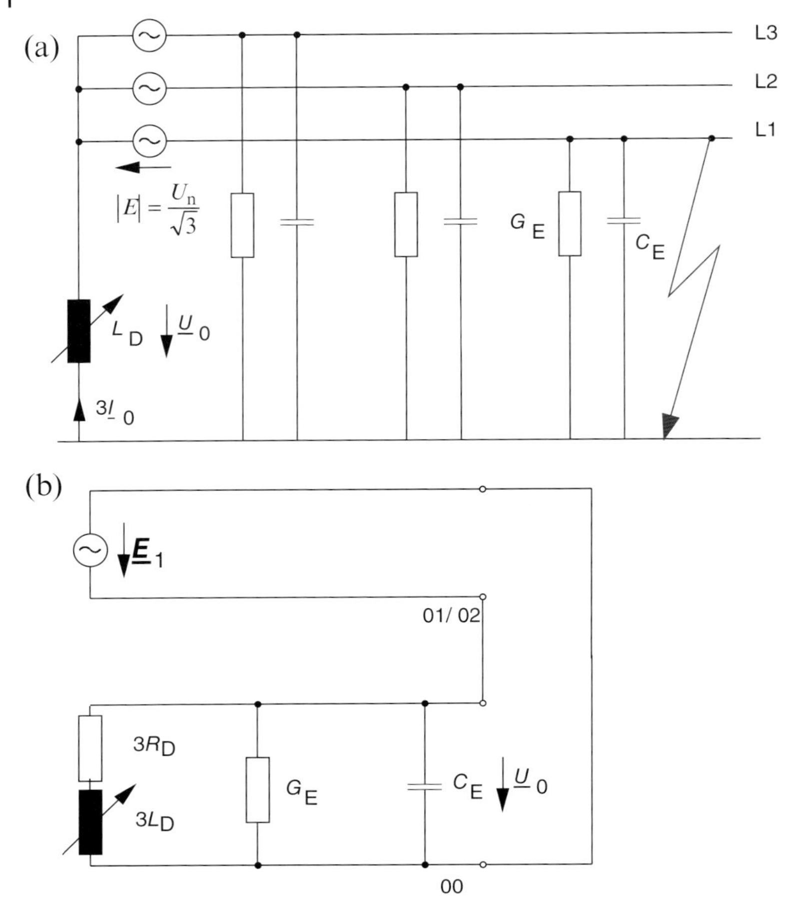

Figure 15.10 System with resonance earthing, earth-fault in phase L1. (a) Equivalent circuit diagram in the three-phase system; (b) equivalent diagram in the system of symmetrical components.

will be set in resonance to the line-to-earth capacitances of the system. The principal of the arrangement of a power system with resonance earthing is outlined in Figure 15.10.

The impedances of transformers and lines of the positive-sequence component can be neglected compared with those of the zero-sequence component due to the order of magnitude of the impedances. The admittance of the zero-sequence component is given by Equation 15.12

$$\underline{Y}_0 = \mathrm{j}\omega \cdot C_\mathrm{E} + \frac{1}{3 \cdot R_\mathrm{D} + \mathrm{j}3 \cdot X_\mathrm{D}} + G_\mathrm{E} \tag{15.12}$$

with

C_E = line-to-earth capacitances of the system
ω = angular frequency of the system
R_D = resistance of the Petersen coil
X_D = reactance of the Petersen coil, $X_D = \omega L_D$
G_E = admittance representing the conductive line losses.

After some conversions, Equation 15.13 follows:

$$\underline{Y}_0 = j\omega \cdot C_E \cdot \left(1 - \frac{1}{3 \cdot \omega^2 \cdot L_D \cdot C_E \cdot \left(1 - j\frac{R_D}{X_D}\right)}\right) + G_E \tag{15.13}$$

The impedance of the Petersen coil appears with its threefold value in the zero-sequence component [80]. It is assumed that $R_D << X_D$ and that the losses of the Petersen coil are summed with the line-to-earth losses and are represented as admittance G_E of the line. The admittance in the zero-sequence component is then given by Equation 15.14a.

$$\underline{Y}_0 = j\omega \cdot C_E \cdot \left(1 - \frac{1}{3 \cdot \omega^2 \cdot L_D \cdot C_E}\right) + G_E \tag{15.14a}$$

The maximal impedance is obtained if the imaginary part according to Equation 15.13 is equal to zero and the current of the Petersen coil I_D is equal to the capacitive current I_{CE} of the system. As indicated in Figure 15.10, the line-to-earth capacitance C_E, the reactance $3L_D$ and the ohmic losses $R_0 = 1/G_E$ form a parallel resonance circuit with resonance frequency according to Equation 15.15.

$$\omega = \frac{1}{\sqrt{3 \cdot L_D \cdot C_E}} \tag{15.15}$$

The resonance frequency in case of resonance earthing is to be the nominal frequency $f = 50$ Hz or $f = 60$ Hz. Defining the detuning factor v according to Equation 15.16a,

$$v = \frac{I_D - I_{CE}}{I_{CE}} = 1 - \frac{1}{3 \cdot \omega^2 \cdot L_D \cdot C_E} \tag{15.16a}$$

and the system damping δ_0 according to Equation 15.16b,

$$\delta_0 = \frac{G_E}{\omega \cdot C_E} \tag{15.16b}$$

the admittance of the zero-sequence component is given by Equation 15.14b.

$$\underline{Y}_0 = \omega \cdot C_E \cdot (jv + \delta) \tag{15.14b}$$

The admittance will be minimal and the impedance will be maximal in case of resonance tuning ($\nu = 0$). The earth-fault current I_{Res} in general is obtained with Equation 15.17a.

$$\underline{I}_{Res} \approx \sqrt{3} \cdot U_n \cdot \omega \cdot C_E \cdot (j\nu + \delta_0) \tag{15.17a}$$

In case of resonance tuning ($\nu = 0$) the earth-fault current is a pure ohmic current and is calculated according to Equation 15.17b.

$$\underline{I}_{Res} \approx \sqrt{3} \cdot U_n \cdot \omega \cdot C_E \cdot \delta_0 \tag{15.17b}$$

The line-to-earth voltages of the non-faulted phases increase to the value of the line-to-line voltage in case of a single-phase earth-fault, which is further increased due to asymmetrical system voltages, resulting in a higher displacement voltage between neutral and earth. In order to avoid the high voltages in case of exact resonance tuning, a small detuning of 8–12% is chosen in practice.

The task of resonance earthing is to reduce the earth-fault current at the fault location to the minimum or nearly to the minimum by adjusting the Petersen coil to resonance or nearly to resonance with the line-to-earth capacitances. The ohmic part of the residual current I_{Res} cannot be compensated by this. If the residual current is small enough, self-extinguishing of the arc in air at the fault location is possible. VDE 0228-2:1987 defines the limits for self-extinguishing of residual currents I_{Res} (and capacitive earth-fault currents I_{CE}) for different voltage levels as outlined in Figure 15.11. It can be seen from Figure 15.11 that the limit for ohmic currents is nearly twice the limit for capacitive currents.

The Petersen coil can only be tuned for one frequency (nominal frequency) in resonance. Harmonics that are present in the system voltage increase the residual current at the fault location.

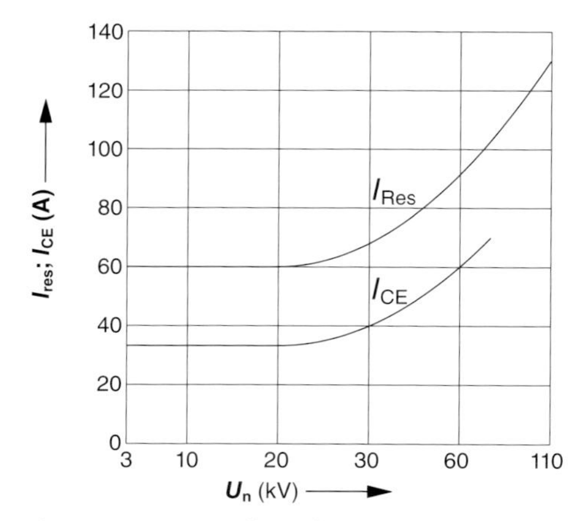

Figure 15.11 Limits for self-extinguishing of ohmic currents I_{Res} and capacitive currents I_{CE} according to VDE 0228-2:1987.

As the line-to-earth capacitances are changing during system operation, for example, due to switching of lines, the reactance of the Petersen coil has to be adjusted to the system conditions. Reliable criteria have to be established to tune the Petersen coil in resonance with the line-to-earth capacitances.

15.5.2 Calculation of Displacement Voltage

In real power systems, the line-to-earth capacitances are unequal: for example, in case of a transmission line due to different clearance of the phase-conductors above ground or in case of cables due to manufacturing tolerances. Under normal operating conditions, a displacement voltage between transformer neutral and earth U_{NE} can be measured. This voltage is equal the voltage U_0 in the zero-sequence component. The calculation of the displacement voltage can be carried out in the three-phase system (Figure 15.12a) as well as with the system of symmetrical components (Figure 15.12b).

Based on Figure 15.12a, the displacement voltage is calculated according to Equation 15.18.

$$\underline{U}_{NE} = \frac{U_n}{\sqrt{3}} \cdot \frac{j\omega \cdot (C_{L1E} + \underline{a}^2 \cdot C_{L2E} + \underline{a} \cdot C_{L3E})}{j\omega \cdot (C_{L1E} + C_{L2E} + C_{L3E}) - j\dfrac{1}{\omega \cdot L_D} + 3 \cdot G_E} \tag{15.18}$$

with

U_n = nominal system voltage
ω = angular frequency of the system
C_{L1E}; C_{L2E}; C_{L3E} = line-to-earth capacitances according to Figure 15.12a
L_D = inductance of the Petersen coil
G_E = admittance representing the conductive line losses.

If the line-to-earth capacitances are different and if the asymmetry is assumed to reside in phases L1 and L2, the capacitances are defined according to Equation 15.19.

$$C_{L1E} = C_E + \Delta C_{L1E} \tag{15.19a}$$

$$C_{L2E} = C_E + \Delta C_{L2E} \tag{15.19b}$$

$$C_{L3E} = C_E \tag{15.19c}$$

with

ΔC_{L1E}; ΔC_{L2E} = asymmetry of the line-to-earth capacitances.

The displacement voltage is given by Equation 15.20.

$$\underline{U}_{NE} = \frac{U_n}{\sqrt{3}} \cdot \frac{\Delta C_{L1E} + \underline{a}^2 \cdot \Delta C_{L2E}}{(3 \cdot C_E + \Delta C_{L1E} + \Delta C_{L2E}) - j\dfrac{1}{\omega \cdot L_D} + 3 \cdot G_E} \tag{15.20}$$

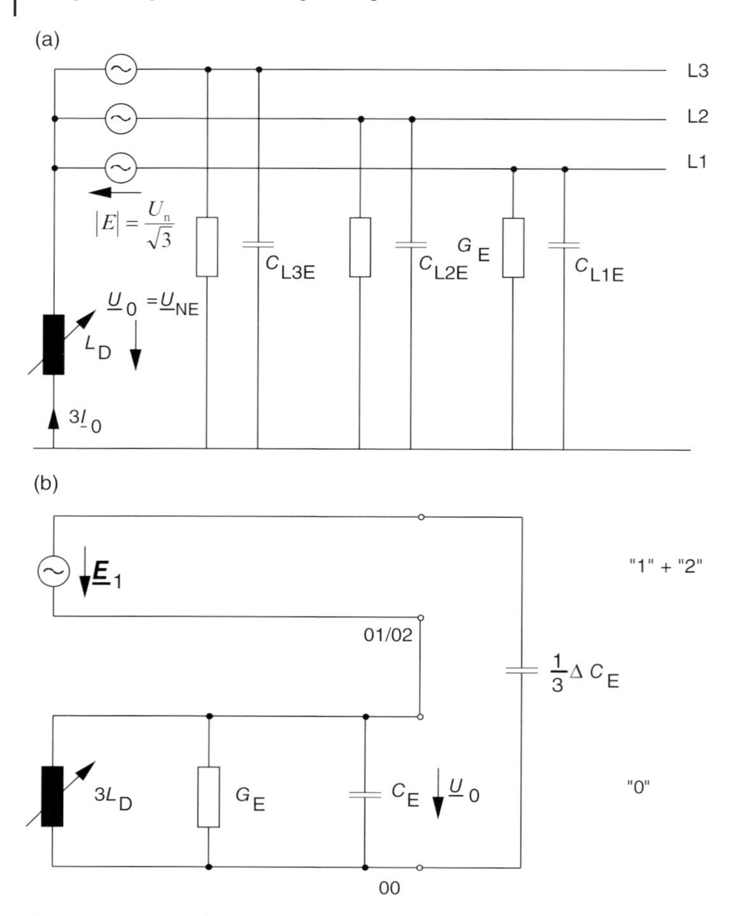

Figure 15.12 Equivalent circuit diagrams of a power system with asymmetrical line-to-earth capacitances. (a) Equivalent circuit diagram in the three-phase system; (b) equivalent circuit diagram in the system of symmetrical components.

Defining the asymmetry factor k according to Equation 15.21a,

$$\underline{k} = \frac{C_{L1E} + \underline{a}^2 \cdot C_{L2E} + \underline{a} \cdot C_{L3E}}{C_{L1E} + C_{L2E} + C_{L3E}} = \frac{\Delta C_{L1E} + \underline{a}^2 \cdot \Delta C_{L2E}}{3 \cdot C_E + \Delta C_{L1E} + \Delta C_{L2E}} \tag{15.21a}$$

the system damping δ_0 according to Equation 15.21b,

$$\delta_0 = \frac{3 \cdot G_E}{\omega \cdot (C_{L1E} + C_{L2E} + C_{L3E})} = \frac{3 \cdot G_E}{\omega \cdot (3 \cdot C_E + \Delta C_{L1E} + \Delta C_{L2E})} \tag{15.21b}$$

and the detuning factor v according to Equation 15.21c,

$$\nu = \frac{\frac{1}{\omega \cdot L_{D}} - \omega \cdot (C_{L1E} + C_{L2E} + C_{L3E})}{\omega \cdot (C_{L1E} + C_{L2E} + C_{L3E})} = \frac{\frac{1}{\omega \cdot L_{D}} - \omega \cdot (3 \cdot C_{E} + \Delta C_{L1E} + \Delta C_{L2E})}{\omega \cdot (3 \cdot C_{L1E} + \Delta C_{L2E} + \Delta C_{L3E})} \tag{15.21c}$$

the displacement voltage $\underline{U}_{NE}$ is calculated from Equation 15.22.

$$\underline{U}_{NE} = \frac{U_{n}}{\sqrt{3}} \cdot \frac{k}{\nu + j\delta_{0}} \tag{15.22}$$

Assuming the asymmetric capacitance ΔC_{E} concentrated in phase L1 ($\Delta C_{E} >> \Delta C_{L1E}$ and $\Delta C_{E} >> \Delta C_{L2E}$),the displacement voltage $\underline{U}_{NE}$, equal to the voltage in the zero-sequence component $\underline{U}_{0}$, is calculated with the system of symmetrical components based on Figure 15.12b according to Equation 15.23.

$$\underline{U}_{0} = \frac{U_{n}}{\sqrt{3}} \cdot \frac{j\omega \cdot \Delta C_{E}}{j\omega \cdot 3 \cdot C_{E}} \cdot \frac{1}{1 - \left(\frac{1}{3 \cdot \omega^{2} \cdot L_{D} \cdot C_{E}} \right) - j\frac{G_{E}}{\omega \cdot C_{E}}} \tag{15.23}$$

The asymmetry factor k, the system damping δ_{0} and the detuning factor ν can be calculated based on these assumptions according to Equations 15.24.

$$k = \frac{\Delta C_{E}}{3 \cdot C_{E}} \tag{15.24a}$$

$$\delta_{0} = \frac{G_{E}}{\omega \cdot C_{E}} \tag{15.24b}$$

$$\nu = 1 - \frac{1}{3 \cdot \omega^{2} \cdot L_{D} \cdot C_{E}} \tag{15.24c}$$

The displacement voltage $\underline{U}_{NE}$, equal to the voltage in the zero-sequence component $\underline{U}_{0}$, is then calculated with Equation 15.25.

$$\underline{U}_{NE} = \underline{U}_{0} = \frac{U_{n}}{\sqrt{3}} \cdot \frac{k}{\nu + j\delta_{0}} \tag{15.25}$$

The polar plot of the displacement voltage $\underline{U}_{NE}$ according to Equations 15.22 and 15.25 outlined in Figure 15.13 indicates a circular plot through the zero point. The phase-angle of the detuning factor $\nu = 0$ is determined by the phase-angle of the capacitive asymmetry. The diameter of the polar plot is defined according to Equation 15.25 as the ratio of capacitive asymmetry k and damping δ_{0}.

The capacitive asymmetry is comparatively high in power systems with overhead transmission lines, resulting in a sufficiently high displacement voltage. Cable systems have a comparatively small asymmetry, resulting for most cable

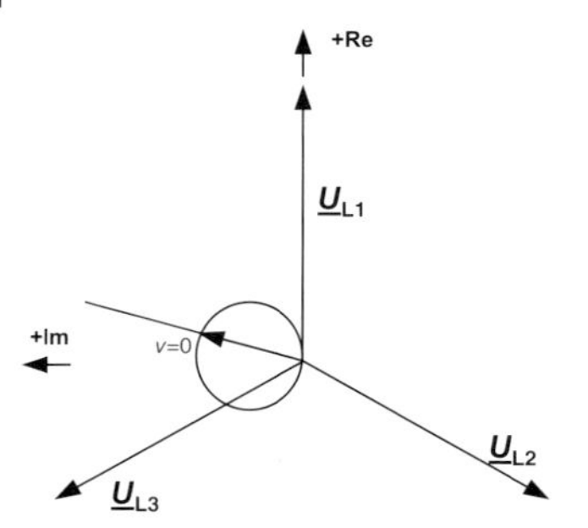

Figure 15.13 Polar plot of the displacement voltage in a power system with resonance earthing.

systems in an insufficiently low displacement voltage and problems while tuning the Petersen coil into resonance. Capacitors between two phases or between one phase and earth will increase the displacement voltage to the required value.

15.5.3 Tuning of the Petersen Coil

The Petersen coil can be constructed as a plunger-coil (tuning-coil) with continuous adjustment of the reactance, which can be tuned into resonance by successive operation. The displacement voltage measured at the Petersen coil is maximal in case of resonance tuning; the value depends on the capacitive asymmetry and on the losses of the reactor. The earth-fault current will be minimal in this case and the power frequency component of the capacitive earth-fault current is compensated by the reactive current of the Petersen coil. Figure 15.14 indicates the displacement voltage and the residual current for different tuning of the reactor.

The displacement voltage is to be limited to $U_{NE} < 10\,kV$ in 110 kV-systems. From Figure 15.14 it can be seen that the residual current is increased if the displacement voltage is reduced. Residual currents above 130 A in 110 kV systems or above 60 A in 10 kV systems are not self-extinguishing; both values define the tuning limits of the Petersen coil as indicated in Figure 15.14. Tuning of the Petersen coil can be done in such a way that the resonance circuit is either capacitive (undertuning; $v < 0$), resulting in an ohmic-capacitive residual current or inductive (overtuning; $v > 0$), resulting in an ohmic-inductive residual current at the earth-fault location. A small overtuning up to $v = 10\%$ is often recommended as the displacement voltage will not increase in case of switching of lines, because the capacitances will be reduced by this and the resonance circuit will be detuned without any further adjustment. The limits for the displacement voltage and the residual current as indicated in Figure 15.14 have to be guaranteed even under outage conditions.

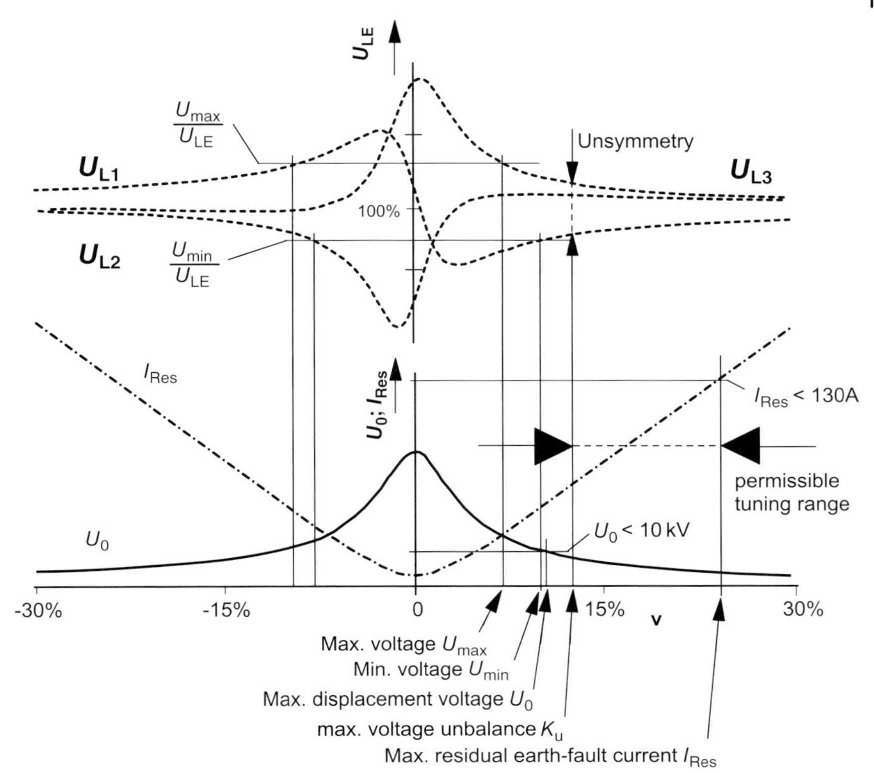

Figure 15.14 Voltages and residual current in case of an earth-fault; displacement voltage without earth-fault; system [87] and [88] parameters are: $U_n = 110\,kV$; $I_{CE} = 520\,A$; $d = 3\%$; $k = 1.2\%$.

Figure 15.14 also indicates the line-to-earth voltages for different tuning factors (system parameters are: $U_n = 110\,kV$; $I_{CE} = 520\,A$; $d = 3\%$; $k = 1.2\%$), which also limit the range of detuning of the Petersen coil. Assuming a minimal permissible voltage of $U_{min} = 0.9U_n/\sqrt{3}$ according to IEC 60038, a maximal permissible voltage according to IEC 60071-1 of $U_{max} = 123\,kV/\sqrt{3}$ and a permissible asymmetry of the three voltages according to DIN EN 50160 of $k_u = 2\%$, it can be seen that the permissible tuning range of the Petersen coil is $v = 12–22\%$.

All considerations so far are based on a linear current–voltage characteristic of the Petersen coil. Figure 15.15 indicates the nonlinear characteristic of a Petersen coil ($U_r = 20\,kV/\sqrt{3}$; $I_r = 640\,A$) for minimal and maximal adjustment.

Due to the nonlinear characteristic, the minimum of the residual current is not achieved at the maximal displacement voltage (adjustment criterion of the Petersen coil). The difference is typically in the range of 3–15% of the rated current as outlined in Figure 15.16.

(a)

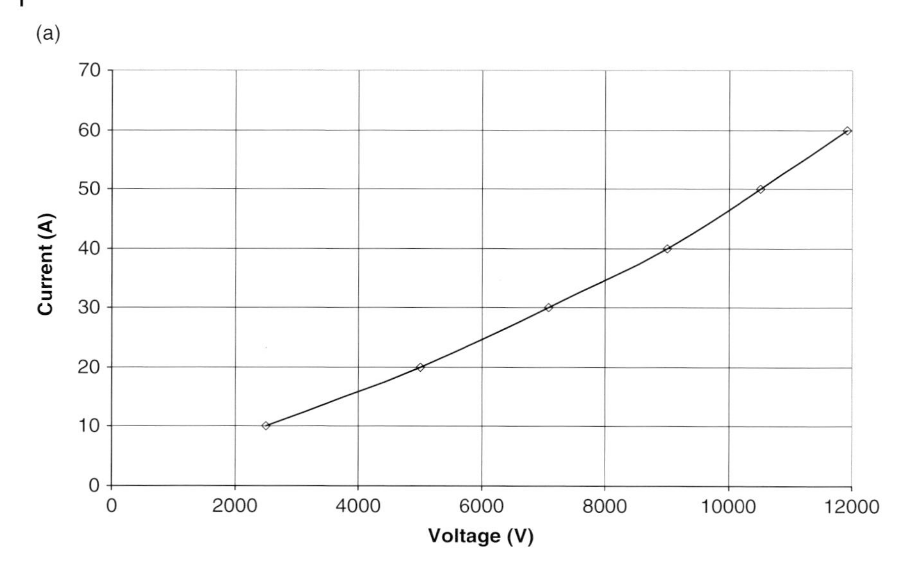

(b)

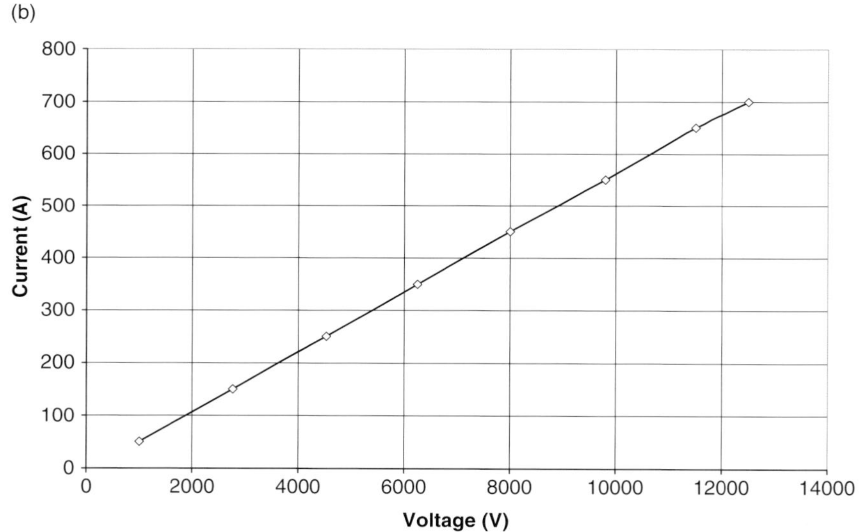

Figure 15.15 Current–voltage characteristic of a Petersen coil; $U_r = 20\,\text{kV}/\sqrt{3}$; $I_r = 640\,\text{A}$. (a) Minimal adjustment (50 A); (b) maximal adjustment (640 A).

15.5.4
Residual Current Compensation

The great advantage of resonance earthing is the self-extinguishing effect of the nearly ohmic residual currents in case of single-phase faults in air; see Figure 15.11. The disadvantage is the possibility of reignition of earth faults, permanent

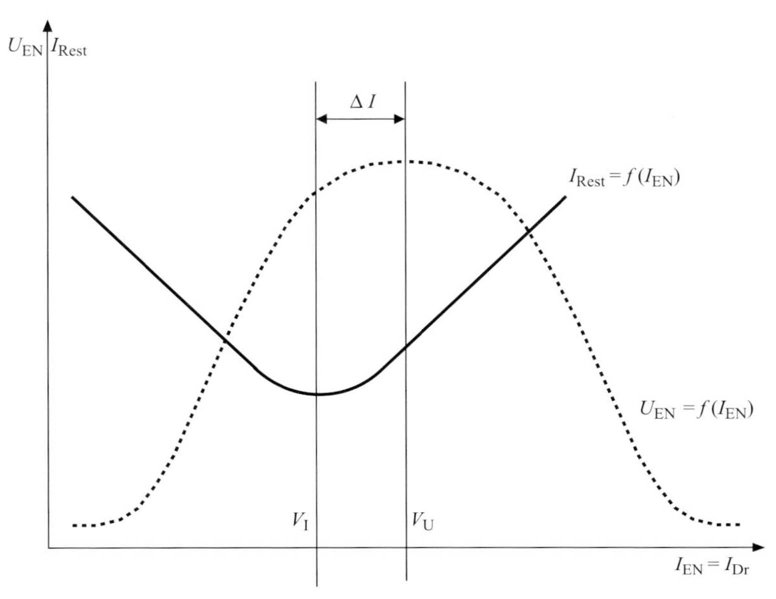

Figure 15.16 Displacement voltage in non-faulted operation and residual current under earth-fault conditions; non-linear characteristic of the Petersen coil.

ground faults, and/or double earth faults. The residual current at the fault location comprises several components:

- Current due to detuning of the Petersen coil as compared with exact resonance tuning.
  Exact resonance tuning should be avoided, in order to limit the voltage increase and/or voltage drop in of the phase voltages; see Figure 15.14.
- Ohmic residual current due to the resistance of the Petersen coil, other resistive components, such as corona losses, and leakage resistances of insulation.
  This part of the residual current cannot be avoided due to the ohmic losses of the coil and the insulation of the line; see Table 14.4.
- Capacitive current due to asymmetrical capacitances of the different line-to-earth capacitances of the three phases of the power system.
  The capacitive asymmetry is comparatively high with overhead lines and comparatively low in cable networks; see Section 15.5.2.
- Currents with higher frequencies due to prevailing harmonic (and interharmonic) voltages.

The harmonic content of the residual current, caused by random occurrence of the harmonic voltages, cannot be determined accurately in advance. Individual studies [93] and [94] indicated, for example, 150-Hz components of residual currents up to 80 A in 110-kV systems. The harmonic components of the residual current are furthermore determined by the resonance frequency of the network. A first estimate of the resonant frequency is provided by Equation 15.26.

$$f_{\text{res}} = \frac{f_1}{U_n} \cdot \sqrt{0.2 \cdot \frac{S''_{k3}}{\omega C_E}} \tag{15.26}$$

With:

f_1 Power system frequency
S''_{k3} Three-phase short-circuit power
C_E Line-to-ground capacitance

Basically, two different methods for the residual current compensation are possible [94–98], as stated in the next section.

15.5.4.1 Residual Current Compensation by Shifting of the Neutral-Point Displacement Voltage

The capacitive asymmetry of the power system is changed by connecting a single-phase capacitor to one of the phase conductors; see Section 15.5.2. By choosing a suitable size and location of the connection of the single-phase capacitor, the capacitive asymmetry is reduced; the neutral-point displacement voltage may be tuned to almost zero [94]. The residual current will then also be near to zero. This method is applicable only for one case of capacitive asymmetry. If the asymmetry changes, for example, due to rearrangements in the system topology, the displacement voltage can only be reduced to a minor extend and the residual current can only be compensated to a reduced amount. Residual currents due to harmonics cannot be compensated. The method however is simple, reliable, and economical.

15.5.4.2 Residual Current Compensation by Injection of Current into the Neutral

The method of injecting current into the neutral provides that a static inverter is connected to the auxiliary winding of the Petersen coil, having a rated power of some percent of the rating of the coil itself [94] and [98]. If necessary, the method can be combined with a switchable single-phase capacitor as mentioned above. Under normal operating conditions, the system damping, asymmetry, and detuning of the Petersen coil are monitored and compared with specified limits. In principle, the control method is based on the measurement of the zero-sequence admittance by injecting a current into the neutral. A measurement at no-fault conditions (normal operating condition) is taken as reference value for the admittance. By slight modulation of the set-point of the static inverter (change of the injected current), the required displacement voltage at the Petersen coil for the measurement can be modified. In case of earth fault, a second measurement is carried out. Magnitude and phase-angle of the injected current are modified by the static inverter, and the measured admittance is changing as well. If the admittance, in case of an earth fault, is equal to the reference value measured under normal operating conditions, the displacement voltage is zero and the residual

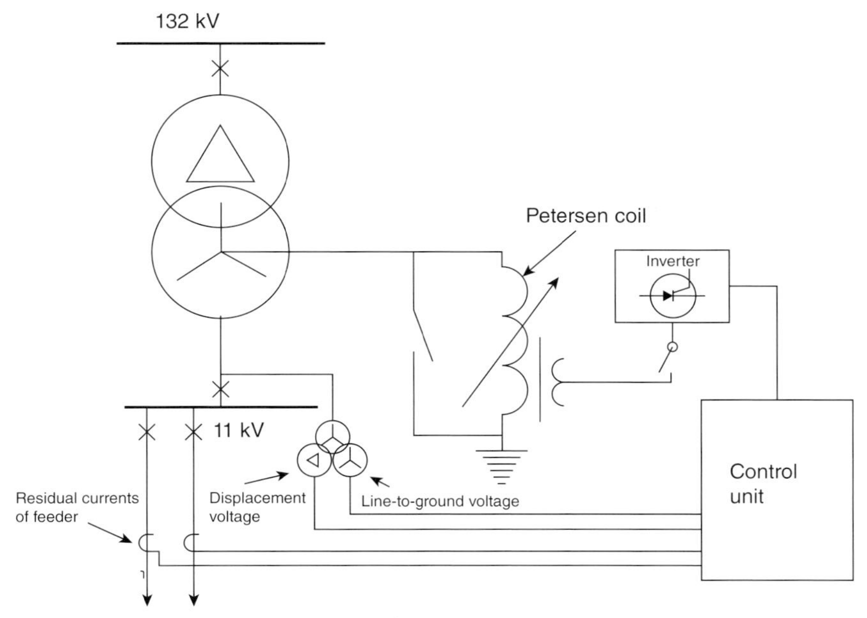

Figure 15.17 Scheme for compensation of residual currents.

current is zero as well. With this method, the faulty feeder and the fault location can be determined with an accuracy of a few percent related to the length of the faulty feeder. The method of compensation by injection of current into the neutral can be applied for any case of asymmetry. Residual earth-fault currents caused by harmonic voltages cannot be compensated using this method. First investigations to improve the method to compensate harmonic currents are described in References [96] and [97]. Figure 15.17 outlines the connection of a residual current compensation unit in a 132/11-kV substation.

15.6 Earthing of Neutrals on HV Side and LV Side of Transformers

The selection of the type of neutral earthing on the HV side and LV side of transformers demands special attention. The neutral earthing on one side of the transformer has an influence on the system performance on the other side in the case of earth-faults or single-phase short-circuits as the voltages in the zero-sequence component are transferred from one side of the transformer to the other. The neutral earthing of a 110/10 kV transformer (vector group Yyd) according to Figure 15.18 is taken as an example.

The impedances $\underline{Z}_{E1}$ and $\underline{Z}_{E2}$ according to Figure 15.18 represent the earthing conditions for the different types of neutral earthing. In case of a single-phase

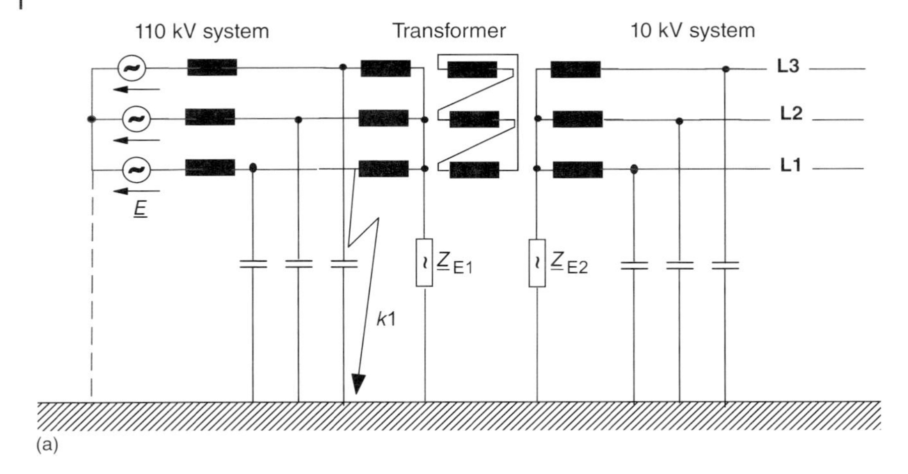

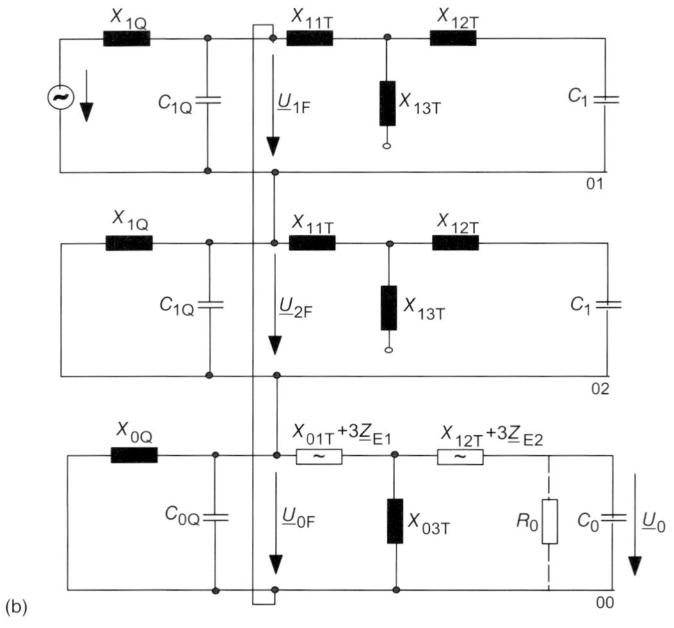

Figure 15.18 Transformation of voltage in the zero-sequence component of transformers in case of single-phase faults. (a) Equivalent circuit diagram in three-phase system; (b) equivalent circuit diagram in the system of symmetrical components.

Table 15.2 Voltages u_0 in the zero-sequence component transferred through a 110 kV/10 kV transformer in the case of single-phase fault in the 110 kV system of Figure 15.17.

10 kV system		**110 kV system and compensation winding of the transformer**		
		Low-impedance earthing $u_0 = U_0/(U_n/\sqrt{3})$		**Resonance earthing**
Z_{E2}	**Limitation I_{k1} (A)**	**With compensation winding**	**Without compensation winding**	**With compensation winding**
Low-resistance earthing. $Z_{E2} = 0\,\Omega$		0.2	0.6	0.03
Current limitation. Z_{E2} inductive	2000	0.2	0.6	0.03
	500	0.25	0.7	0.04
Current limitation. Z_{E2} ohmic	2000	0.2	0.6	0.03
	500	0.2	0.6	0.03
Resonance earthing		<7	>10	<0.3
Isolated neutral		Voltage transfer through stray capacitances		

U_n = nominal system voltage.

fault in the high-voltage system (110 kV), the voltage U_0 in the zero-sequence component is transferred to the medium-voltage system (10 kV) in the same amount. Similar considerations indicate that the voltage in the zero-sequence component is transferred to the HV side in the case of a single-phase fault in the LV system. In both cases, a fault current is measured in the system, which has no fault. Table 15.2 indicates the results of a fault-analysis [89] and [90] with the voltages transferred through the transformer in case of faults.

The 110 kV/10 kV transformer can be operated with low-impedance earthing on both sides if a third winding (compensation winding, delta-connection) is available, as can be seen from Table 15.2. If the transformer is not equipped with compensation winding, the voltages in the zero-sequence component may reach values up to 70% of the line-to-earth voltage.

Low-impedance earthing on the 110 V side and resonance earthing on the 10 kV side should be avoided due to high voltages in the zero-sequence component, which furthermore depend on the tuning of the Petersen coil. The maximal voltage in this case is not reached for resonance tuning but depends on the ratio of the capacitive reactance to the resistance in the zero-sequence component X_{C0}/R_0. The strategy of limiting the displacement voltage under normal operating conditions according to Section 15.5.2 may result in an increased displacement voltage in the 10 kV system (resonance earthing) in case of an earth-fault in the 110 kV system (low-impedance earthing).

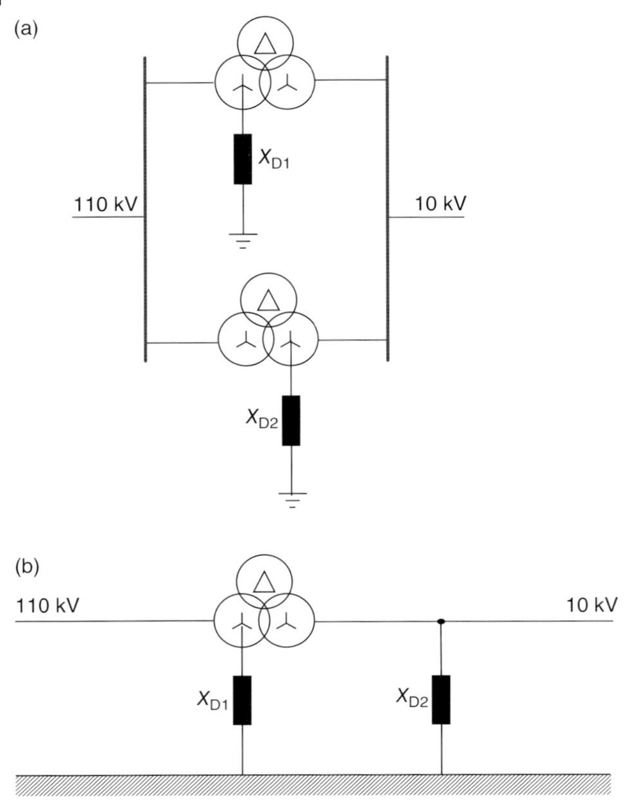

Figure 15.19 Alternate earthing of transformer neutrals by Petersen coils. (a) Two parallel transformers; (b) earthing at artificial neutral with reactor X_{D2}.

Resonance earthing in the 110 kV system can be combined with all types of neutral earthing in the 10 kV system if the transformer is equipped with a compensation winding. The connection of Petersen coils to both neutrals (110 kV and 10 kV) has to be investigated for special cases and is not generally recommended. The voltage transfer by stray capacitances in case of isolated neutral in the 10 kV system can be reduced by installing capacitances in the 10 kV system. If the earthing of both neutrals of transformers by Petersen coils cannot be avoided in the same substation in case of two parallel transformers, the neutral earthing should be carried out as indicated in Figure 15.19a. If only one transformer is installed, one Petersen coil X_{D1} can be connected directly to the transformer, the second one X_{D2} should be connected at an artificial neutral according to Figure 15.19b.

If the feeding system (e.g. 110 kV) is operated with low-impedance earthing and the medium voltage system (e.g. 20 kV) is earthed through Petersen coils or by fault-limiting impedance, fault currents will occur in the medium-voltage system

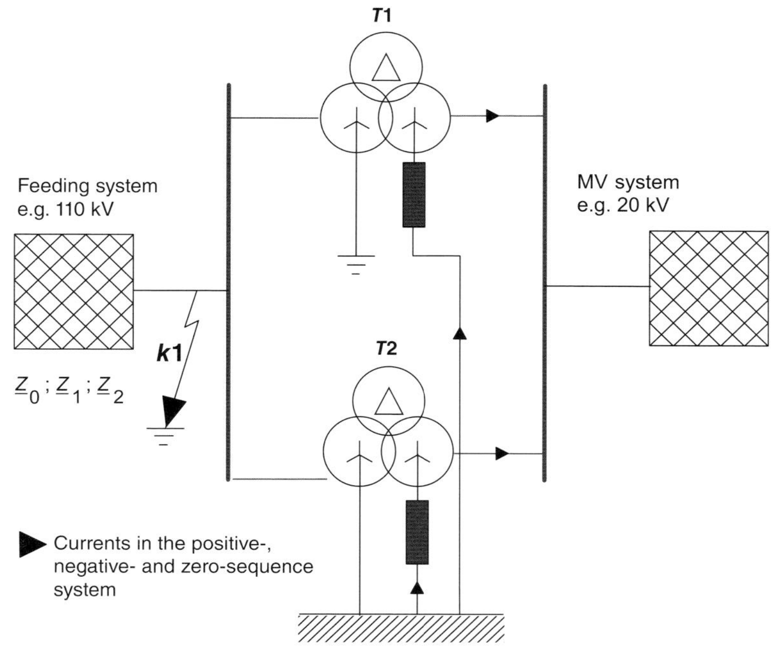

Figure 15.20 Fault current in the MV system in case of a short-circuit in the HV system.

in the case of a single-phase short-circuit in the high-voltage system, as outlined in Figure 15.20. The magnitude of the fault current depends on the impedance of the earthing in the medium-voltage system. In some cases, this current may exceed the rated current of the transformer, thus causing operation of power system protection on the MV side [91] and [92].

16
Tendering and Contracting

16.1
General (Project Definition)

The preceding chapters have dealt with the needs and basics of power system planning and principles as well as focusing on basic design aspects of main power system components. Based on recommendations resulting from various studies, the project engineering phase, as the next step, mainly covers the design, tendering and contracting/project implementation phases.

In this chapter the focus is set upon the general procedures and activities during the project engineering phase up to contracting/project implementation, after the project has been approved for realization and execution.

The creation of projects in the field of power and energy may be required for various reasons (refer also to Chapter 3), such as:

- Construction of new substations, overhead transmission lines or power cable systems as a result of power system planning
- Extension of existing power systems due to expansion of the supply areas, increase in power demand, addition of new supply point(s)
- Re-configuration of existing power network(s) to cope with developments and requirements in cities, regions or countries
- Interconnection between power systems
- Development of new power supply network(s) for new industrial complexes, newly created cities, resorts
- Addition of new or extension of existing power plants
- Improvement of existing power supply or electrical networks to increase reliability, operation flexibility, reduce network losses
- Rehabilitation or refurbishment of existing plants, substations, lines or components to meet the increasing requirements of power system development and expansion.

The engineering activities required to realize the planned project are carried out by the engineering divisions of the client (e.g. a utility), or are assigned to an engineering company.

Power System Engineering: Planning, Design and Operation of Power Systems and Equipment, Second Edition.
Jürgen Schlabbach and Karl-Heinz Rofalski.

The prior studies and planning work provide the basis for the definition of parameters and design criteria for the project material and components. Further engineering activities include development and preparation of technical concepts, specifications, layout plans, route plans, project implementation schedules and estimation of project costs. All the data and plans are compiled for the client's and, if required, the authority's (e.g. the government ministry of power and energy) review and approval to proceed with the next step.

After the decision is made by the client/authority to set up the project as part of its development plan under a national economic and social policy, the necessary steps are initiated for the project start by the client preparing the terms of reference. In Figure 16.1 a scheme of main activities during the tendering/contracting stage is shown.

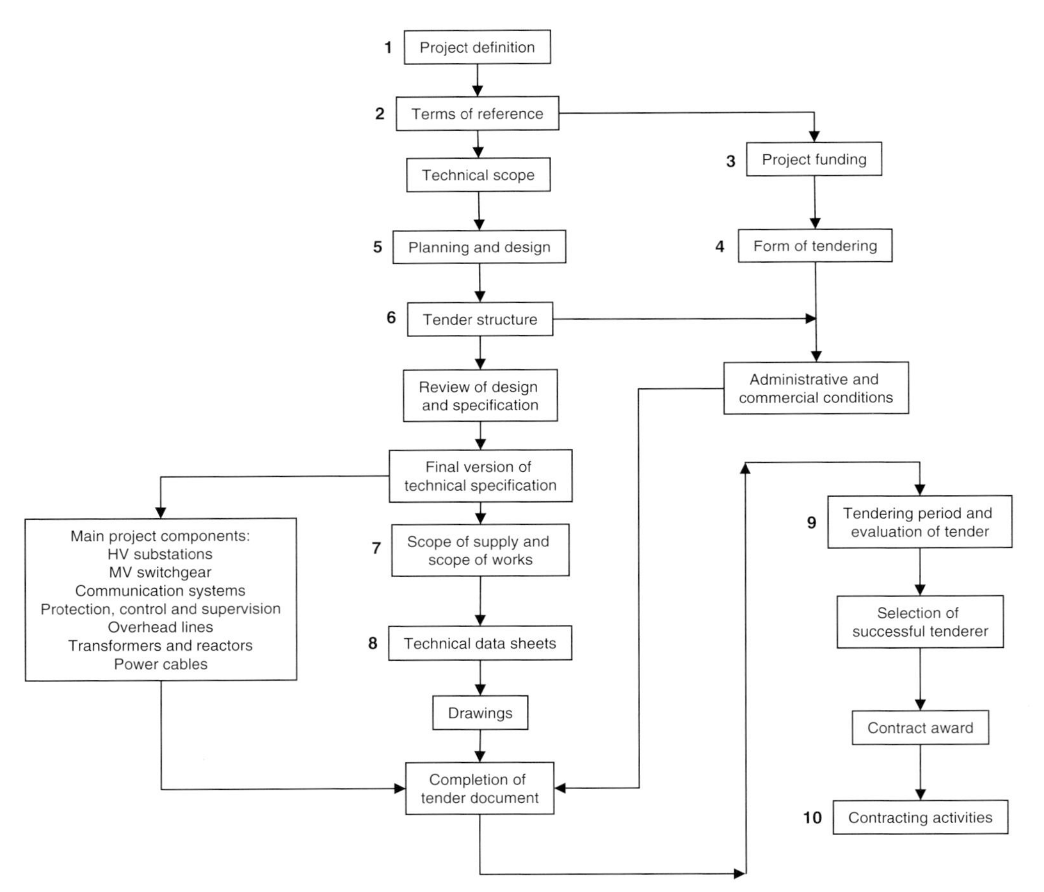

Figure 16.1 Scheme of main activities during the tendering and construction stages. The numbering refers to section numbers in the text.

16.2
Terms of Reference (TOR)

Terms of reference for a defined project are issued by clients/authorities and describe the services, supplies and work requested for the execution of the project under its terms and relevant regulations. Eligible or short-listed companies are invited to offer the work and services.

The terms of reference (TOR) are normally prepared by the client. The TOR serve to give a comprehensive overview of the client and its organization, the nature and status of the project, the requirements for services, engineering services, hardware and software, implementation schedule and commercial conditions and to outline the objective (such as to construct new high-voltage substations and overhead lines to meet the continuing growth in power demand, or to investigate the power interconnection with an industrial company to improve the reliability of power supply).

Formally the terms of reference are structured as detailed below.

16.2.1
Background

The functions and areas of responsibility of the client and the organization of the power sector are outlined, for example, responsibility for generation and transmission of bulk power throughout the country, or responsibility for distribution of electric power. Data and information relevant to the power system and system composition are stated, for example, radial distribution and transmission system.

16.2.2
Objective

Under this section three examples of projects are selected which are defined e.g. by utilities or power supply companies under development or power system expansion programs, or part thereof, with the aim to cope with the requirements of increasing power demand or changes in the power network structure.

- The client (e.g. in south-east Asia) plans to expand the existing 115/22 kV supply system with new 115/22 kV substations and switching stations in order to cope with the continuing growth in demand. Engineering services are needed to design the substations and switching stations and prepare turnkey solicitation followed by the tendering procedures.
- The client (e.g. in a Middle Eastern country) intends to carry out conceptual and engineering design, including preparation of tender packages for the reinforcement of electrical power supply at a large industrial complex. The engineering tasks include the conceptual and front end engineering design, preparation of tender packages, and identification of connection options with the required power system studies to support the selected option.

- The objective is strengthening of power distribution networks (e.g. in Africa) to improve quality, efficiency and reliability of power supply to different supply areas. New networks and extensions of existing 66/33 kV networks, new 33 kV overhead lines on concrete poles or lattice steel towers, and extension of the telecommunications network are the main aspects of the project. Engineering activities shall cover the engineering and design as well as the preparation of tender documents followed by the tendering procedures.

16.2.3
Scope of Engineering Activities

The required engineering activities range from feasibility studies to project monitoring during the implementation stage. For project execution, two main project phases are usually defined – the preconstruction phase and the construction phase. The activities in the preconstruction phase include planning and design as well as technical studies as may be required for determination of design parameters or to confirm power system configurations. The preconstruction phase generally covers the engineering and design work until award of a contract to the successful contractor(s). The main engineering activities required and defined by the client or utility company are of the sort described below.

- Collection of data and review of previous studies/reports (design parameters and requirements, standards, environmental aspects, technical risks)
- Review of design practices, preparation of design for the network components (e.g. substation and associated equipment, overhead lines, communication equipment)
- Preparation of tender documents with all associated administrative, commercial and technical sections, schedules and drawings: for example, invitation, general conditions, technical specifications, data sheets, price schedules and drawings
- Preparation of prequalification documents as required
- Assistance during the tendering period
- Tender evaluation work
- Assistance during the contract award procedure
- Detailed power system studies, for example, power system optimization studies for an industrial power system.

The main engineering activities during the construction phase focus on the following aspects.

- Approval of the contractor's drawings and documents, submitted for the defined project and forming the basis for manufacturing, testing and execution of works
- Tests on equipment in the contractor's factory or at independent testing sites for type tests
- Construction supervision and monitoring
- Site tests, commissioning and taking over of completed facilities
- Review and compilation of "as-built" drawings and documents; review of operations and maintenance manuals.

16.3 Project Funding

When planning a power system project, for example, power transmission/power distribution, interconnection of power systems, extension of telecommunication system, apart from the technical and economic feasibility it is of great importance for successful implementation and timely completion to secure the necessary funding and financing for the technical equipment, work and services in connection with the project. Following careful research on the funding possibilities, the plan for project financing is established. In most cases various sources of financing are applied, typically:

- The client's government, covering the local cost component
- Development banks and funding agencies covering the cost component for procurement of equipment, work and services in foreign currency
- The involvement of multilateral, regional banks and financing agencies is required for large, diversified or interconnection projects

The individual development banks were mostly founded as regional development banks, operating mainly in their respective regions (e.g. Africa, Asia, Pacific) to promote power projects and economic and social development through loans and technical assistance in those regions. But they also participate in projects worldwide. Loans are offered by the development banks under favorable conditions relating to such aspects as interest rate, service charge, long re-payment period, initial grace period. From among the ranks of development banks of good repute a few are mentioned below as examples (in alphabetical order).

- African Development Bank (**AfDB**) Financing Institute, main office in Abidjan, Ivory Cost
- Asian Development Bank(**ADB**), main office in Manila, Philippines
- European Investment Bank (**EIB**), main office in Brussels, Belgium
- Kreditanstalt für Wiederaufbau (**KfW**), main office in Frankfurt.
- **World Bank Group**, main office in Washington DC, USA, comprising:
  The International Bank for Reconstruction and Development (IBRD)
  The International Development Association (IDA)

16.4 Form of Tendering

After the design parameters and equipment specifications have been established in the preceding stages, preparation of the tender documents follows as the next step. The whole project may be divided into component packages or lots as may be defined between the client and the funding agency if this is required because of the project's size and varying project financing or to allow contractors to quote in their special fields only (for example, switchgear, substations or overhead lines), or for the whole project as general contractor.

For assistance and support in the engineering activities, in the majority of cases the services of external engineering firms or institutes are engaged by the client according to its own policy or/and in line with the regulations of the funding agency.

Upon approval of the tender documents by the client and higher authorities, the tender procedure can commence. A tendering period of up to 3 months for normal size overhead line or substation projects is the standard time allowed for the contractor to prepare the tender. Further steps are described in the following subsections.

16.4.1 International Tendering

The procedures for tendering and awarding of contracts for international construction projects are generally based on guidelines and recommendations by FIDIC (Federation Internationale des Ingenieurs-Conseils (International Federation of Consulting Engineers), Lausanne, Switzerland [99] to [101]. Provisions are made to assist the client or the assigned engineering company to receive complete, sound and competitive tenders in line with the tender documents. At the same time, tenderers are given the opportunity to respond easily to invitations to tender for projects for which they are qualified.

The main activities of the tendering and contract awarding stage are as follows:

- Project objectives
- Establishment of procurement method and form of tendering
- Preparation of program (if applicable)
- Prequalification of tenderers (if applicable)
- Preparation of tender documents
- Issue of tender documents
- Site visit by tenderer
- Tenderers' queries; addenda to tender documents
- Submission and receipt of tenders
- Evaluation of tenders
- Review of tenders; tenders containing deviations
- Adjudication of tenders
- Issue of letter of acceptance
- Award of contract, issue of letter of acceptance
- Performance security
- Preparation of contract agreement.

16.4.2 Prequalification

Experience has shown that for projects involving international tendering, prequalification is desirable since it enables the client to establish the competence of companies subsequently invited to tender. It is also in the interest of contractors since, if prequalified, they will know that they are competing against a limited number of other firms, all of whom possess the required competence and capability. The prequalification procedure includes the following steps:

- Preparation of prequalification documents
- Invitation to prequalify
- Issue and submission of prequalification documents
- Analysis of prequalification applications
- Selection of tenderers; notification of applicants.

16.4.3 Short Listing

If the client applies the short-listing method, the number of tenderers is confined to the companies that have satisfied the preceding qualification procedures and criteria for the particular field of engineering, design, manufacturing and project execution. From the list, short-listed companies and manufacturers are selected and invited for tendering at the discretion of the client/authority in cooperation with the bank. Qualification criteria include:

- Sound commercial and financial background, legal status, certification according the relevant parts of ISO 9000 or equivalent
- Organization, management, personnel
- The tenderer shall prove that he strictly works in line with the requirements of the applicable standards and codes, giving evidence of his ability and qualification by references of similar projects successfully completed
- Quick response to queries during maintenance and guarantee periods
- Experience and references in projects nature of the same nature, project-related references, particular references in the region.

16.5 Planning and Design

Diligent planning, design and dimensioning of electrical power systems and their related components are important engineering activities and form the basis for the technical specifications of equipment, layouts and work to be prepared for the tender documents.

The activities include the review and updating of existing specifications, data, layouts, routing for overhead lines and cables, power system configurations, and determination of parameters in connection with the use in the defined project. Relevant system and network studies are to be performed, where necessary, to confirm the planned configuration, design data or parameters, or to recommend measures for improvement or selection of higher ratings for equipment and so on.

Specialized engineers from the different engineering disciplines are assigned to the specific tasks: for example, electrical and mechanical engineers for power system planning, substation design, including switchgear and associated components and equipment, protection design, auxiliaries, power transformers, power cables and overhead lines; civil engineers for the design and construction of buildings, foundations and structures.

The main scope of planning and design work and aspects to be considered are outlined in the context of this book's coverage. Detailed design specification is not the intention here.

The aim of the planning and design work is to reach the best technical standard and coordination between requirements for the equipment and work under the definition of the project and integration into existing power systems such as to ensure maintenance and operational compatibility.

Design requirements and activities to be dealt with are presented for the main components, such as substations, overhead lines and distribution network.

Substations

- General technical substation requirements
- Basic electrical data, insulation coordination
- Layout plan and general arrangement of complete substation
- Drawings including electrical single-line diagrams
- Detailed description and requirements for indoor/outdoor type high-voltage substations, including switchgear, circuit breaker, isolator, current and voltage transformers, earthing switch and control units
- Detailed description and requirements for MV metal-clad switchgear, including withdrawable circuit-breaker, instrument transformers, metal-clad cubicles and earthing switch
- Control equipment, including interlocking, measuring and metering instruments, alarm indication and voltage regulation
- Protective and control systems
- Station supply, power and auxiliary cables and lighting system.

Overhead lines

- Line route profile
- Configuration/design of towers and tower earthing in view of insulation coordination
- Applicable standards
- Line design regarding conductor size, loading parameters, maximum/minimum and average spans and foundation types
- Mechanical and electrical requirements for all line components such as phase-conductors, earth-conductors, insulator strings and hardware
- Soil classification and foundation design standards
- Requirements for construction, erection and maintenance
- Environmental aspects of line routing.

Electrical distribution network

- Review of existing loads
- Review of actual network topology with view to adaptation to new requirements
- Review of design philosophy; for example, conductor cross-section, concrete/wooden poles in MV/LV network
- Review of technical network data with view to adaptation to new requirements, for example, the need for upgrading to overcome constraints
- Updating of the single-line diagrams, layout plans, tables with revised data.

Environmental aspects

- Environmental regulations and requirements need to be considered. In some cases the tender documents ask for an environmental impact study to clarify the impact on hydrological conditions, dewatering, animal life, noise-emission, and so on, and suggest suitable countermeasures.

16.6 Tender Structure

16.6.1 General

The design and specifications developed or updated during the engineering phase, especially the design phase, for the equipment and work involved in the implementation of the defined project form the basis for the preparation of the tender documents. The tender documents contain all necessary information, instructions, commercial procedures, technical specifications, scope of supply and work, data and drawings enabling the tenderer to offer the desired equipment and services in line with the client's and the bank's guidelines.

The tender documents are made up of technical parts, containing specifications, technical requirements regarding the engineering, supply of material, installation and putting into operation of equipment and plants, and the administrative, commercial part.

In case of large projects or as determined by the authorities or for reasons of control, the tender documents often are split into lots, also called packages, to differentiate individual project components or groups or equipment. For example:

- LOT 1 HV (400 kV) transmission line from location A to substation B
- LOT 2 Substations
- LOT 2A New HV (400/132 kV) grid station(s) (with conventional control, and supervision or computerized control and supervision)
- LOT 2B Extension of HV (400 kV) grid station(s)
- LOT 2C1 New HV/MV (132/33 kV) substation(s)

- LOT 2C2 Extension of HV/MV (230/33 kV) substation
- LOT 3 Underground cables (HV and MV)
- LOT 4 Substations and overhead lines (33 kV).

Generally the tender documents are prepared for international competitive bidding taking into consideration local regulations/stipulations as well as the guidelines of funding agencies (e.g. development banks, see Section 16.3).

16.6.2 Tender Set-Up

A standard has been developed regarding set-up and arrangement of the tender documents. Typically, the tender documents consist of documents or sections common to all lots or packages and documents specific to each individual lot or package subject to the nature of the project as described in the sections below.

16.6.2.1 General, Common Sections

The administrative, commercial and financial aspects are compiled in four sections as indicated below:

- Section 1 Tender invitation and instructions to tenderers
- Section 2 Tender forms
- Section 3 Price tabulation sheets and time schedule
- Section 4 General conditions of contract.

In Section 1, guidelines and instructions are laid down for tenderers concerning information: for example, project location, routes, limits of work, climatic and environmental conditions, preparation and submission of tenders, time schedule, bid opening and evaluation procedure and criteria, set-up of the tender documents, filling-in of form sheets, currency, certificates, references and documents from the tenderers to prove their eligibility, qualification and experience to carry out the work according to international standards.

In Section 2, instructions are given to be observed by the tenderer as regards acknowledgement of contract conditions, confirmation of prices, completion of works, tender and performance bond to be filled or provided by the tenderer, company profile and form of contract agreement.

Section 3 contains the price tabulation sheets and time schedule. The price sheets are prepared for the tenderer to quote for supply of material and equipment as well as for erection. Foreign and local currencies are differentiated. Key dates are to be indicated in the time schedules for the main activities, such as design work, production, transport, erection work and commissioning. Table 16.1 shows a sample price sheet.

Section 4 deals with the general conditions to be observed in the execution of a contract. In general the "Conditions of contract for electrical and mechanical works" prepared by FIDIC [99] to [101] are taken as a basis. The main topics include:

Table 16.1 Sample price sheet for an HV grid-station.

		C + F price		Erection price/foreign		Erection price/local	
	Estimated	(Currency:)		(Currency:)		(Currency:)	
	Quantity	Unit rate	Total	Unit rate	Total	Unit rate	Total
	a	b	a × b	c	a × c	d	a × d
Lot No							
New HV gridstation at location A							
(Computerized control and supervision – SCMS)							

Note: The Technical Specifications, the description of the Scope of the Work/Scope of Supply and the subsequently prepared Price Tabulation Sheets are intended to give the frame of the equipment needed for the implementation of the work. It is understood that the work(s) include everything requisite and/or necessary to finish the entire work properly and the equipment has to be complete in every respect, notwithstanding the fact that every item may not be specifically mentioned.

- Definitions and interpretations
- General instructions, such as language, approval procedures and standards
- Documents to be submitted
- General obligations, such as performance bond, customs duties and taxes
- Quality of material, equipment and workmanship
- Transport
- Erection work, completion and acceptance of work
- Responsibility and liability of contractor, insurance
- *Force majeur*, settlement of disputes
- Terms of payment
- Damage penalties.

16.6.2.2 Sections Specific to Each Lot or Package

The technical specifications, scope of work and supply, technical data sheets, tender drawings are specific for each individual project, depending on the nature of the project: for example, substation project, overhead line project, communication project. The required papers are compiled in the sections detailed below.

For the individual project lot or package, the "technical" sections are combined with the common sections (Sections 1–4 above), to form the complete tender document.

- Section 5 General technical specifications
- Section 6 Particular technical specification, scope of work and supply
- Section 7 Technical data sheets
- Section 8 Tender drawings.

16.6.3 General Technical Specifications

In Section 5, the technical specifications as applicable for the electromechanical equipment, material and work are compiled and included in the tender documents. It must be assured that the last valid versions of the technical specifications and data are being used. The relevant specifications are to be applied subject to the nature of the power project. For example:

- General criteria for the design, including climatic and ambient conditions such as temperature, rainfall, storms, relative humidity, and so on
- High-voltage metal-clad SF_6-insulated switchgear
- Medium voltage switchgear (gas-insulated type)
- Control, supervision and protection
- Overhead line
- Power and auxiliary cable
- Communication system
- Transformers and reactors
- Auxiliary equipment
- Spare parts, tools and test equipment
- Site services and civil works.

Taking a substation project as an example, the following component specifications are applicable, as expanded upon in the subsequent text.

- General criteria for the design
- High-voltage metal-clad SF_6-insulated switchgear or high voltage outdoor switchgear
- Medium-voltage switchgear (gas insulated type) or air-insulated metal-enclosed switchgear
- Control, supervision and protection
- Auxiliary equipment and spare parts
- Site services and civil works for substations

16.6.3.1 General Rules and Provisions Related to the Design

General parameters and information are outlined related to prevailing local conditions and practice to be observed in the design work and for tendering. The equipment must be suitable in all respects for use and operation within the defined power systems. For example:

- Location, site levels, climatic conditions (data and information on wind velocity, storms, etc.)
- Ambient temperatures, relative humidity and meteorological data
- Soil conditions (soil thermal resistivity in $km\,W^{-1}$)
- Power system data and characteristics, such as system operating voltages; HV level, for example, 380/420 kV, 220/245 kV; rated voltage, HV and MV levels, for example, 132/145 kV, 33/36 kV, 11/12 kV; low-voltage system, for example, 400/240V (+10–15%), number of phases, 3 (3-phase, 5-wire system); frequency 50 Hz; neutral earthing.

A guideline for the preparation and scope of the tender document is given in the context of this book's coverage.

The technical specifications, as part of the tender document, are the basis for the technical design of the equipment and work required for the defined project. The client's standard technical specifications may have to be adapted to the individual project needs in the course of completing the tender documents. Tenderers are responsible for including everything required and/or necessary to complete the entire work properly, irrespective the fact that not every item may be specifically mentioned in the specifications.

16.6.3.2 High-Voltage Metal-Clad SF_6-Insulated Switchgear

In the case of a high-voltage substation project, the high-voltage switchgear is the main project component. From the two basic designs, air-insulated outdoor switchgear and gas-insulated switchgear, the metal-clad SF_6-insulated switchgear is taken as an example for the switchgear specification.

The specification of the gas-insulated switchgear and switchgear components include principal aspects such as common features, circuit-breakers, isolators, high-speed earthing switches, measuring transformers and transducers, interlocking and control and the gas system itself.

Common Features General arrangement of switchgear and accessories for installation in switchgear rooms shall consider the following, *inter alia*.

- The switchgear construction shall be of suitable material and thickness to withstand the mechanical and thermal stresses due to short-circuits. The rated duration of short-circuit is 3 seconds.
- Rupture diaphragms shall be provided in each compartment to allow for pressure relief.
- Future extension shall be enabled.
- The latest modern engineering practice shall be followed.
- The switchgear shall be supplied complete with all auxiliary equipment necessary for operation, routine maintenance, repairs or extensions.
- The design shall be compact, fully metal-clad and of the sulfur hexafluoride (SF_6)-insulated type.
- Equipment shall be constructed for the indicated busbar system and include all necessary switches, current and voltage transformers, as indicated in the single-line diagrams (Section 8 of TOR).
- Components shall have interchangeability as far as possible.
- The arrangement of the switchgear shall be such that any part can be removed without interruption or disturbance to adjacent feeders or circuits.

Circuit-Breakers Generally, the circuit-breakers shall fulfill the service operation conditions, for example, making and breaking of fault current. The circuit-breaker operating mechanism must be capable of storing energy for the operation sequence as specified in the technical data sheets. The three-phase circuit-breakers shall incorporate SF_6 gas as an insulating as well as an arc-quenching medium (re-strike-free switching must be guaranteed). Switching conditions are defined.

Isolators Isolators shall switch under zero-current condition and keep the switch position during short-circuit.

High-Speed Earthing Switches High-speed earthing switches are required at the outgoing ends of every feeder and for the busbars. They shall be constructed to withstand an accidental switching onto a live part, that is, they shall be of the make-proof type.

Measurement Transformers Measurement transformers include current transformers (CT) and voltage transformers (VT) according to the applicable standard IEC 60044. Main design particulars include:

Current Transformers

- Ring-core design of current transformer, secondary windings embedded in cast resin.
- Requirements as to short-time primary rating to be not less than that of the associated switchgear.

- Thermal rating of the current transformer such as to allow, under site conditions, a 20% continuous overloading referred to nominal rating of the current transformer.
- Requirements regarding CT-cores for measuring and protection, magnetizing curves, rated output (30 VA minimum), class of accuracy, rated accuracy limit factors, rated primary current, turns ratio, knee-point voltage and resistance of the secondary windings.

Voltage Transformers

- SF_6-insulated voltage transformers shall be of the inductive type, encapsulated, the gas compartment to be segregated from the adjacent compartments.
- Minimum rated output shall be 100 VA.
- Busbar-VT shall be connected through hand–operated isolators.
- Construction and testing in accordance with IEC 60066.

Interlocking and Control The interlock system is defined, including circuit-breakers, isolators, earthing switches and bus-couplers, to prevent any incorrect operation of the circuit-breakers and switches and to at least fulfill the general requirements according to the detailed specification. In case of key-operated switches, the operator is responsible for all switching operations.

Gas System Principal features are as follows.

- Due to the requirements for maintenance, the switchgear requires individual compartments, each having its own overpressure relief device. Extension of the busbar system shall be possible without de-gassing the existing part; that is, gas-tight bushings shall be provided at each busbar end.
- The individual compartments shall be supervised via gas-density monitors with temperature compensated pressure gauges.
- Gas losses shall be guaranteed less than 1% per year.
- Further components form part of the specification, such as SF_6-terminals, bus ducts, outdoor wall-mounted bushings, metal-oxide surge arresters.
- SF_6-gas losses shall be capitalized.

The maximum gas losses per switchgear compartment are defined not to exceed 1% per year. This value is to be guaranteed by the manufacturer.

16.6.3.3 **Medium-Voltage Switchgear**

For to medium-voltage switchgear, standard designs and models from the manufacturer's switchgear program are generally specified under the condition that the equipment meets the requirements of the specifications and serves the intended purpose. The minimum quality and performance requirements must be fulfilled. The manufacturer must prove at least 5 years of successful service in the field.

The specification covers the design, ratings, testing, shipping, installation and commissioning of factory-assembled, type-tested switchgear of different characteristics, for example, air-insulated switchgear, metal-clad type, single or double busbar system, draw-out section, vacuum circuit-breaker and SF_6-insulated switchgear, triple-pole or single-pole metal-clad type, single or double busbar system, draw-out section, vacuum circuit-breaker.

The voltage levels extend up to 36 kV; the maximum busbar rated current at this voltage level is stated as 2500 A and the maximum rated short-time current 31.5 kA. Detailed and specific data required are contained in the drawings and data sheets of the tender documents.

16.6.3.4 Control, Supervision and Protection

The specification covers the control, supervision and protection of the substation. For control and supervision, the microprocessor implementation termed SCMS (substation control and monitoring system) is the preferred technique rather than the conventional technique. General features and design requirements related to the various components are specified in detail. For the protection systems, transformer tap changers, transformer supervision, meters, recorders and alarm indication panels shall be provided and installed in the control room. The panels shall be arranged in at least the following sections:

- Switchgear control
- Transformer on-load-tap-changer (OLTC) control/parallel interlocking
- Alarm annunciation
- Fault monitoring system (FMS)
- Event recorder
- Protection panel(s)
- Synchronization
- Communication system
- Load dispatch center (interface)
- Air conditioning and ventilation
- Fire protection
- AC-supply system
- DC-supply system.

16.6.3.5 Overhead Lines

For high-voltage and extra-high-voltage overhead transmission lines, a large number of different configurations is available, using self-supporting lattice steel towers. For low- and medium-voltage lines, concrete, steel or wood poles are used. The specification for the overhead line is set up according to the nature of the project and covers the design and requirements concerning manufacture, factory testing, delivery, transport, installation, site testing of towers, insulators, phase-conductors, earth-conductors, fittings, and so on as well as the associated civil works, access roads, foundations.

All required equipment, material and work for the overhead line, whether specified or not, shall be included to achieve a safe and reliably designed system. The main components and aspects dealt with in the specifications include the following.

Soil Investigations and Tests Soil investigations and tests are done to determine the necessary soil mechanical parameters for the foundation design of towers. The soil conditions and characteristics shall be ascertained, using approved methods. After determining the specific soil characteristics, the applicable type of foundation shall be defined.

Foundations Foundation types are, for example, normal foundations and pile foundations for towers, which may be employed where special ground conditions exist. A number of other types of foundation can be required depending on the tower or pole type, the soil investigations and results of laboratory tests. Prior to the selection of foundation types, all relevant calculations, data (design criteria, uplift criteria, safety factors) and drawings are to be submitted by the contracting firm for approval.

Towers Towers for a high-voltage transmission line project are assumed in the following. The specification covers the main features and design data for construction and materials as well as for phase- and earth-conductors.

The different tower types, such as suspension, tension and angle towers, shall be of standard construction. Single-circuit, double-circuit or multi-circuit towers are defined. The tower shall be designed as self-supporting lattice type steel frame with square base. Tower outlines shall be as shown on the tender drawings. The members of the lattice structure are to be of hot-rolled steel angle sections, factory made and hot-dip-galvanized. Existing proven tower design may be used, if equaling or exceeding the design loading and clearances required by the specification.

Criteria such as design loading and design unbalanced loading (broken wire conditions), safety factor and overload capacity are to be considered in the design, as well as wind conditions. Permissible values according to standards must not be exceeded. No damage or permanent distortion of any members, bolts, connections of fittings or elongation of bolt holes shall be permitted for these design conditions.

Other aspects include:

- Tower grounding
- Solar-powered aviation obstruction lighting
- Workmanship equal to the best modern practice in the manufacture and fabrication of materials covered by this specification
- Tower locations based on detailed survey work performed by the contracting firm to determine the tower locations in the map and longitudinal profiles and prepare tower lists with main data for each tower including wind span and weight span
- Tower testing by load tests to be applied to the tower to specify that each tower shall withstand the test loads for at least five minutes without failure or permanent distortion of any member, fitting, bolt or part and without elongation of bolt holes.

Phase-Conductors and Earth-Conductors Particulars of the conductors to be supplied and of standards are set out in the schedule. The conductors shall be manufactured and tested in accordance with standards as indicated in the schedules. The conductor is supplied on reels. Further specifications include optical fiber ground wire (OPGW) to comprise an optical unit integral with the earth-conductor. The OPGW design shall be mechanically and electrically compatible with design of the transmission line.

16.6.3.6 Power and Auxiliary Cable

The specification covers the design, manufacture, factory testing, supply, transport, laying and installation, and site testing of power cables, associated pilot, telephone and optical fiber cables, auxiliary cables and control cables, including all civil works, cable terminals, cable racks, cable fixing material, and so on. All equipment shall be covered, irrespective whether specified or not, to form a complete and reliable system. Typical power, control and communication cables used in electrical power systems are:

- High-voltage XLPE cables
- Oil-filled cables of the low pressure oil-filled type
- Medium-voltage XLPE cables
- Low-voltage power cables (nominal AC-distribution voltage, 3-phase/5-wire system, PVC insulation)
- Protection and telephone cables (e.g. 17-pair cable, 5 pairs for protection, 12 pairs for telephone usage)
- Optical fiber cables (shall have fibers of the single-mode type, suitable for transmitting light signals).

Requirements are laid down for the different types of power, control and communication cables, including standards to be complied with, cable design and construction, sealing and drumming, tests at independent institutes (type tests) or manufacturer's premises and at site, laying and routing.

16.6.3.7 Telecommunication System

The specification covers the design, manufacture, testing, delivery, transportation and erection as well as the commissioning of all material and equipment required for the telecommunication systems and system extension. Requirements are defined such as extension of existing communication systems for telephony, teleprotection signaling and data/alarm transmission by use of fiber-optic transmission media. Compatibility of equipment to cater for any upgrading without limitation shall be guaranteed.

16.6.3.8 Transformers and Reactors

The specifications and requirements cover the design, manufacture, factory testing, delivery, transport, erection and commissioning related to:

- Power transformers
- Distribution transformers (power rating up to 2.5 MVA)

- Reactors, including shunt reactors, current limiting reactors and neutral grounding reactors.

Detailed specifications cover the mechanical and electrical design of the transformers and reactors, including the aspects listed below (the applicable standards are included in the list of standards and norms).

- Ambient temperature, operation condition
- Number and type of windings
- Magnetic core
- Transformer/reactor tank, oil conservator, cooling (natural air ONAN, forced air ONAF)
- Tap changer (on-load, off-load), voltage regulation
- Control and monitoring device
- Terminals, bushings
- Condition of parallel operation
- Test requirements.

16.6.3.9 Auxiliary Equipment

Specifications of the auxiliary equipment are part of the tender documents for substations. The features and requirements regarding construction and design, parameters and arrangement (e.g. panels) are detailed For example:

- **AC-supply** from two independent power sources with specified voltage, for example, 400/230 V, with defined tolerance, including distribution panels of a self-standing cubicle type, constructed equivalent to the control and protection panels of the switchgears. The continuous and short-time/short-circuit ratings of the switchgears shall be according to the specified transformer ratings and the expected short-circuit rating. Operation conditions are defined.

- **DC-supply** with battery and charger, including two battery chargers, two battery banks, two voltage-control units for regulating the output voltage, two main switchboards 110 V for distributing power to the various loads. Batteries shall be either of the nickel–cadmium type or of the long-life, sealed lead–acid type. The distribution panels shall be of a self-standing cubicle type, constructed equivalent to the control and protection panels of the switchgears. The continuous and short-time/short-circuit ratings of the switchgears shall be according to the specified transformer ratings and the expected short-circuit rating. Operation conditions are defined.

Station Lighting A complete station lighting system, indoor as well as outdoor, is to be included.

Lightning Protection and Earthing System It is required to determine by calculations and measurements whether impermissible touch and step voltages occur at any place of the station that may be endangered. Design principles are stated in the specification (e.g. the HV and MV systems are solidly earthed at the neutral of the transformers).

16.6.3.10 Civil Works for Substations

The civil works are part of the tender documents and cover all civil work in connection with the related substation project, including for example, design, manufacture, testing, delivery, transport, storage at site, erection, installation, commissioning, performance testing, and handing over in satisfactory operating condition of all civil work, such as switchgear building, control building and miscellaneous work.

Regarding the buildings, air-conditioning system, fire-fighting system, station lighting system for indoor and outdoor as well as lightning protection and earthing for the entire station shall be provided.

16.7 Scope of Work and Supply

Section 6, specific to each lot or package, deals with the determination of the scope of supply of equipment and material and associated work required under the defined project. The main objective of the project, the nature of the project and general requirements are introduced. For details of technical requirements and descriptions of the complete work, reference is made to the technical specifications and the related particular sections of the tender documents. An example is given below for a substation project (only one lot is assumed).

16.7.1 General

The tender documents call for the supply, delivery, erection, commissioning and handing over of a new 380/123 kV grid-station at location A and the extension of the 380 kV/123 kV substations at location B.

Within the framework of supply and wark the following shall be outlined.

16.7.2 380 kV Switchgear

The gas-insulated switchgear (GIS) shall consist of a double busbar system for five feeders, current rating 3150 A, short-circuit current 63 kA – 3 s, general single-line diagram as per the tender drawing. Space for one future feeder shall be allowed for. Each feeder bay for overhead line connection to be equipped with

- Three-pole circuit-breaker
- Three-pole maintenance earthing switches
- Three-pole high-speed earthing switch
- Three-pole line-disconnecting switch
- Three-phase current transformers with separate cores for different protection and metering purposes as per tender drawings
- Set (three-phase) of voltage transformers, SF_6-insulated

- Set (three-phase) of gas-insulated type surge arresters
- Set (three-phase) of outdoor terminations, consisting of SF_6-bus-ducts with outdoor bushings
- Related control cubicle for each feeder.

The other bays are equipped analogously according to the specification.

16.7.3 123 kV Switchgear

The GIS shall consist of a two-busbar system for ten feeders, current rating 3150 A, short-circuit current 40 kA – 3 s, general single-line diagram as per the tender drawing.

The feeders each shall be equipped with

- Two three-pole busbar selection isolators
- One three-pole circuit-breaker
- Two three-pole maintenance earthing switches
- One three-pole high-speed earthing switch
- One three-pole line-disconnecting switch
- Two three-phase current transformers with separate cores for different protection and metering purposes as per the tender drawings
- One set (three-phase) of voltage transformers, SF_6-insulated
- One three-pole cable end unit suitable for connection of XLPE cable sealing ends.

The related local control cubicle for each feeder shall be equipped with the necessary interlocking unit, one amperemeter with selector switch or three amperemeters, one voltmeter with selector switch, one selector switch with positions: off-local-remote, transducer and annunciator block.

16.7.4 Transformers and Reactors

This section covers the design, manufacture, factory testing, transport, erection, installation, commissioning and handing over in satisfactory operating condition of transformers/reactors, including all panels, auxiliary equipment and accessories.

The design of the transformers, for example 300 MVA, shall comply with the following main design data:

- Rated power — 300 MVA
- No-load voltage ratio — 400 kV ± 15%/142 kV
- Vector group symbol — YNyn0(d)
- Type of cooling — ONAN/ONAF/ODAF
- Rated frequency — 50 Hz
- Voltage regulation — on-load tap changer.

16.7.5
Telecommunication System

The scope of work and supply covers the design, production, supply, transport, installation, cable laying, wiring, testing, commissioning and handing over in satisfactory operating condition of the telecommunication system, including all auxiliary equipment and accessories for multiplex equipment (optical terminals), telephone-alarm system, teleprotection, signaling equipment, power supply for communication system, fiber-optic cables, radio equipment and closed-circuit television systems.

The equipment shall be accommodated in cubicles/distribution racks. Further components included within the scope are power cables, control cables and communication cable, auxiliaries and civil works.

16.8
Technical Data Sheets

The technical data sheets form an integral part of the tender documents and shall be diligently completed by the tenderer. The tenderer shall be bound to adhere to the design data and criteria as stated in the technical data sheets. By comparing the data documented by the tenderer against the requirements it can be evaluated whether the quality and design of the tenderer's equipment are in accordance with the specifications or whether deviations from the specifications may lead to the rejection of the equipment. Technical data sheets of all major and significant project components of the project in question shall be prepared. For evaluation of the equipment and work they form an essential part of the specifications and tender document. An example of a technical data sheet is given in Table 16.2 for a 400 kV overhead line tower and associated insulator.

Only columns headed "Tendered" or "Data required by the tenderer" of the data sheets must be filled in by the tenderer, without omission. The manner and breakdown of the data sheets must not be changed, that is, no changes or additions are acceptable within "Required" columns. The values stated by the tenderer shall be guaranteed limit values, allowing for a margin on its "safe side." If there are deviations from the specifications, the tenderer shall give explanations on a separate paper for assessment by the client.

There are standards and general rules for filling in of the data sheets and providing complete information on the equipment:

- They must be typewritten.
- If a particular item is not applicable or not quoted, the letters **NA** (Not Applicable) or **NQ** (Not Quoted) shall be typed in the space provided.
- For any data not duly inserted in the data sheets, the least favorable data stated by any competitor shall be used in the evaluation process.
- In case of deviations from the technical specifications and the technical data sheets, further explanations must be stated on extra sheets only.

Table 16.2 Sample technical data sheet of a 400 kV overhead line tower and associated insulator.

400 kV Overhead line	Unit	Min. data required	Tenderer
Steel tower weights			
The Tenderer hereby states that in accordance with the design conditions the estimated net weight of the steel towers will be as follows:			
Type of tower (net weight in kg)			
– S + 0 m Basic tower	kg		
– S – 3 m	kg		
– S + 6 m	kg		
– AT 30 + 0 m Basic tower	kg		
– AT 30 – 3 m	kg		
– AT 30 + 6 m	kg		
– BAT 60 + 0 m Basic tower	kg		
– BAT 60 – 3 m	kg		
– BAT 60 + 6 m	kg		
– BAT/DE + 0 m Basic tower	kg		
– Marine crossing tower	kg		
– GANTRY (double)	kg		
Insulators			
–Insulator Units			
Manufacturer			
Standards	IEC/VDE/DIN		
Type		HTV silicone rubber	
Creepage path: 16800	mm		
Maximum working load: min. 68	kN		
Specified mechanical load: min. 310	kN		
Routine test load: min. 176	kN		
Core diameter	mm		
Number of sheds			
Shed diameter	mm		
Shed spacing	mm		
Shed inclination			
– Upside	deg.		
– Downside	deg.		
– Mass	kg		
– Lightn. impulse withstand voltage (dry): 1425	kV		
– Switch. impulse withstand voltage (wet): 1050	kV		
– Power frequency withstand voltage (wet): 630	kV		
– Electrical Values of insulator strings	kV		
– Lightn. impulse withstand voltage (dry): 1425	kV		
– Switch. impulse withstand voltage (wet): 1050	kV		
– Power frequency withstand voltage (wet): 630	kV		
– RIV above 1 V across 300 ohm at 1 MHz and 320 kV: max. 46	dB		
– Corona extinction voltage: min. 260	kV		

Table 16.3 Sample technical data sheet of 110 kV switchgear.

110 kV Switchgear	Unit	Min. data required	Tenderer
General			
Rated voltage	kV	123	
Rated frequency	Hz	50	
Number of Phases		3	
Busbar system			
Type		Tubular	
Rated short-time withstand current (3 s)	kA	31.5	
Rated peak short-circuit current	kA	80.0	
Rated normal current:	A	2500	

- In case the tenderer is able to offer better values than required (e.g. test voltage levels, ratings at ambient temperature, etc.), he is invited to offer this equipment.
- Catalogues and further descriptive information on all equipment quoted and the pertinent data shall be submitted to give sufficient details of the equipment offered.

Further examples of technical data sheets are given in Tables 16.3– 16.8 for other applications.

16.9 Tendering Period and Evaluation of Tender

16.9.1 Tendering Period

For the preparation of the tender by the tenderer, a period of about two months up to three months is considered adequate for a normal-size power transmission or distribution project. For large complex power projects, with power transmission, distribution and generation, a longer period may be allowed the tenderer for coordination of the various project components and suppliers to assure good quality of equipment and work.

During the tender period, queries from bidders have to be clarified and subsequently correlative circulars are to be issued to all bidders.

The tenders have to be submitted to the client's office at the specified date. In most cases the tenderer has to prepare the offer in two separate packages, for example, the technical part and separately the price/commercial offer. This is to conform to the evaluation procedure by international bidding. This means that only after completion of the evaluation of the technical offers will the price and commercial tender be opened. However, only the best three technically evaluated tenderers will be part of the commercial evaluation.

Table 16.4 Sample technical data sheet of 110 kV circuit-breaker and disconnector.

110 kV Circuit-breaker	Unit	Min. data required	Tenderer
Name of manufacturer			
Country of manufacture			
Type			
Type tested	Yes/No	Yes	
Standards		IEC/VDE	
Arc-quenching medium	Type	SF6	
Number of phases		3	
Impulse withstand voltage (1.2/50 μs)	kV (peak)	550	
Power frequency withstand voltage (1 min):	kV (r.m.s)	230	
Rated short-time withstand current (3 s)	kA	31.5	
Rated peak short-circuit current	kA	80.0	
Rated normal current	A	1250	
Rated short-circuit breaking current	kA		
Rated short-circuit making current	kA		
Maximum capacitive breaking current	A		
Rated operating sequence:		O-t-CO-t′-CO	
No. of making coil		2	
No. of tripping coil		1	
Operating mechanism			
For closing		Spring	
For opening		Spring	
Motor			
Voltage	V (DC)	110	
Power	W		
Autoreclosure		Three phase	

110 kV Disconnector (busbar)	Unit	Min. data required	Tenderer
Name of manufacturer			
Country of manufacture			
Type		Rotary	
Type tested	Yes/No	Yes	
Standards		IEC/VDE	
Rated current	A	1250	
Rated short-time withstand current (1 s)	kA	31.5	
Rated peak short-circuit current	kA	80	
Impulse withstand voltage (1.2/50 μs)	kV (peak)	550	
Power frequency withstand voltage (1 min):	kV (r.m.s.)	230	
Auxiliary contacts			
Voltage rating	V (.c.)	110	
Current rating	A	10	
Operating mechanism			

Table 16.5 Sample technical data sheet of 110 kV current transformer.

110 kV Current transformer	Unit	Min. data required	Tenderer
Name of manufacturer			
Country of manufacture			
Type			
Type tested	Yes/No	Yes	
Standards		IEC/VDE	
Arrangement		Outdoor	
Number of phases			
Rated primary current:			
Type A	A		
Rated secondary current			
Protection core 1	A	1	
Protection, if required core 2	A	1	
Instruments/measuring core 3	A	1	
Measuring cores	VA		
Accuracy class		0.5M5	
Burden		20	
Protection cores	VA		
Accuracy class		5P10/Cl.X	
Burden		30	
Rated short-time withstand current	kA	31.5	
Rated peak short-circuit current	kA	80	
Impulse withstand voltage (1.2/50 μs)	kV (peak)	550	
Power frequency withstand voltage (1 min)	kV (r.m.s.)	230	
Insulation material			
Limits continuous primary current	A		
Kneepoint voltage (Cl. X)	V		

Table 16.6 Sample technical data sheet of 110 kV inductive voltage transformer.

110 kV Inductive voltage transformer	Unit	Min. data required	Tenderer
Name of manufacturer			
Country of manufacture			
Type			
Type tested	Yes/No	Yes	
Standards		IEC/VDE	
Type of voltage transformer		Inductive	
Arrangement		Outdoor	
Number of phases			
Impulse withstand voltage (1.2/50 μs)	kV (peak)	550	
Power frequency withstand voltage (1 min)	kV (r.m.s.)	230	
Rated Voltage Factor		$1.9 \times U_n$, 30 s	
Rated thermal burden	VA	1500	
Winding 1			
Rated primary voltage	kV	$110/\sqrt{3}$	
Rated secondary voltage	V	$110/\sqrt{3}$	
Accuracy class		Cl. 0.5	
Rated output	VA		
Winding 2			
Rated primary voltage	kV	$110/\sqrt{3}$	
Rated secondary voltage	V	$110/\sqrt{3}$	
Accuracy class		Cl. 0.5	
Rated output	VA		

Table 16.7 Sample technical data sheet for 33 kV cable–electrical data.

33 kV Cable	Unit	Min. data required	Tenderer
Rated voltage (U/U_r)	kV	19/33	
Highest voltage for equipment (U_{max})	kV	36	
Frequency	Hz	50	
Standard switching impulse withstand voltage	kV		
Impulse withstand voltage (+90 °C)	kV	170	
Power frequency withstand test voltage (4 × U_r) for 4 h	kV	76	
Maximum partial discharge at 1.5 U_r	pC	3	
Dielectric stress at power frequency voltage	$kV\,mm^{-1}$		
Loss current to earth at three-phase system at 90 °C and U_r	$mA\,km^{-1}$		
Maximum positive-sequence capacitance under full load condition	$\mu F\,km^{-1}$		
Positive-sequence inductance	$mH\,km^{-1}$		
Relative permittivity of XLPE insulation			
Maximum value of dielectric loss angle of cable at U_n, 50 Hz, conductor temperature of 20 °C	$\times 10^{-3}$		
Dissipation factor at 90 °C conductor temperature	$\times 10^{-3}$		
Maximum charging current at U_r per conductor	$A\,km^{-1}$		
Charging capacity of three-phase system at U_r	$kVar\,km^{-1}$		
Relative permittivity of outer covering			
Positive- sequence impedance (complex value) at 90 °C	$\mu\Omega\,m^{-1}$		
Zero-sequence impedance (complex value) at 90 °C for defined cable arrangement as per tender drawing	$\mu\Omega\,m^{-1}$		

Table 16.8 Sample technical data sheet for 33 kV cable – permissible currents, temperatures, and so on.

33 kV Cable	Unit	Min. data required	Tenderer
Maximum continuous current carrying capacity at: Rated voltage U_r; soil temperature 43 °C; $\lambda = 1.0\,\text{km W}^{-1}$; 100%-load factor as per IEC 60287; cable arrangement as per tender drawings	A		
Conductor temperature at the above mentioned conditions	°C		
Maximum emergency current rating at 60% pre-load as per IEC 60853	A		
Copper loss at full load per phase	W m^{-1}		
Maximum dielectric loss per phase at U_r, 50 Hz and maximum conductor temperature	W m^{-1}		
Metallic sheath loss per phase at U_r, 50 Hz at defined load conditions	W m^{-1}		
Maximum loss of steel wire armor per phase at normal conductor temperature and at maximum conductor temperature	W m^{-1}		
Total losses per phase at normal operating conditions and at maximum conductor temperature	W m^{-1}		
Permissible asymmetrical conductor fault current for 0.5 s	kA		
Minimum asymmetrical fault current in metallic sheath for 0.5 s	kA	31.5	
Screening factor (ground current/total earth fault current)			
Maximum permissible conductor temperature (continuous operation)	°C	90	
Maximum permissible temperature of outer sheath	°C		
Maximum permissible conductor temperature for emergency conditions as specified	°C	105	
Maximum permissible short-circuit temperature	°C	250	
Maximum DC-conductor resistivity at 20 °C and at 90 °C	$\text{m}\Omega\,\text{km}^{-1}$		
Maximum DC-metallic sheath resistivity at 20 °C	$\Omega\,\text{km}^{-1}$		
Minimum insulation resistivity at 20 °C	$\text{G}\Omega\,\text{km}^{-1}$		
Maximum DC-resistivity of steel wire Armour of cable at 20 °C	$\mu\Omega\,\text{m}^{-1}$		
Loss current to earth at three-phase system at 90 °C and Ur	mA km^{-1}		

16.9.2 Bid Evaluation

Evaluation of the tenderer's technical data, specifications, calculations and drawings as well as the submitted time and implementation schedules will be performed carefully in the first step for all bids received in order to determine compliance and conformity with the specifications and standards. Relevant comparative calculations and tabulations to support the findings in evaluating the bids will be prepared.

As stated above, tender prices in the commercial part of the tender will be evaluated after completion of the technical evaluation and generally only the first three technically qualified tenderers are considered for the commercial evaluation.

The technical evaluation includes the following criteria:

- Compliance with tender requirements
- Completeness of bid
- Compliance with general and specific conditions
- Compliance with technical specifications
- Technical competence of the tenderer and any associated manufacturer, supplier or subcontractor
- Experience of the tenderer in similar projects and in the country concerned or in neighboring countries
- General engineering quality of equipment and materials offered
- Reliability of design
- Compliance with or deviations from the data sheets
- Completion time of the different bidders.

The financial evaluation is to include the following criteria:

- Arithmetical errors contained in the bids
- Deviation from prices and terms of payment
- Evaluation of shortcomings in the performance data.

All data specified in the tender documents have to be compared with the quoted data of the tenderers in such a way as to enable easy identification of deviations and comparison between the different tenderers. Price comparison sheets are to be prepared showing the evaluated prices of the bidders, resulting in a ranking of bids based on their financial attractiveness. Deviations detected in those bids considered acceptable are then subject to subsequent negotiation. The findings shall be compiled in the tender evaluation report which is required by the client's management as well as by the funding agency, if required.

16.10 Contracting

Under the term "contracting" or "contracting phase" the implementation of the defined project is considered, for example, from award of contract and selection of the contractor up to the commissioning of the project. For standard (e.g. 132 kV or 115 kV) high-voltage overhead line projects with associated substations in south-east Asia or in the Gulf region, the execution phase was given as about 30 months, including the initial engineering work, manufacturing, erection and installation, plus some lead time for the contractor to organize the work. Of course the implementation time can vary depending on the project nature.

In the context of this book, the main procedures and engineering activities during the contracting phase are touched upon with view to illustrating the main aspects during the contracting phase. An example of an activity plan for an implementation project comprising construction of a HV overhead line and substations is shown in Figure 16.2.

During the contracting stage, close cooperation between client, authorities, contractor and engineering company, if assigned, is necessary for efficient execu-

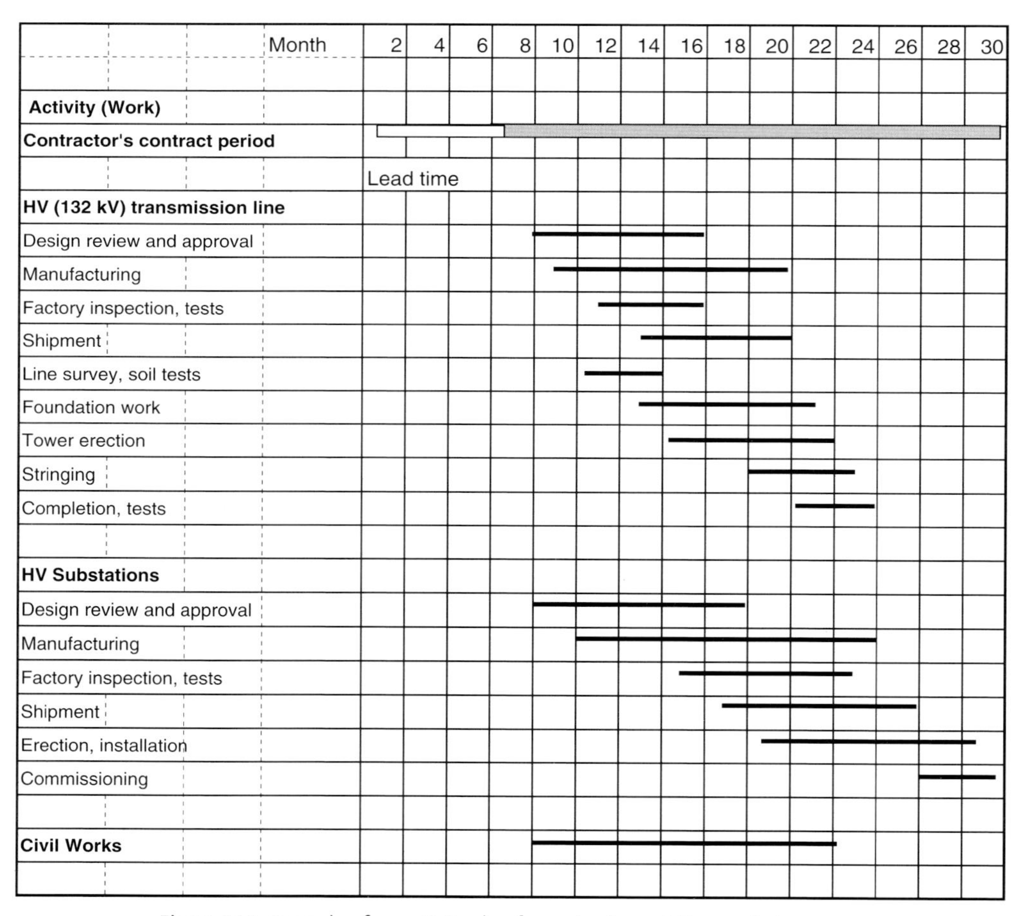

Figure 16.2 Example of an activity plan for an implementation project.

tion of the project. Key activities during the contracting stage include the following.

- At project start, set-up of procedures and project schedule and time schedules to be adhered to and intensively monitored during execution of the work.
- Design review and approval of documents submitted by the contractor for all project components; for example, overhead lines, substations, complex interconnection or industrial projects. The documents include drawings, layouts, single-line diagrams, operation and maintenance manuals, design calculations, literature and data.
- Factory inspections and tests of contractors' equipment and material at the premises of the contractor or type tests at independent test institutions, if required by the contract, and quality assurance.
- Supervision of construction works to assure compliance with the specifications and standards.
- Monitoring of work and activities during the entire construction period to achieve timely completion of the project within the planned budget.
- Site testing and commissioning of equipment and work to prove satisfactory quality and operation, also with interfacing components or projects.

Further aspects cover environmental conditions and training.

Appendix

Norms and Standards

As far as norms and standards are mentioned in this book, the respective citation number of IEC or DIN EN is given. If IEC-standards or DIN EN-norms are not available, the European norms (EN number), VDE-norms or DIN-norms are given. A reference is given for some standards to corresponding ones; amendments and corrigenda are mentioned if available.

Engineering works always requires the application of the latest editions of standards, norms, and technical recommendations, which can be obtained from the IEC secretariat or from the national standard organization.

IEC citation number	*Issue*	*Title*	*Remarks*
IEC 60017	2008	Graphical symbols for use on equipment	ANSI (IEEE C37.2) Electrical Power System Device Function Numbers, Acronyms, and Contact Designations See IEC 61082
IEC 60038	2009-06	IEC standard voltages	
IEC 60044	Different issue dates	Instrument transformers	See IEC 61869
IEC 60071-1	2011-03	Insulation coordination – Part 1: Definitions, principles, and rules	VDE 0111-1

Continued

Power System Engineering: Planning, Design and Operation of Power Systems and Equipment, Second Edition.
Jürgen Schlabbach and Karl-Heinz Rofalski.

IEC citation number	*Issue*	*Title*	*Remarks*
IEC 60071-2	1996-12	Insulation coordination – Part 2: Application guide	VDE 0111-2
IEC/TR 60071-4	2004-06	Insulation coordination – Part 4: Computational guide to insulation coordination and modeling of electrical networks	
IEC 60076-2	2011-02	Power transformers – Part 2: Temperature rise for liquid-immersed transformers	VDE 0532-76-2
IEC 60076-7	2005-12	Power transformers – Part 7: Loading guide for oil-immersed power transformers	
IEC 60076-12	2008-11	Power transformers – Part 12: Loading guide for dry-type power transformers	
IEC 60085	2007-11	Electrical insulation – Thermal evaluation and designation	
IEC 60287-1-1	2006-12	Electric cables – Calculation of the current rating – Part 1-1: Current rating equations (100% load factor) and calculation of losses – General	See IEC 60287-1-2 IEC 60287-1-3
IEC 60354	1991	Loading guide for oil-immersed transformers	Replaced by IEC 60076-7
IEC 60617-DB-12M	2011-01	Graphical symbols for diagrams – 12-month subscription to online database comprising parts 2 to 13 of IEC 60617	DIN 40101 DIN EN 60617

IEC citation number	Issue	Title	Remarks
IEC 60853-1	1985-01	Calculation of the cyclic and emergency current rating of cables. Part 1: Cyclic rating factor for cables up to and including 18/30(36) kV	Amendment
IEC 60853-2	1989-09	Calculation of the cyclic and emergency current rating of cables. Part 2: Cyclic rating of cables greater than 18/30 (36) kV and emergency ratings for cables of all voltages	Amendment
IEC 60853-3	2002-02	Calculation of the cyclic and emergency current rating of cables – Part 3: Cyclic rating factor for cables of all voltages, with partial drying of the soil	
IEC 60865-1	2011-10	Short-circuit currents – Calculation of effects – Part 1: Definitions and calculation methods	VDE 0103
IEC/TR 60890	1987-07	A method of temperature-rise assessment by extrapolation for partially type-tested assemblies (PTTA) of low-voltage switchgear and controlgear	VDE 0660
IEC 60909-0	2001-07	Short-circuit currents in three-phase a.c. systems – Part 0: Calculation of currents	VDE 0102
IEC 60947	2011-03	Low-voltage switchgear and control gear	
IEC 60949	1988-11	Calculation of thermally permissible short-circuit currents, taking into account non-adiabatic heating effects	Amendment

Continued

IEC citation number	***Issue***	***Title***	***Remarks***
IEC 60986	2008-11	Short-circuit temperature limits of electric cables with rated voltages from 6 kV up to 30 kV	Amendment
IEC 61000-3-2	2009-04	Electromagnetic compatibility (EMC) – Part 3-2: Limits – Limits for harmonic current emissions (equipment input current ≤16 A per phase)	Amendment Corrigendum VDE 0838-2
IEC 61000-3-3	2013-05	Electromagnetic compatibility (EMC) – Part 3-3: Limits – Limitation of voltage changes, voltage fluctuations and flicker in public low-voltage supply systems, for equipment with rated current ≤6 A per phase and not subject to conditional connection	VDE 0838-3
IEC/TS 61000-3-4	1998-04	Electromagnetic compatibility (EMC) – Part 3-4: Limits – Limitation of emission of harmonic currents in low-voltage power supply systems for equipment with rated current greater than 16 A	
IEC/TS 61000-3-5	2009-07	Electromagnetic compatibility (EMC) – Part 3-5: Limits – Limitation of voltage fluctuations and flicker in low-voltage power supply systems for equipment with rated current greater than 75 A	Corrigendum
IEC/TR 61000-3-7	2008-02	Electromagnetic compatibility (EMC) – Part 3-7: Limits – Assessment of emission limits for the connection of fluctuating installations to MV, HV and EHV power systems	

IEC citation number	Issue	Title	Remarks
IEC 61000-3-11	2008-08	Electromagnetic compatibility (EMC) – Part 3-11: Limits – Limitation of voltage changes, voltage fluctuations and flicker in public low-voltage supply systems – Equipment with rated current ≤75 A and subject to conditional	VDE 0838-11
IEC 61000-3-12	2011-05	Electromagnetic compatibility (EMC) – Part 3-12: Limits – Limits for harmonic currents produced by equipment connected to public low-voltage systems with input current >16 A and ≤75 A per phase	VDE 0838-12
IEC 61082	2007-03	Preparation of documents used in electrotechnology – Part 1: Rules	VDE 0040
IEC 61660-1	1997-06	Short-circuit currents in d.c. auxiliary installations in power plants and substations – Part 1: Calculation of short-circuit currents	Corrigendum VDE 0102-10
IEC 61660-2	1997-06	Short-circuit currents in d.c. auxiliary installations in power plants and substations – Part 2: Calculation of effects	VDE 0103-10
IEC/TR 62095	2003-06	Electric cables – Calculations for current ratings – Finite element method	

DIN number DIN EN number	Issue	Title	Remarks
DIN EN 805	2000-03	Water supply – Requirements for systems and components outside buildings	
DIN 1998	1978-05	Placing of service conduits in public areas; directives for planning	
DIN 4124	2012-01	Excavations and trenches – Slopes, planking, and strutting breadths of working spaces	In German
DIN EN 42504-1	1983-05	Transformers; three-phase oil immersed type power transformers with off-circuit tapchanger or with on-load tap-changer 2000 to 10000 kVA for 50 Hz and U_m up to 123 kV; characteristics	Replaced by DIN 42508
DIN EN 42508	2009-08	Transformers – Oil-immersed power transformers from 3150 kVA up to 80000 kVA and U_m up to 123 kV	
DIN EN 50052	2008-06	Cast aluminum alloy enclosures for gas-filled high-voltage switchgear and controlgear	Amendment In German See VDE 0670-801
DIN EN 50064	1995-10	Wrought aluminum and aluminum alloy enclosures for gas-filled high-voltage switchgear and control gear;	Amendment In German VDE 0670-803
DIN EN 50160	2011-02	Voltage characteristics in public distribution systems	
DIN EN 50182	2001-12	Conductors for overhead lines. Round wire concentric lay stranded conductors	

DIN number DIN EN number	Issue	Title	Remarks
DIN 50267	1999-04	Common test methods for cables under fire conditions – Tests on gases evolved during combustion of materials from cables	In German VDE 0482-267
DIN EN 50341-1	2013-11	Overhead electrical lines exceeding AC 1 kV – Part 1: General requirements – Common specifications	VDE 0210-1
DIN EN 50423-1	2005-05	Overhead electrical lines exceeding AC 1 kV up to and including AC 45 kV – Part 1: General requirements – Common specifications	VDE 0210-10
DIN EN 50482	Different issue dates	Instrument transformers	VDE 0414 See DIN EN 61869
DIN EN 60076-2	2012-02	Power transformers – Part 2: Temperature rise for liquid-immersed transformers	VDE 0532-76-2
DIN EN 60282	2010-08	High-voltage fuses – Part 1: Current-limiting fuses	Amendments VDE 0670
DIN EN 60617	Different issue dates	Graphical symbols for diagrams	
DIN EN 60947	2011-10	Low-voltage switchgear and control gear	VDE 0660
DIN EN 61400-21	2009-06	Wind turbine generator systems – Part 21: Measurement and assessment of power quality characteristics of grid-connected wind turbines	VDE 0127-21
DIN EN 61869	2010-04	Instrument transformers	See also DIN EN 50482 VDE 0414

VDE classification	*Issue*	*Title*	*Remarks*
VDE 0141	2000-01	Earthing of special high-voltage installations with nominal voltages above 1 kV	
VDE 0271	2007-01	Definitions for cables with nominal voltages above 0.6/1 kV with specila applications [Festlegungen für Starkstromkabel ab 0,6/1 kV für besondere Anwendungen]	
VDE 0276-1000	1995-05	Electric cables current carrying capacity, general conditions, calculation factors [Strombelastbarkeit, Allgemeines; Umrechnungsfaktoren]	
VDE 0289-1	1988-03	Definitions for cables and isolated lines – General conditions [Begriffe für Starkstromkabel und isolierte Starkstromleitungen – Allgemeine Begriffe]	
VDE 0414	Different issue dates	Instrument transformers	DIN EN 50482 DIN EN 61689
VDE 0532-76-2	2012-02	Power transformers – Part 2: Temperature rise for liquid-immersed transformers	DIN EN 60076-2

References

References to Chapter 2

1 Dittmann, A. and Zschernig, J. (eds) (1998) *Energiewirtschaft [Economics of Energy Trade]*, B.G. Teubner-Verlag, Stuttgart, ISBN 3-519-06361-1.

2 Mende, W. and Albrecht, K.-F. (1994) Das prognostische Potential des Evolon-Modells [The Potential of Evolon-model for load forecasting], in *Neue Ansätze der Prognostik* (ed. M. Härter), Verlag TÜV-Rheinland, Köln, p. 127.

3 Kaufmann, W. (1995) *Planung öffentlicher Elektrizitätsverteilungs-Systeme [Planning of Public Power Supply Systems]*, VDE-Verlag, Berlin und Offenbach, ISBN 3-8022-0469-7.

4 Lebau, H. (2002) Nutzen und Notwendigkeit von Lastprofilen aus Sicht der Netzbetreiber [Advantages and needs of load-profiles], in Lastprofile in der leitungsgebundenen Energieversorgung. VDI-GET-Seminar 24.04.2002, VWEW-Energieverlag, Frankfurt am Main.

5 Meyer, J., Fünfgeld, C., Adam, T. and Schieferdecker, B. (1999) Repräsentative VDEW-Lastprofile [Representative load-profiles of the VDEW], in VDEW-Materialien M-28/1999, VWEW-Energieverlag, Frankfurt am Main.

6 Fünfgeld, C. and Tiedeman, R. (2000) Anwendung der repräsentativen VDEW-Lastprofile step-by-step [Application of representative load-profiles step-by-step], in VDEW-Materialien M-05/2000, VWEW-Energieverlag, Frankfurt am Main.

7 Fünfgeld, C., Fiebig, C., Hofmann, A. and Hofmann, J. (2002) Lastprofile für Geschäftskunden – belastbare Einzelkundenplanung [Load-profiles for business customers for planning of bulk load customers]. Elektrizitätswirtschaft, **101** (8), 307–311.

8 Stamminger, R. (ed.) (2009) *Synergy Potential of Smart Domestic Appliances in Renewable Energy Systems*, Shaker Verlag, Aachen.

Reference to Chapter 4

9 VWEW (ed.) (1978) *Netzverluste, Eine Richtlinie für ihre Bewertung und Verminderung [Losses in Power Systems – Regulations for Assessment and Reduction]*, VWEW-Energieverlag, Frankfurt am Main, ISBN 3-8022-0007-1.

References to Chapter 5

10 ENTSO (2011) Statistical Yearbook 2011, European Network of Transmission System Operators for Electricity. ENTSO, Brussels.

References to Chapter 6

11 Kaufmann, W. (1995) *Planung öffentlicher Elektrizitätsverteilungs-Systeme [Planning of Public Power Supply Systems]*, VDE-Verlag, Berlin und Offenbach, ISBN 3-8022-0469-7.

Power System Engineering: Planning, Design and Operation of Power Systems and Equipment, Second Edition.
Jürgen Schlabbach and Karl-Heinz Rofalski.

12 Schlabbach, J. (2009) *Elektroenergieversorgung [Electrical Energy Supply]*, 3rd revised edn, VDE-Verlag, Berlin und Offenbach, ISBN 3-8007-3108-4.

13 Müller, L. and Matla, W. (2003) *Selektivschutz in elektrischen Anlagen [Protection of Electrical Installations and Equipment]*, 3rd edn, VWEW-Energieverlag, Frankfurt am Main, ISBN 3-8022-0658-4.

References to Chapter 7

14 Oeding, D. and Oswald, B. (2011) *Elektrische Kraftwerke und Netze [Power Stations and Electrical Power Systems]*, 7th edn, Springer-Verlag, Berlin, Heidelberg, New York, ISBN 3-642-19246-3.

15 Schlabbach, J. (2002) *Sternpunktbehandlung in elektrischen Netzen [Neutral Earthing in Electrical Power Systems]*, VWEW-Energieverlag, Frankfurt am Main, ISBN 3-8022-0677-0.

16 Vosen, H. (1997) *Kühlung und Belastbarkeit von Transformatoren [Cooling and Permissible Loading of Transformers]*, VDE-Schriftenreihe No. 72, VDE-Verlag, Berlin und Offenbach.

17 McShane, C.P.; Rapp, K.J., Corkran, J.L. et al. (2002) Aging of paper insulation in natural ester dielectric fluid, IEEE/PES Transmission and Distribution Conference, IEEE 0-7803-7257-5/01.

18 VDE 0536 (2008) *Belastbarkeit von Öltransformatoren [Permissible Loading of Oil-immersed Transformers]*, VDE-Verlag, Berlin und Offenbach.

19 IEC 60354 (1991) *Loading Guide for Oil-immersed Transformers*, IEC.

20 IEC 60076-7 (2005) Loading guide for oil-immersed transformers. IEC.

21 IEC 60076-12 (2008) Loading guide for dry-type transformers, IEC.

22 IEC 60905 (1987) Loading Guide for Dry-type Power Transformers, IEC.

23 Schlabbach, J. (2004) *Short-Circuit Currents*, IEE Power and Energy Series Nr. 51, IEE, London/UK, ISBN 8-5296-514-8.

References to Chapter 8

24 Heinhold, L. and Stubbe, R. (1999) *Kabel und Leitungen für Starkstrom [Cables and Conductors for Power System Installations]*, 5th revised edn, Wiley VCH, Berlin und Weinheim, ISBN.

25 Brakelmann, H. (1984) *Belastbarkeiten der Energiekabel [Permissible Loading of Power Cables]*, VDE-Verlag, Berlin und Offenbach, ISBN 3-8007-1406-X.

26 IEC 60287 (1993–2002). *Calculation of Continuous Current Rating of Cables (100% Load Factor)*, IEC.

27 Rheydt, K. (ed.) (1989) *Strombelastbarkeit von Hochspannungskabeln [Current Carrying Capacity of High-Voltage Cables]*, KABEL RHEYDT, Mönchengladbach.

28 Hitchcock, J.A. and Preece, R.J. (1979) Simultaneous diffusion of heat and moisture around normally buried cable systems, IEE Conf. Publ. 176, S. 262–67.

29 Donazzi, F., Occhini, E. and Seppi, A. (1979) Soil thermal and hydrological characteristics in designing underground cables. *Proceedings of the IEE*, S. 506–16.

30 Schlabbach, J. (2003) *Optimized Design of Multiple-supply Service Trench*, IEEE-Conference EUROCON, Ljubljana/Slowenien, ISBN 078037763X, Paper 713.

31 Brakelmann, H. (1991) CAE bei der Planung von Kabeltrassen [Application of CAE for the design of cable trenches], *Elektrizitätswirtschaft*, S. **90**, 384–93.

32 IEC 60853 (1985–2008). *Calculation of Cyclic and Emergency Rating of Cables.*

33 DIN 1998 (1978). *Unterbringung von Leitungen und Anlagen in öffentlichen Flächen, Richtlinien für die Planung [Arrangement of Cables and Installations in Public Areas. Recommendations for Planning and Design].*

34 DVGW (ed.) (1976) *Errichtung von Gasleitungen bis 4 bar Betriebsdruck aus Stahlrohren [Installation of Gas-Pipelines for Operating Pressure up to 4 Bar Made from Steel Pipes]*, Arbeitsblatt G 462/I des DVGW, Bonn.

35 DVGW (ed.) (1985) *Gasleitungen aus Stahlrohren von mehr als 4 bar bis 16 bar Betriebsdruck [Gas-Pipelines with*

Operating Pressure 4 Bar up to 16 Bar], Arbeitsblatt G 462/II des DVGW, Bonn.

36 DVGW (ed.) (1988) *Planungsregeln für Wasserleitungen und Wasserrohrnetze [Rules for Planning and Design of Water-Pipelines]*, Merkblatt W 403 des DVGW, Bonn.

37 DIN EN805 (2000). *Anforderungen an Wasserversorgungssysteme und deren Bauteile außerhalb von Gebäuden [Requirements of Water Supply Systems and Their Installations outside Buildings]*.

38 Schlabbach, J. (2004) *Short-Circuit Currents*, IEE Power and Energy Series Nr. 51, IEE, London, ISBN 8-5296-514-8.

References to Chapter 9

39 Webs, A. (1963) Dauerstrombelastbarkeit von nach DIN 48201 gefertigten Freileitungsseilen [Continuous loading of overhead conductors manufactured according to DIN 48201]. *Elektrizitätswirtschaft*, **62** (*23*), 861–872.

40 Kirn, H. (1988) Die Bestimmung der Temperatur von Freileitungsseilen [Determination of the temperature of overhead line conductors]. *Elektrizitätswirtschaft*, **87** (*21*), 1055–1065.

41 Schlabbach, J. (2009) *Elektroenergieversorgung [Electrical Energy Systems]*, 3rd revised edn, VDE-Verlag, Berlin und Offenbach, ISBN 3-8007-3108-4.

42 Schlabbach, J. (2004) *Short-Circuit Currents*, IEE Power and Energy Series Nr. 51, IEE, London, ISBN 8-5296-514-8.

43 D'heil, F. and Herzig, K. (1977) Auswirkungen des Seiltyps auf die Qualität der Mastausteilung von Freileitungen [Dependency of the conductor type on tower arrangement of overhead lines]. *Elektrizitätswirtschaft*, **76** (*19*), 664–672.

44 CIGRE WG 22.09 (ed.) (1991) Parameter studies of overhead transmission costs, *Electra*, **136**, 31–67.

45 VWEW (ed.) (1978) *Netzverluste, Eine Richtlinie für ihre Bewertung und Verminderung [Losses in Electrical Power System – Regulations for Assessment and Reduction]*, VWEW-Energieverlag, Frankfurt am Main.

46 FGH (ed.) (1977) Elektrische Hochleistungsübertragung und -verteilung in Verdichtungsräumen [Transmission and distribution of electrical energy in urban areas], Mannheim, BMFT-ET 4042.

47 Berndorf (ed.) (2002) *TAL – Temperaturbeständige Aluminiumlegierung für Freileitungsseile [TAL – Temperature Stabilised Aluminium Alloys for Overhead Line Conductors]*, Berndorf GmbH, Austria.

48 Niemeyer, D. and Grohs, A. (2008) *Overhead Lines [Freileitungen]*, 2nd edn, EW Medien und Kongresse, Frankfurt am Main.

49 Cigre (2004) Conductors for the uprating of overhead lines. Cigre SC B2 WG B2.12, Paris.

References to Chapter 10

50 Schlabbach, J. (2009) *Elektroenergieversorgung [Electrical Energy Systems]*, 3rd revised edn, VDE-Verlag, Berlin und Offenbach, ISBN 3-8007-3108-4.

51 Oeding, D. and Oswald, B. (2011) *Elektrische Kraftwerke und Netze [Power Stations and Electrical Power Systems]*, 7th edn, Springer-Verlag, Berlin, Heidelberg, New York, ISBN 3-642-19246-3.

52 Song, Y.H. and Johns, A.T. (1999) *Flexible AC Transmission Systems*, IEE, London, ISBN 8-5296-771-3.

53 Fahland, P., Schlabbach, J., Umulu, E. and Tarkan, O. (1985) Einsatz von Reihenkondensatoren bzw. quergeregelten Transformatoren zur Erhöhung der Übertragungsfähigkeit paralleler 110-kV-Leitungen [Application of series capacitors and phase-shift transformers to increase the power transfer through parallel 110-kV overhead lines]. *Elektrizitätswirtschaft*, **84**, 933–35.

54 Große-Gehling, M., Just, W., Reese, J. and Schlabbach, J. (2013) *Reactive Power Compensation [Blindleistungskompensation]*, EW Medien und Kongresse, Frankfurt am Main.

References to Chapter 11

55 Laible, W. (1968) Abhängigkeit der Wirk- und Blindleistungsaufnahme passiver Netze von Spannungs- und Frequenzschwankungen [Dependency of active and reactive power from voltage and frequency fluctuations]. *Bull SEV*, **59**, 49–65.
56 Buse, G., Bopp, R. and Junkermann, H. (1983) Das Verhalten der Netzleistung bei veränderlicher Netzspannung [Changing of power and changing voltage]. *Elektrizitätswirtschaft*, **82**, 829–32.
57 Funk, G. (1971) Der Einfluß von Netzimpedanzen auf die Spannungsabhängigkeit von Drehstromlasten [Influence of load impedances on the voltage dependency of three-phase system load]. *Wiss. Ber. AEG Telefunken*, **44**, 1–5.
58 Schlabbach, J. (2009) *Elektroenergieversorgung [Electrical Energy Supply]*, 3rd revised edn, VDE-Verlag, ISBN 3-8007-3108-4.
59 Schlabbach, J. (2004) *Short-Circuit Currents*, IEE Power and Energy Series Nr. 51, IEE, London, ISBN 8-5296-514-8.
60 Jenkins, N., Allan, R., Crossley, P. *et al.* (2000) *Embedded Generation*, in Power and Energy Series, No. 31, IEE, London/UK, ISBN 8-5296-774-8.
61 Oeding, D. and Oswald, B. (2011) *Elektrische Kraftwerke und Netze [Power Stations and Electrical Power Systems]*, 7th edn, Springer-Verlag, Berlin, Heidelberg, New York, ISBN 3-642-19246-3.
62 Schlabbach, J., Blume, D. and Stephanblome, Th. (1999) *Voltage Quality in Electrical Power Systems*, IEE Power and Energy Series Nr. 36, IEE, London/UK, ISBN 8-5296-975-9.
63 Schlabbach, J. (2002) *Sternpunktbehandlung [Handling of Neutrals in Power Systems]*, VWEW-Energieverlag, Frankfurt, ISBN 3-8022-0677-0.

References to Chapter 12

64 Oeding, D. and Oswald, B. (2011) *Elektrische Kraftwerke und Netze [Power Stations and Electrical Power Systems]*, 7th edn, Springer-Verlag, Berlin, Heidelberg, New York, ISBN 3-642-19246-3.
65 ENTSO (2012) Network code for requirements for grid connection applicable to all generators. ENTSO, Brussels.
66 VDN (2007) Transmission Code 2007, Network and system rules for the operation of the German transmission system [Transmission Code, Netzsystemregeln für den Betrieb des Deutschen Übertragungsnetzes]. EW Medien und Kongresse, Frankfurt am Main.
67 VDN (2004) Technical conditions for connection of renewable generating plants to high and extra-high voltage systems [EEG-Erzeugungsanlagen am Hoch- und Höchstspannungsnetz]. EW Medien und Kongresse, Frankfurt am Main.
68 E VDE-AR-N 4130 (2013) *Technical Conditions for Connection of Generation Units to EHV-Systems, [Technische Anschlussbedingungen für Erzeugungsanlagen am Höchstspannungsnetz]*, VDE-Verlag, Berlin und Offenbach.
69 E VDE-AR-N 4120 (2012) *Technical Conditions for Connection of Customer Equipment to High-Voltage Systems (110 Kv), [Technische Bedingungen für den Anschluss von Kundenanlagen an das Hochspannungsnetz (TAB Hochspannung)]*, VDE-Verlag, Berlin und Offenbach.
70 BDEW-technical guideline (2008) Connection of generation units to medium-voltage systems, [Erzeugungsanlagen am

Mittelspannungsnetz. Supplements February 2011, BDEW, Berlin.

71 VDE-AR-N 4105 (2011) *Connection of Generating Units to Low-Voltage Systems. [Erzeugungsanlagen am Niederspannungsnetz]*, VDE-Verlag, Berlin und Offenbach.

72 VDN (2007) DA-CH-CZ Technical Rules for the Assessment of Power System Perturbations [Technische Regeln zur Beurteilung von Netzrückwirkungen]. EW Medien und Kongresse, Frankfurt am Main.

73 Schlabbach, J., Blume, D. and Stephanblome, Th. (1999) *Voltage Quality in Electrical Power Systems*, IEE Power and Energy Series Nr. 36, IEE, London/UK, ISBN 8-5296-975-9.

74 Hormann, W., Just, W. and Schlabbach, J. (2008) *Netzrückwirkungen [Power System Disturbances]*, 3rd edn, VWEW-Energieverlag, Frankfurt am Main, ISBN 3-8022-0917-8.

References to Chapter 13

75 Schossig, W. (2013) *Netzschutztechnik [Protection of Power Systems]*, 4th edn, EW Medien und Kongresse, Frankfurt am Main, ISBN 3-8022-1060-0.

76 Müller, L. and Matla, W. (2003) *Selektivschutz elektrischer Anlagen [Selective Protection of Power Installations and Equipment]*, 3rd edn, VWEW-Energieverlag, Frankfurt am Main, ISBN 3-8022-0658-4.

77 Schlabbach, J. (2009) *Elektroenergieversorgung [Electrical Energy Systems]*, 3rd edn, VDE-Verlag, Berlin und Offenbach, ISBN 3-8007-3108-4.

78 Schlabbach, J. (2002) *Sternpunktbehandlung [Handling of Neutrals in Power Systems]*, VWEW-Energieverlag, Frankfurt am Main, ISBN 3-8022-0677-0.

79 Doemeland, W. (1997) *Handbuch Schutztechnik [Handbook of power system protection]*, 6th revised edn, VDE-Verlag, Berlin und Offenbach, ISBN 3-8007-2259-3.

References to Chapter 14

80 Berger, K., Anderson, R.B. and Kröninger, H. (1975) Parameters of lightning flashes. Report of CIGRE study committee 33. Report ELT 041-1.

81 Garbagnati, E. and Lo Piparo, G.B. (1982) Parameter von Blitzströmen [Parameters of lightning strokes]. *Elektrotechnische Zeitschrift etz*, **2**, 103.

82 Eriksson, A.J. and Weck, K.-H. (1991) Simplified procedures for determining representative substation impinging lightning overvoltages. Report of CIGRE technical committee 33-01.02.

83 Eriksson, A.J. (1987) An improved electrogeometric model for transmission line shielding analysis. *IEEE PWRD-2*, **3**, 271–282.

84 Oeding, D. and Oswald, B. (2011) *Elektrische Kraftwerke und Netze [Power Stations and Electrical Power Systems]*, 7th edn, Springer-Verlag, Berlin, Heidelberg, New York, ISBN 3-642-19246-3.

85 Electromagnetic Transients Program (EMTP) (1986) *Theory Book*, Bonneville Power Administration, Portland, Oregon/USA.

86 Alternative Transients Program (ATP) (1987–2003) *Rule Book*, Canadian/American EMTP User Group.

References to Chapter 15

87 Schlabbach, J. (2002) *Neutral handling [Sternpunktbehandlung]*, VWEW-Energieverlag, Frankfurt am Main, ISBN 3-8022-0677-0.

88 Schlabbach, J. (2009) *Electrical power system engineering [Elektroenergieversorgung]*, 3rd edn, VDE-Verlag, Berlin, Offenbach/Germany, ISBN 3-8007-3108-4.

89 Balzer, G. (1988) Beidseitige Sternpunktbenhandlung von Transformatoren [Double-side earthing of transformers], in *Sternpunktbanendlung in 10-kV-bis 110-kV-Netzen, ETG-Fachbericht*, Vol. 24

(ed. VDE), VDE-Verlag, Berlin and Offenbach/Germany.

90 Gröber H. and Komurka, J. (1973) Übertragung der Nullspannung bei zweiseitig geerdeten Transformatoren [Transformation of zero-sequence voltage through transformers]. Technische Mitteilung FGH Mannheim/ Germany.

91 Niemand, T. and Kunz, H. (1996) *Erdungsanlagen [Earthing equipment in power systems]*, VWEW-Energieverlag, Frankfurt am Main, ISBN 3-8022-0362-3.

92 Arbitrary agency of VDEW (1987) Technical recommendation No. 1 – Induced voltages in telecommunication circuits, VWEW-Energieverlag, Frankfurt am Main.

93 Koetzold, B., Gauger, V., and Winter, K.M. (1997) *Residual Current Compensation, [Erdschlußschutzsystem mit Reststromkompensation]*, ETG-Fachbericht 66, VDE-Verlag, Berlin und Offenbach.

94 Swedish Neutral (1999) Installation for earth-fault protection, [Erdschlußschutzanlage SN-RCC98-A1]. Swedish Neutral, S-19631 Kungsängen.

95 Kurosawa, Y., *et al.* (2008) Effect and influence of residual current compensation. IEEE transmission and distribution conference.

96 Huang, S.F., Chen, Z.H. and Bi, T.S. (2005) Adaptive residual current compensation for robust fault-type selection in mho elements. *IEEE Transactions on Power Delivery*, 20, (2).

97 Wiede, H. (2009) Active Neutral Handling [Aktive Sternpunktbehandlung], DE-Patent Nr. DE 102008017927.

98 Raisz, D. and Dan, A. (2011) *Innovativ Measures for Fault Loacting and Residual Current Compensation in Power Systems with Resonace Earthing [Innovative Lösung für die Fehlerortung und Reststromkompensation in gelöschten Netzen]*, VDE-Conference STE 2011, VDE-Verlag, Berlin und Offenbach, ISBN 978-3-8007-3370-5.

References to Chapter 16

99 FIDIC (ed.) (1987) *Conditions of Contract for Electrical and Mechanical Works*, 3rd edn, International Federation of Consulting Engineers, Geneva, ISBN 2-88432-000-8.

100 FIDIC (ed.) (1997) *Conditions of Contract for Electrical and Mechanical Works*. Supplement to the 3rd edition, 1st edn, International Federation of Consulting Engineers, Geneva.

101 FIDIC (ed.) (1988) *Guide to the Use of FIDIC Conditions of Contract for Electrical and Mechanical Works*, International Federation of Consulting Engineers, Geneva, ISBN 2-88432-003-2.

Index

Note: Page numbers in *italics* refer to figures; those in **bold** refer to tables.

Power System Engineering: Planning, Design and Operation of Power Systems and Equipment, Second Edition.
Jürgen Schlabbach and Karl-Heinz Rofalski.

t